D1742085

Indonesia's Technological Challenge

The **Institute of Southeast Asian Studies (ISEAS)** was established as an autonomous organisation in 1968. It is a regional research centre for scholars and other specialists concerned with modern Southeast Asia, particularly the many-faceted problems of stability and security, economic development, and political and social change.

The Institute's research programmes are the Regional Economic Studies Programme (RES, including ASEAN and APEC), Regional Strategic and Political Studies Programme (RSPS), Regional Social and Cultural Studies Programme (RSCS), and the Indochina Programme (ICP).

The Institute is governed by a twenty-two-member Board of Trustees comprising nominees from the Singapore Government, the National University of Singapore, the various Chambers of Commerce, and professional and civic organisations. A ten-man Executive Committee oversees day-to-day operations; it is chaired by the Director, the Institute's chief academic and administrative officer.

The **Research School of Pacific and Asian Studies (RSPAS)** at the Australian National University is home to the Indonesia Project, the organiser of the annual *Indonesia Update* conference. The *Update* offers an overview of recent economic and political developments, and also devotes attention to a theme of particular importance in Indonesia's development. Financial support for the conference series is provided by the Australian Agency for International Development (AusAID). The *Indonesia Assessment* series contains papers presented at these conferences, and also sometimes includes other specially commissioned pieces.

The **Indonesia Project** is a major international centre of research and graduate training on the economy of Indonesia. Established in 1965, it is well known and respected in Indonesia and in other places where Indonesia attracts serious scholarly and official interest. The Project obtains its core funding from the Australian National University; in addition, the Australian Department of Foreign Affairs and Trade has provided an annual grant since 1980. A major activity is producing and distributing three times a year an internationally recognised journal on the Indonesian economy, the *Bulletin of Indonesian Economic Studies*, each issue of which contains a Survey of Recent Economic Developments.

The **Department of Political and Social Change** in the Research School of Pacific and Asian Studies focuses on research in domestic politics, social processes and state-society relations in Asia and the Pacific. It has a long established interest in Indonesian affairs. Each year staff members of the department work with the Indonesia Project to plan and organise the Indonesian Update Conference.

Indonesia's Technological Challenge

Edited by

Hal Hill
Thee Kian Wie

Research School of Pacific and Asian Studies
AUSTRALIAN NATIONAL UNIVERSITY, Canberra
INSTITUTE OF SOUTHEAST ASIAN STUDIES, Singapore

Published jointly by

Institute of Southeast Asian Studies Research School of Pacific and Asian Studies
30 Heng Mui Keng Terrace   Australian National University
Pasir Panjang Road   Canberra ACT 9299
Singapore 119614   Australia

Internet e-mail: publish@iseas.edu.sg
World Wide Web: http://www.iseas.edu.sg/pub.html

*The responsibility for facts and opinions expressed in this publication rests
exclusively with the authors and their interpretations do not necessarily reflect the
views or the policy of ISEAS or the RSPAS.*

Cataloguing in Publication Data

Indonesia's technological challenge / edited by Hal Hill and Thee Kian Wie.
   1. Technological innovations--Indonesia.
   2. Technology and state--Indonesia.
   3. Labor supply--Effect of technological innovations on--Indonesia.
   I. Hill, Hal, 1948-
   II. Thee, Kian Wie, 1935-
HC450 T4I42   1998   sls98-30211

ISBN 981-230-013-9 (soft cover)
ISBN 981-230-019-8 (hard cover)

Printed and bound in Singapore by Prime Packaging Industries Pte Ltd

CONTENTS

PART V TECHNOLOGY: INTERNATIONAL DIMENSIONS

LIST OF TABLES

LIST OF FIGURES

CONTRIBUTORS

Colin Barlow — Research School of Pacific and Asian Studies, Australian National University, Canberra

Rony M. Bishry — BPPT (Agency for the Assessment and Application of Technology), Jakarta

William Cole — Asia Foundation, San Francisco

Geoff Forrester — Research School of Pacific and Asian Studies, Australian National University, Canberra

Murman Hidayat — BPPT (Agency for the Assessment and Application of Technology), Jakarta

Hal Hill — Research School of Pacific and Asian Studies, Australian National University, Canberra

Sri Mulyani Indrawati — Faculty of Economics, University of Indonesia, Jakarta

Sanjaya Lall — Queen Elizabeth House, University of Oxford

Chris Manning — Research School of Pacific and Asian Studies, Australian National University, Canberra

Ross H. McLeod — Research School of Pacific and Asian Studies, Australian National University, Canberra

Saadillah Mursjid — Minister/Cabinet Secretary, Republic of Indonesia; President Director, PT Dua Satu Tiga Puluh

Leslie O'Brien — Department of Industry, Science and Tourism, Canberra

Robert C. Rice	Faculty of Business and Economics, Monash University, Melbourne
Remy Rohadian	Department of Industry, Science and Tourism, Canberra
Samaun Samadikun	Bandung Institute of Technology; former Chair of LIPI
Yuri Sato	Overseas Senior Research Fellow in Jakarta, Institute of Developing Economies, Tokyo
Don Scott-Kemmis	Department of Industry, Science and Tourism, Canberra
Shannon Luke Smith	Research School of Pacific and Asian Studies, Australian National University, Canberra
Hadi Soesastro	Centre for Strategic and International Studies, Jakarta; Research School of Pacific and Asian Studies, Australian National University, Canberra
Adam Szirmai	Department of Technology and Development Studies, Eindhoven University of Technology, Eindhoven
Thee Kian Wie	Centre for Economic and Development Studies, Indonesian Institute of Science, Jakarta; Research School of Pacific and Asian Studies, Australian National University, Canberra
Marcel Timmer	Department of Technology and Development Studies, Eindhoven University of Technology, Eindhoven
Rick van der Kamp	School of Oriental and African Studies, University of London

FOREWORD

I am delighted to formally open the 1997 Indonesia Update Conference, and to write the Foreword to the proceedings volume, *Indonesia's Technological Challenge*.

This is my second visit to the beautiful and peaceful city of Canberra, and to the Australian National University (ANU). On my first visit to Canberra, in 1993, I was astonished to see the great depth and breadth of academic interest in Indonesia to be found at the ANU. I was able to meet with many world-class experts on Indonesia from a wide range of disciplines. I do not know of any other place in the world outside Indonesia where it is possible to find such an intense concentration of academic interest in my country.

I was informed that, among the younger generation, there are postgraduate students from many disciplines doing research on Indonesia, and that in the Faculty of Asian Studies, Indonesian language is very popular among undergraduates. I am especially pleased to learn that arrangements have been made for undergraduates studying Bahasa Indonesia to live in Indonesia for a year, taking intensive language courses and studying other subjects in Indonesian classrooms alongside Indonesian university students as part of their degree.

I have also learned that Australia is now one of the most important countries of destination for Indonesians studying abroad, and that there are currently some 16,700 studying here in schools, universities, and vocational and technical colleges. This is an increase of about 50% in just three years! From my own experience, I know that living and studying in another country is an extremely effective way of learning to appreciate and understand other cultures.

We are all aware that our two countries have at times disagreed on various issues. That is probably inevitable between two large and very different neighbouring countries. But I am convinced that the increasing flow of students, from Indonesia to Australia and from Australia to Indonesia, has already paid ample dividends in allowing us to view our differences calmly and from a perspective that is based on mutual respect.

The size of the audience at the Update Conference shows clearly the depth of interest in Indonesia to be found in Australia. I understand that the conference usually has around 300–400 participants, drawn not only from the universities, but also from government departments, the business sector, the media and the general public. It is very gratifying to me, as a representative of the Indonesian government and the Indonesian people, to be present.

This is the fifteenth in the series of Indonesia Update Conferences, and the papers presented at them have since 1988 been published in the now well-established *Indonesia Assessment* series. The *Assessment* volume has grown to include additional commissioned papers alongside those presented at the conference. It is no longer published in-house at the ANU, but as a joint venture with the Institute of Southeast Asian Studies in Singapore. This allows the volume to reach a far wider international readership than was possible previously. The *Assessment* series has therefore become an important complement to the ANU's *Bulletin of Indonesian Economic Studies*, which has been in continuous publication for over three decades.

The topics for discussion at the Update Conference in recent years have ranged widely over such matters as political scenarios for the future; the state of higher education; manpower issues; the financial sector; the development of Eastern Indonesia; and population and human resources. I do not know of any other conference or publication series of this kind anywhere in the world that examines development in my country in such depth and from so many perspectives.

There has been a nice balance between Indonesians and non-Indonesians in the list of contributors to the various *Assessment* volumes. Looking over this list of names, as an Indonesian, I cannot help but be impressed by the organisers' ability to attract Indonesian speakers and contributors of the highest calibre—including, of course, Dr Thee Kian Wie, the co-organiser of the 1997 conference.

At the 1997 Indonesia Update Conference, the special theme for discussion is 'Indonesia's Technological Challenge'. Clearly, technological improvement is an essential part of the development process, and we have seen enormous change in the technologies applied throughout the Indonesian economy. One aspect of this is the Green Revolution, in which technological improvement made possible great increases in rice production, so that we were able to become self-sufficient in rice more than a decade ago. Another is the reform of our trade policies, which made it feasible for us to expand vastly our use of manufacturing technology to produce manufactured goods for sale on world markets.

The services sector has also witnessed vast changes in technology. Many years ago, our government was quick to act when it saw the potential benefits of satellite telecommunications for a far-flung archipelagic nation, by launching the Palapa satellite. More recently, the younger generation has been equally quick to embrace mobile phone technology—although sometimes I wonder if the young people I see talking on their mobile phones are really having a conversation, or just demonstrating the fact that they own one! Indeed, technological change has moved so far in this field that even I can now boast of having my own e-mail facility!

Our people once had to travel by foot and by horse-drawn cart. Then they began to be able to afford bicycles, and to ride in mini-vans. These days, more and more use motorcycles, cars and air-conditioned buses. Such changes have their own technological implications. For example, roads need to be widened, straightened and strengthened; busy intersections need to be controlled with traffic lights; and overpasses have to be constructed. Also, it becomes feasible to operate toll roads in many locations so that the costs of providing the roads are borne by those who use them rather than by the government.

Another field in which new technology brings its own technological challenges is fishing. As you know, our fishing industry has been growing rapidly in recent years, aided by the use of bigger and more sophisticated fishing vessels and satellite technology to locate shoals of fish. The problem that arises here, however, is the danger of overexploitation. This creates the need for monitoring the fishing resource and managing it in a sustainable fashion. Incidentally, this is a field in which I believe Australia could provide considerable assistance to Indonesia, since it is a problem you are already facing yourselves and a field in which you have accumulated a great deal of expertise.

Indonesia is a country in a hurry to catch up to the developed world. This urgency stems not from some inappropriate pride, but from the very strong realisation that we need to grow rapidly in order to provide a better standard of living for our people. Even with the reductions in poverty that have been achieved, millions of Indonesians still live at income levels that leave them classified as 'near poor'. Rapid growth is not a luxury; for us it is a necessity.

Access to technology, as well as the capacity to adopt technology, are important variables in our efforts to increase the rate of growth. Yet we are also aware of the pitfalls of encouraging inappropriate technology acquisition and development. Achieving a balance between these two extremes involves a difficult process of weighing conflicting strategies and opinions. As with all policy choices, the answers are never clear, and so choices can always be criticised. But I am convinced that further research and analysis of these issues will help narrow the range of choices, reducing the chances that major policy errors will be made. The purpose of a conference such as this is to contribute to the process of policy dialogue, to learn from the experience and insights of others, and to draw appropriate lessons for the unique political, social and economic entity that we know as Indonesia.

SAADILLAH MURSJID

PREFACE

This volume comprises a selection of papers presented to the 1997 Indonesia Update Conference, held at the Australian National University (ANU) on 19–20 September. The Indonesia Update is an annual conference devoted to the study of contemporary Indonesia. It is designed to inform the broad academic, government, business, media and international affairs communities in Australia, Indonesia and elsewhere of recent developments in the country, and to provide an in-depth treatment of topical public issues.

This was the fifteenth consecutive Update Conference, and this volume is the ninth to be published in the *Indonesia Assessment* proceedings series. The Update and the *Assessment* can thus be said to have 'grown up', evolving since the early experimental days into major enterprises featuring strong participation from Australia, Indonesia and other parts of the world.

The conference and book format is the same each year, namely a survey of recent political and economic developments, followed by a more detailed examination of a particular issue. We are very pleased to have assembled a team of highly qualified contributors, including a member of the Indonesian cabinet, together with academics, researchers and officials from Australia, Indonesia, Japan, the Netherlands, the UK and the US.

The theme on this occasion has been 'Indonesia's Technological Challenge'. We have interpreted this topic broadly, examining the major analytical and policy issues against a backdrop ranging from detailed micro-level case studies to international dimensions. Our aim, as in all publications in this series, has been to produce an analytically informed volume which is nevertheless accessible to a general readership. To our knowledge, this is the first English-language volume to investigate in any depth technology issues in the Indonesian context.

We wish to place on record our thanks to the conference speakers, discussants and chairs, some of whom made the long journey to Canberra from Indonesia and elsewhere (in search of a promised Spring which did not eventuate!). We are especially grateful to Bapak Saadillah Mursjid for taking time off from a demanding ministerial schedule to be with us, both to officially open the conference and to present a paper. We are also indebted to our paper writers, who met tight deadlines and who tolerated innumerable requests for further information and editorial modifications.

As always it is a pleasure to thank the many organisations and individuals who have supported our endeavours. The Australian Agency for International Development, AusAID, has kindly supported the Update and the *Assessment* series for a number of years, while the Research School of Pacific and Asian Studies (RSPAS) at the ANU has provided a congenial home since their inception. We also thank the Asia Foundation, the Institute of Developing Economies, Tokyo, the Soedarpo Corporation and PT Indomobil for funding participants' international travel expenses.

The conference was organised by a cheerful and capable team which included Bev Fraser, Kim Kelo and Alison Ley, assisted by Liz Drysdale, Lynn Moir and Trish van der Hoek. Several ANU students provided much useful logistical assistance at the conference.

In addition, a number of colleagues at the ANU provided friendly assistance and advice, both in developing the program and in facilitating the smooth functioning of the conference. These included in particular recent Update organisers Colin Barlow, Harold Crouch, Terry Hull, Gavin Jones, Ross McLeod and Chris Manning. Ross Garnaut, Ben Kerkvliet and Merle Ricklefs were, as always, supportive of the enterprise.

The quality of the production owes a great deal to the deft and highly professional touch of Beth Thomson of Japan Online, who efficiently produced the manuscript for publication. This is now the fourth *Assessment* to be published in conjunction with the Institute of Southeast Asian Studies, Singapore, and we are very pleased to acknowledge again our close association with Triena Ong, the Institute's Managing Editor, and her expert production team.

This year's Update and *Assessment* occur at a difficult and troubling time for Indonesia as the country experiences its most serious economic and political crises since 1965–66. Inevitably, in such a fast changing environment, political and economic 'updates' date rapidly, and topics not directly connected to the current crises— such as the one examined in this book—are pushed into the background. However, we believe that the contributors to the first section of the book have provided an important analysis of the period leading up to the dramatic events of late 1997 and early 1998, while the papers which focus on the major theme of technology address an issue of enduring analytical and policy importance. When Indonesia resolves its current economic, political and social crises, the government and the policy-oriented research community will again need to face the country's 'technological challenge'.

HAL HILL AND THEE KIAN WIE

Canberra and Jakarta

February 1998

GLOSSARY

AAECP	ASEAN–Australia Economic Cooperation Program
ACCI	Australian Chamber of Commerce and Industry
ACIAR	Australian Centre for International Agricultural Research
ACRES	Australian Centre for Remote Sensing
ACTEW	ACT Electricity and Water
ADB	Asian Development Bank
AFTA	ASEAN Free Trade Area
AGSO	Australian Geological Survey Organisation
AIPI	Indonesian Academy of Sciences
AIPO	Australian Intellectual Property Office
ANSTO	Australian Nuclear Science and Technology Organisation
ANU	The Australian National University
ARC	Australian Research Council
ASEAN	Association of Southeast Asian Nations
ATBM	*alat tenun bukan mesin* (partly mechanised handloom)
AusAID	Australian Agency for International Development
Bakosurtanal	National Coordinating Agency for Survey and Mapping
Bapepam	Badan Pelaksana Pasar Modal (Stock Exchange Supervisory Agency)
Bappenas	National Development Planning Board
Batan	National Atomic Energy Agency
BI	Bank Indonesia (the central bank)
BIDA	Batam Island Development Authority

Bimas	Bimbingan Massal (former nationwide rice intensification program, including subsidised loans)
BKPM	Badan Koordinasi Penanaman Modal (Investment Coordinating Board)
BPIS	Badan Pengelola Industri Strategis (Agency for the Management of Strategic Industries)
BPPI	Agency for Industrial Research and Development
BPPT	Badan Pengkajian dan Penerapan Teknologi (Agency for the Assessment and Application of Technology)
BPS	Biro Pusat Statistik, recently changed to Badan Pusat Statistik (Central Bureau of Statistics)
Bulog	Badan Urusan Logistik (national food logistics agency)
chaebol	Korean business conglomerates
COSTAI	Collaboration on Science and Technology between Australia and Indonesia
CRC	Cooperative Research Centre
CSIRO	Commonwealth Scientific and Industrial Research Organisaton
Dati II	Daerah Tingkat II (second level region, that is, *kabupaten/kotamadya*)
DEET	Department of Employment, Education and Training (Australia) (now DEETYA)
DEETYA	Department of Employment, Education, Training and Youth Affairs (Australia)
Depkop	Departemen Kooperasi dan Pembinaan Pengusaha Kecil (Department of Cooperatives and Development of Small Entrepreneurs)
DIFF	Development Import Finance Facility
DIST	Department of Industry, Science and Tourism (Australia)
DKI	Daerah Khusus Ibukota (Special Capital Region)
DRN	Dewan Riset Nasional (National Research Council)
DSN	Dewan Standardisasi Nasional (National Standards Council)

DSP	Daftar Skala Prioritas (Investment Priority List)
DSTO	Defence Science Technology Organisation (Australia)
DSTP	PT Dua Satu Tiga Puluh (PT Twenty-one Thirty, manufacturer of the N-2130 aircraft)
EDB	Economic Development Board (Singapore)
EFIC	Export Finance Insurance Corporation
FAA	Federal Aviation Administration (US)
FDI	foreign direct investment
fob	free on board
GBHN	Garis-garis Besar Haluan Negara (Broad Guidelines of State Policy)
GDP	gross domestic product
Golkar	Golongan Karya (government political party under the New Order)
golput	*golongan putih* (protest non-participation vote)
GSLP	Government Sector Linkages Program (operated by AusAID)
HIA	Housing Industry Association (Australia)
hp	horsepower
HRD	human resource development
ICMI	Ikatan Cendekiawan Muslimin Indonesia (Indonesian Association of Muslim Intellectuals)
IMF	International Monetary Fund
IPTN	Industri Pesawat Terbang Nusantara (the state-owned aircraft manufacturing factory in Bandung)
ISTP	International Science and Technology Program, operated by DIST
IT	information technology
ITD	industrial technology development
kabupaten	subprovincial administrative region
KADI	Komite Anti Dumping Indonesia (Indonesian Committee for Anti-dumping)
Kadin	Kamar Dagang dan Industri (Chamber of Commerce and Industry)
kecamaten	subdistrict

kemitraan	partnership schemes, strategic alliances
kepala desa	village head
Kepres	Presidential Decree
Kodam	Regional Military Command
Kopassus	Special Forces Command
Kostrad	Army Strategic Command
KUK	Kredit Usaha Kecil (Small Business Loans)
Lapan	National Institute of Aeronautics and Space
Lemhannas	Armed Forces National Resilience Institute
Lemtek	Lembaga Teknologi (Technology Institute, University of Indonesia)
LIPI	Lembaga Ilmu Pengetahuan Indonesia (Indonesian Institute of Sciences)
LNG	liquefied natural gas
LPND	Lembaga Pemerintah Non-Departemen (non-departmental government institute)
M1	narrow money
M2	broad money
MBA	Master Builders Association (Australia)
Menristek	Office of the Minister of State for Research and Technology
MNC	multinational corporation
MOIT	Ministry of Industry and Trade
MPR	Majelis Perwakilan Rakyat (People's Consultative Assembly)
MREP	Marine Resource Environment Project
MSTQ	metrology, standards, testing and quality assurance
NAM	Non-aligned Movement
NIE	newly industrialising economy
NGO	non-government organisation
NU	Nahdatul Ulama (major, mainly Java-based Islamic organisation)
OECD	Organisation for Economic Cooperation and Development
OEM	original equipment manufacturing

Pancasila	the five guiding principles of the Indonesian state
PDI	Partai Demokrasi Indonesia (Indonesian Democracy Party)
PET	Perusahaan Eksportir Tertentu (registered export companies)
PLN	Perusahaan Listrik Negara (State Electricity Company)
PMDN	Penanaman Modal Dalam Negeri (domestic investment)
PP	Peraturan Pemerintah (Government Decree)
ppm	picks per minute
PPN	*produktivitas prestasi nasional* (national performance productivity)
PPP	Partai Persatuan Perbangunan (United Development Party, the Islamic political grouping)
PRD	Partai Rakyat Demokrasi (People's Democratic Party)
pribumi	indigenous ethnic groups
PSLP	Private Sector Linkages Program (operated by AusAID)
Puspiptek	National Centre for Science and Technology Development
R&D	research and development
RDE	research, development and engineering
Repelita	Rencana Pembangunan Lima Tahun (Five-Year Development Plan)
rpm	revolutions per minute
RUK	Program Riset Unggulan Kemitraan (Priority Partnership Research Program)
RUSNAS	Riset Unggulan Kemitraan Nasional (National Priority Research Partnership)
SBI	Sertifikat Bank Indonesia (Bank Indonesia Certificates)
SBPU	Surat Berharga Pasar Uang (Money Market Securities)
Sekneg	Sekretariat Negara (Minister/State Secretary)
SITC	Standard International Trade Classification

SMEs	small and medium-sized enterprises
SPRU	Science Policy Research Unit
S&T	science and technology
STAID	Science and Technology for Industrial Development
STP	Science and Technology Policy (Project)
TFP	total factor productivity
TIPSE	Tim Pengkajian Strategi Ekspor (Export Strategy Review Team)
TISA	technology-based international strategic alliance
UNCTAD	United Nations Conference on Trade and Development
UNESCO	United Nations Educational, Scientific and Cultural Organisation
unit desa	village unit
UU	Undang-Undang (Law)
WTO	World Trade Organisation
Yantek	Pelayanan Jasa Teknologi (Technological Service Provision)

CURRENCY

$	US dollar
A$	Australian dollar
DM	Deutschmark
Rp	rupiah
S$	Singapore dollar
¥	yen

1

INTRODUCTION

Hal Hill*

THE YEAR IN SUMMARY: TURBULENCE AND UNCERTAINTY

The year 1997 will undoubtedly go down in history as Indonesia's most difficult since 1966 and in some respects since Independence. The twin pillars of the New Order—political stability and strong economic growth—have been challenged as never before. It is now almost certainly the case that the regime has changed irrevocably: there is little prospect of a quick return to strong economic growth, and the probability of a smooth and painless leadership transition is much diminished. It is precisely the conjunction of serious economic and political problems, together with escalating social tensions, which is at the nub of the current malaise. A strong, credible and decisive political leadership would be able to tackle the foreign exchange and financial crisis; the external shocks of the 1980s were in some respects almost as severe, and these were overcome quite quickly. Alternatively, a buoyant economy would smooth the path of political succession in the sense that all major parties with a stake

---

*In preparing this chapter, I wish to acknowledge in particular the assistance of my co-editor, Dr Thee Kian Wie, together with the contribution of our writers, on whose papers I have freely drawn, and of participants at the Indonesia Update Conference, who contributed to several high-quality discussions. I have also drawn on my own writings on technology and industrialisation (Hill 1997) and those of Thee Kian Wie (Chapter 6 in this volume, and 1997).

in the outcome would want to preserve the essential ingredients of success. Instead, Indonesia has to deal with a nasty cocktail of economic decline, rising impoverishment and political uncertainty. Scenarios which only a year ago would have been considered very much 'worst case' outcomes now appear increasingly likely.

In 1996–98 Indonesia has been the focus of world attention to an unprecedented extent. This period of unwelcome publicity began with the ruthless manipulation of the opposition Indonesian Democracy Party (PDI) and the subsequent crushing of political dissent in June–July 1996. In late 1996 and early 1997, a series of nasty incidents with unpleasant ethnic undercurrents occurred in several mid-sized Javanese cities. The unrest then continued with extremely violent ethnic conflicts in the province of West Kalimantan, mostly involving the indigenous Dayak community and long-term settlers from the island of Madura. In May the formal electoral process got under way. The predictable outcome, of a powerful government party overwhelming an emasculated opposition, was accompanied by unprecedented violence. In his comprehensive analysis of these events, Geoff Forrester in Chapter 2 characterises the 1997 campaign as the 'most turbulent ever' of the six that have been conducted during the New Order period. Forrester also chronicles the meticulous preparation for the March 1998 meeting of the People's Consultative Assembly (MPR), which meets every five years to elect the President and Vice-President.

By mid 1997 attention began to shift to the catastrophic forest fires engulfing much of Sumatra and Kalimantan. For a period the resulting haze spilled over to neighbouring Malaysia and Singapore. The government was very slow to respond—it was not until late October that the licences of several offending timber companies were revoked. Less publicised, though perhaps more serious in its social impact, the extended dry season not only exacerbated the problem of indiscriminate forest burning and clearing, but also disrupted agricultural production. The effects of the fires on the health of local communities, particularly children, have yet to be carefully documented, but they will almost certainly be extremely adverse. In distant, predominantly agrarian communities with poor distribution systems, daunting problems of widespread food shortages, and in some cases starvation, have emerged. In addition to all this, the country experienced its worst ever civil aviation disaster in August; there was a serious kidnapping in Irian Jaya earlier in the year; the government crushed youth dissent with heavy-handed prison sentences; simmering discontent continued in outlying provinces such as

East Timor and Irian Jaya; cronyism centred around the country's 'first family' became ever more visible and unpopular; the notion of a widening gap between rich and poor, though without clear empirical support one way or the other, was almost universally held to be true; anecdotal information suggested that corruption may have become even more prevalent; there was an apparent increase in crime and violence in both large cities and smaller towns; and there was widespread frustration that the much discussed trend towards *keterbukaan* [political liberalisation] of the early 1990s had failed to materialise.

However, the lull in political activity for several months after the election appeared to confirm the view of many that the political bases of the regime were invincible, and that it would be business as usual in a heavily managed MPR meeting. Moreover, these mounting political, social and environmental problems, while extremely serious, were nevertheless set against a backdrop of vigorous economic performance. As Sri Mulyani Indrawati documents in Chapter 3, while there was a large unfinished policy reform agenda, most economic indicators through to the middle of 1997 were sound: growth and investment were strong; inflation was declining; fiscal policy (notwithstanding serious off-budget problems) was relatively tight; there was a modest current account deficit; and international debt (including some prepayment of liabilities) was manageable. Critics of the government, both at home and abroad, recognised that the regime continued to generate reasonably broad-based improvements in living standards. As long as this process held up, it was widely assumed that President Soeharto would remain in power for as long as he wished, and that the transition to a similar sort of leadership and system of governance would be a smooth one. Optimists hoped that an expanding middle class and the growing internationalisation of virtually all facets of the Indonesian economy and society would gradually lead to a more open and responsive political system, a more independent press and judiciary, and a cleaner, more accountable bureaucracy.

The initial effects of Southeast Asia's financial crisis on Indonesia appeared to be quite mild. In response to persistent speculation, the Thai government finally floated the baht on 2 July, but the ripple effects elsewhere were, for a short period, modest. Tubagus (1997, p. 8) expresses the generally prevailing sentiment that Indonesia was 'not Thailand' for a number of reasons: Indonesia had liberalised its financial sector earlier and more completely; its political system appeared more stable; and its current account deficit was lower. Even when the ripples became larger, a sense of cautious optimism continued to prevail. In response to speculative pressure, on 14 August the government announced that the rupiah would be floated, albeit with continuing heavy indirect intervention through interest rate policy. The government was also quick to announce that it was inviting the International Monetary Fund (IMF) for detailed policy consultations and advice. Through to late September, notwithstanding rising apprehension in some circles, the cautious optimism persisted. The devaluation at this stage, of the order of about 33%, was not large by Indonesian standards of the past 20 years and, as Ross McLeod shows in Chapter 4 of this volume, past devaluations of this magnitude have invariably been associated with a lagged acceleration in economic growth. Moreover, many subscribed to the notion that the buffeting of the rupiah would constitute a 'wake-up call' for the government, entrenching the power of the technocrats, reversing the drift in microeconomic reform, curbing palace-based cronyism, and ensuring that conservative fiscal policy remained in place.

On 23 September, the Finance Minister announced that several major planned infrastructure projects would be shelved or reviewed. On 8 October the IMF and the World Bank were formally approached, and Professor Widjojo Nitisastro became actively involved in coordinating the crisis management effort. Following these negotiations, a package of assistance and reform was released on 31 October, including the provision of a stand-by facility of up to $43 billion. The next day it was stated that some 16 'unsound' banks were to be liquidated immediately. During most of October the rupiah remained in the range Rp 3,500–4,000/$1, which at that stage seemed to be the lower limit of any likely currency movement. In early November, buoyed by these announcements, and with Bank Indonesia intervention, the rupiah began to appreciate slightly, and for a period it seemed as though the worst might be over.

However, the situation began to deteriorate quite sharply from this point onwards.[1] As foreign debt became due, there was a scramble for dollars. The government appeared to backtrack on its commitments to the IMF: one of the President's children whose bank had been closed down was permitted to resume operations under another guise, and some new politically connected megaprojects were

---

[1]For a comprehensive analysis of events over this period, see Soesastro and Basri (1998).

approved. As Korea's crisis worsened, there were renewed region-wide concerns. Thus the rupiah began its free-fall from about mid November. Rumours about the President's ill health—a possible stroke—in early December caused panic. On 6 January, the government's budget was greeted negatively, in particular its seemingly unrealistic growth, inflation and exchange rate targets (of 4%, 9% and Rp 4,000 respectively). Shortly afterwards, a report in the *Washington Post* newspaper quoted anonymous senior IMF officials as being unhappy with the Indonesian government's handling of the crisis. The two events pushed the rupiah below the psychological threshold value of 10,000, which in turn triggered widespread panic buying of essential goods. The further deterioration resulted in a second agreement with the IMF, signed on 15 January, which committed the country to comprehensive macroeconomic stabilisation and microeconomic reform. This too was greeted with scepticism by the markets, especially when followed shortly afterwards by reports that B.J. Habibie would be a strong contender for the position of Vice-President. On very thin trading, the rupiah plunged to a record low of Rp 17,000/$1 on 22 January. Since then there has been modest, though erratic, improvement, towards the range 7,000–10,000. A banking rehabilitation package was announced on 22 January, together with a proposal aimed at encouraging debt renegotiations, while in mid February the government announced its intention to establish a controversial currency board system, in the face of stiff opposition not only from the IMF and international donors, but also from most informed Indonesian commentators.

It is too early to provide an authoritative assessment of the socioeconomic dimensions of the crisis. For the year through to 28 January 1998, the rupiah's decline against the US dollar was the greatest among the 25 major low and middle-income comparator countries listed by *The Economist* (31 January 1998), and well above that of the other two Asian cases of IMF intervention, South Korea and Thailand. The decline in the country's stock market over this period, measured in US dollar terms, was also the greatest in the comparison. Some companies, especially those known to be in extreme financial distress or tainted politically, have seen their stock market capitalisation slashed dramatically. Indonesia's debt ratings have fallen below 'junk' status. As the devaluations feed through to the domestic economy, and the government is unable to sustain food subsidies, inflation is rising sharply; it may even approach 100% or more on an annualised basis for a short period. There are no very recent, reliable estimates of economic growth for early 1998, but previous forecasts of zero or slightly negative numbers now seem excessively optimistic. Several, admittedly very approximate, estimates suggest that unemployment rates are increasing sharply, although the most serious manifestation of the problem will almost certainly be declining incomes as workers displaced from the formal sector attempt to eke out an existence in the informal sector. One million construction workers, and possibly many more, are thought to have lost their jobs. Much of the manufacturing sector has come to a standstill. Compounding these problems, rice production is declining because of below average rainfall; in cumulative terms, for the years 1997–98 the decline may be as much as 7–8%.

Thus Indonesia's problems are exceptionally serious by any historical or contemporary standard. Its financial and economic crisis, as noted, is worse than that of either South Korea or Thailand, and certainly much deeper than in any other Southeast Asian nation. At this stage, its problems even overshadow those of Mexico during the Mexican crisis of 1994–95, at least as measured by the exchange rate collapse. Such an outcome is contrary to all expectations, since by almost any indicator Indonesia appeared in better shape than Mexico prior to their crises.

The events of 1997–98 also differ sharply from those of 1985–86, when Indonesia experienced its most recent serious external shock, albeit of different origins. Then, a powerful and decisive president led an able and united team of technocrats, unencumbered by any significant vested commercial interests, in close cooperation with the international donor community and facilitated by a debt profile dominated by long-term official loans at concessional interest rates, through a process of economic reform which quickly triggered recovery. Every one of these key variables is either not present or much weaker in 1997–98; Indonesia now has a less secure president, powerful cronies opposed to reforms that might damage their own commercial interests, a less cohesive and effective team of key economic advisers, and a huge short-term, unhedged commercial debt.

This is not the place to attempt to explain in any detail the events since mid 1997. They are highly complex, virtually without precedent, and were foreseen by no serious analyst. But it is worth drawing attention briefly to the major factors. First, there was the contagion effect of the Thai crisis. Even if this had not occurred, it is possible that nervousness in the international financial community might have triggered a speculative attack on the rupiah. But it is by no means certain, and therefore Thailand must be regarded as the initial causal factor. Once Indonesia became the subject of speculative attack, three sets of financial factors rendered the country vulnerable. One was the existence of large, unhedged borrowings in international currencies, a situation which arose from the substantial differential between rupiah and dollar lending rates, and the managed stability of the rupiah–dollar rate (which in real effective terms was almost constant) for about a decade. The second was the composition of recent capital inflows, and in particular the growing proportion of highly mobile short-term foreign debt. The third was the conduit for a significant proportion of the loans—a banking system which, while substantially deregulated, was poorly monitored and supervised, and characterised by substantial intra-conglomerate transactions; it was thus a system whose financial reports were not highly credible. In addition to these four factors, there was the more general consideration of regime credibility. Here Indonesia has scored highly for long, but at the most crucial of periods it began to falter seriously. There were serious doubts about President Soeharto's commitment to far-reaching reform, his physical capacity to remain in power, his sources of economic policy advice, and likely presidential successors. All these problems were exacerbated by a number of secondary factors—such as falling rice production, low growth in East Asia, particular nervousness in the Chinese business community and continuing problems of corruption. But the five points adumbrated above do seem to be the central explanatory variables.

What of the future? In such a volatile situation, in which social tension, economic decline and political turbulence are interacting in a potentially explosive manner, it is impossible to forecast likely outcomes. It is certain that the 30 years of near-uninterrupted high growth has come to an end, at least for the medium term, and that Soeharto's legitimacy as a powerful leader capable of delivering strong growth has been damaged beyond repair. Thus, in unexpectedly rapid fashion, we have witnessed the demise of the New Order. One hesitates to conclude on an entirely negative note, however. Modern Indonesia has endured, and recovered from, deep traumas in the past: the struggle for Independence and the sometimes painful birth of the nation state; massive political upheaval and hyperinflation in 1964–66; the Pertamina debacle (involving debts that at one stage were equivalent to about 30% of GDP) in 1974–75; and the collapse of oil prices in 1985–86. On each occasion, national resilience eventually triumphed over despair. There is no technical reason why the current economic disaster cannot be overcome. But, as

in the previous crises, the key is a credible political regime committed to recovery and willing and able to put aside narrow sectional
interests in the pursuit of a common good. The new cabinet, including
most of all the President and Vice-President, will have to measure
up to these formidable challenges or risk terrible socioeconomic
consequences, not to mention the harsh judgement of history.

TECHNOLOGY ISSUES

The country's current economic and political crisis has overshadowed
practically all other issues, including the theme of this book,
'Indonesia's technological challenge'. It is important to bear this
point in mind: topics such as these are of 'second order' importance,
deserving examination only in an environment of political stability
and economic growth. If there is one clear lesson to be learned from
the current crisis, it is this fundamental yet frequently overlooked
proposition. One of the dangers of a long period of sustained economic
development, such as Indonesia experienced for 30 years prior to the
middle of 1997, was that it had engendered a sense of complacency in
which it was tacitly assumed in many quarters that rapid growth
was the natural order of things regardless of policy mistakes, political disturbance or even gross corruption. 'Single issue' advocates—
whether it be of the promotion of technology, investment in infrastructure, education and health policy, industry policy, rural and
regional development, or poverty alleviation—ignore the larger
macroeconomic and political picture at their peril, however important these micro issues may be.

With this caveat in mind, why study technology? First, and
most important, technological innovation and diffusion is a significant component of Indonesia's socioeconomic development. A study of
the changes in the technologies employed, and how, why and for
what purpose, sheds light on broader processes of change in the
country. Perhaps more than any other facet of development and
public policy, technology requires an understanding of the interaction between long-term economic development, the internationalisation of an economy, the mechanics of public policy and the
microeconomics of firm-level innovation. These are the major themes
to be studied in this volume.

Technology is important obviously because it is central to the
process of raising productivity and therefore living standards. One
only has to consider the state of production and distribution

technologies in Indonesia now and 30 years ago to appreciate the importance of this point. Although, as will be argued below, it is very difficult to measure this process of change, it is clear that a technological revolution has swept through the country in recent decades. In this context, one has to pose the question of whether these changes have been the outcome of the country's rapid economic growth—an incidental by-product of the policies which propelled the high growth—or whether they have been in some way the result of deliberate technology policy interventions.

The latter question leads on to a third justification for examining the issue. Technology has been one of the most widely debated questions in Indonesia over the past decade. Professor B.J. Habibie, Indonesia's Minister of State for Research and Technology, has been one of the most influential cabinet members for nearly two decades, and his views and programs are highly controversial. In Indonesia, as in nearly all countries, 'technology policy' remains an ill-defined area. Within the narrow confines of the Ministry there is arguably a clearly defined strategy, although even here the gap between rhetoric and implementation remains considerable. But the range of policies that impinge on technology development, broadly defined, go well beyond these confines, and in this sense, in Indonesia as in many countries, it is difficult to discern any systematic national strategy. This conundrum reflects in part differences in policy approaches (the much discussed debate between 'technologs' and 'technocrats', which seems to have lost some of its sting in recent years) and in part the analytical foundations concerning the presence of market failure and, as a corollary, the case for government intervention.

A related policy-based justification for studying the topic is that Indonesia's budget for technology, though quite small ($400 million at the pre-crisis exchange rate), is projected to rise quite fast. Unless the current crisis derails the strategy, the country could be spending 2% of GDP on technology programs 20 years from now, while in toto the effective reach of such a strategy into other areas of public policy could be much greater still. Programs of this scale warrant attention.

It would be obvious even to the most casual observer of Indonesia that there has been enormous technological development, broadly defined, over the past 30 years. Consider the following snapshot profiles of daily life then and now (at least pre-crisis). In the mid 1960s, motorised transport was not commonly used by the mass of the people, who travelled around cities and towns by bicycle, by *becak* or on foot. Few people had travelled beyond their province of birth; far fewer yet had ever travelled by air. A small percentage of households, mostly confined to the larger cities, had electricity. The reach of newspapers and modern forms of communication was restricted in the main to metropolitan areas. Most Indonesians ate unprocessed food and wore cloth produced on handlooms; foreign consumer goods were a rarity. Most left school before completing their primary education. If they fell ill, it was unlikely that they would have access to modern pharmaceuticals and hospital services; indeed, about one-eighth of Indonesians did not live to see their first birthday and one-quarter did not celebrate their fifteenth birthday.

Production technologies were predominantly simple, mostly involving manual labour. There was a pervasive 'technological' pessimism, on the basis of which originated the belief that the country was unlikely ever to enter an era of modern technological development. In the food sector, the green revolution, which was then beginning to sweep through Asia, had barely caused a ripple in Indonesia. The long-established cash crop sector had displayed little capacity for technological innovation. Indeed, a leading agricultural researcher of that period, David Penny (1969, p. 263), argued that '... the reluctance of farmers to buy fertiliser, modern tools, etc. is still so great that it is unlikely that any substantial modernisation of Indonesian peasant agriculture will take place in the next decade or two'. In manufacturing industry, the country had been starved of new technology, partly because in the low-growth environment there was little inducement to invest in fixed capital, and partly because there had been only a trickle of new foreign investment outside the petroleum sector since the late 1930s. What little new technology there was came mainly from the technologically backward Comecon block, for investment projects which were in any case highly politicised (Gibson 1966a, 1966b; Siahaan 1996).

Indonesia attracted international scholarly attention over this period, not only because of its poor economic performance, but also because it was the laboratory from which much thinking about 'dualism' had emerged. Boeke had pioneered the concept in the 1930s, invoking cultural and social explanations for the phenomenon. In post-Independence Indonesia these were regarded as unsatisfactory, and by the late 1950s an alternative model, of 'technological dualism', had emerged, particularly associated with the work of Benjamin Higgins (see, for example, Higgins 1956). He accepted the notion of dualism, particularly the presence of a foreign enclave, differing in its technology, factor proportions, ownership and market orientation, alongside the indigenous sector. However, he contested Boeke's formulation, and saw the explanation as lying rather in the fixed technical coefficients and capital-intensive production modes of the modern sector, in contrast to the flexible coefficients found in the traditional sector.

The comparisons between the 1960s and the pre-crisis 1990s are stark indeed. Though Indonesia is still a relatively poor nation, the daily lives of its people have been transformed beyond recognition, and 'technology' has played a key role in this process. People's consumption patterns and modes of dress have changed greatly. Personal mobility has risen dramatically: all major cities can be reached from Jakarta in a day; highways have been carved through the main islands; intra and intercity motorised road transport has expanded exponentially. Information networks have grown so fast that the point has now been reached where the government has effectively lost its capacity to control much of what its urban citizens read. Indonesia is a much more cosmopolitan society than it was at the beginning of the New Order, and not just because one can buy pizza or sushi even in small cities. More Indonesians travel abroad, and more foreigners visit and reside in Indonesia than ever before. The achievement of universal mid-secondary education is within reach, and people are living longer.

Openness to the international economy and high levels of investment have transformed production frontiers. The introduction of high-yielding varieties has meant that the country has generally been able to feed itself, while most of the products now produced by its factories were simply unavailable 30 years ago. Technology has brought the country together, spatially and economically, if not necessarily politically. The revolution in transport and communication has meant that an entity which was not much more than a political construct now functions as a national economy for goods, services and labour. Enormous differences in technology persist, but the overarching divide implicit in the concept of technological dualism is no longer present. Instead, one observes a continuum, with considerable mobility along it.

It would be a mistake to argue that these rapid technological developments are necessarily a good thing. Some—indeed most—are, as revealed in the significant improvements in living standards over this period. But obviously not all are. The air quality in Indonesia's cities is undoubtedly poorer than it was in the 1960s, the forest cover is much thinner, and soil erosion and fertility are more serious challenges than ever before. Here technology is both part of the problem and part of the solution. Rapid industrialisation and urbanisation have had serious negative side-effects. But technology, which is one of the keys to rising affluence, may also provide the capacity to overcome many of these problems. One of the challenges for Indonesia, as in all countries, is to employ technology to generate sustained improvements in living standards, the latter defined to take account of both the 'goods' and the 'bads' resulting from rapid growth.

The issue of technology development and policy in Indonesia, and the contributions to this book, may usefully be examined under the following broad headings.

First, what are 'technology', 'R&D' and 'technology policy'? How do we define them, and are there suitable empirical proxies for these analytical and conceptual constructs?

Second, and related to the first issue, is there a case for government intervention in technology markets, and if so, what are the analytical underpinnings of market failure and what sort of intervention might be suggested by this approach? What light does East Asian and other countries' experiences shed on these questions for latecomer countries such as Indonesia?

Third, what is the Indonesian government's approach to technology policy, both in the realm of philosophical foundations and practical implementation strategies? What institutions, in both the public and private sectors, have been established to sustain the country's technological effort?

A fourth topic for consideration concerns quantitative indicators of Indonesia's technological development in comparative international perspective.

Fifth, there are the international dimensions: the mechanisms and magnitudes of technology flows, the special role of multinational corporations (MNCs) in these flows, and the growing globalisation of technology flows.

A sixth issue focuses on the microeconomics of technology development and innovation. This is an important but neglected area of research; many of the really interesting issues are micro—industry or even firm-level—in nature, and secondary data often tell us rather little about these micro processes.

Finally, there is an examination of Indonesia's policy environment, to place technology policies in a broader political economy context and to evaluate interventions in education, international trade and investment, as well as those with an explicit technology mission.

Some Definitions

There is no universally accepted definition of technology and technology progress. The most common approaches define technology as 'a collection of physical processes that transform inputs into outputs and knowledge and skills that structure the activities involved in carrying out these transformations' (Kim 1997, p. 4). More formally, one may regard the production isoquant as embodying the technological options available at a given point of time, based on the existing stock of scientific knowhow. Technology progress may then be regarded as a 'better way of doing things' or of 'producing more from less', by employing new technologies and generating new products and processes. All the contributors to this volume work within a framework that accepts a broadly defined notion of technological progress according to a range of quantitative or qualitative indicators.

Partly because of these definitional difficulties, there is no widely used summary measure of technological competence analogous to, for example, GDP as a single measure of a country's level of economic activity. All technology indicators have serious conceptual and empirical limitations, which explains why a basket of such measures is typically employed in making comparisons over time and across countries, and also why some of the really important insights have originated from case study material. Indeed, the best studies of technology—in a tradition to which this book aspires—are those which combine analytical insights from both macro (economy-wide) and micro studies.

The usual approaches, adopted also in this volume, include both 'input' and 'output' measures. The former focus on resources devoted to the development of technological capacity, such as expenditure on R&D and investment in human capital. The latter concentrate on outcomes of this capacity, and include such indicators as patents and the proportion of output or exports originating from 'technology-intensive' activities. Many of these indicators are empirically slippery, however. Their limitations are well known, and are discussed in several chapters of this volume. But it may be useful to illustrate the common empirical pitfalls immediately by way of a few illustrations.

- Total factor productivity (TFP) is often used as a proxy for technology progress. Yet technologically dynamic Singapore consistently records TFP growth that is very low by East Asian (and global) standards.

- Foreign direct investment (FDI) flows and stocks are often used as an indicator of technological capability. But Korea, a very modest recipient of FDI, has one of the strongest technical bases among developing countries, while Malaysia, with a much more modest technological base, has been one of the largest recipients of FDI.

- Export intensity indicators are often used to gauge technological rankings (based on the percentage of exports considered 'technology intensive'). But these sometimes result in Malaysia being ranked ahead of Japan, and Thailand close to the US.

- R&D expenditure (as a percentage of GDP) is still probably the most widely used indicator of technical sophistication. But it ignores the international and domestic context in which R&D activities take place. A strong domestic R&D base in a closed, distorted economy (such as the former USSR) does not provide a durable and dynamic technological advantage.

The point of these illustrations is not to criticise the use of such indicators. (Apart from anything else, this would entail a healthy dose of self-criticism!) Rather, it is to illustrate the importance of a diverse range of macro and micro indicators in building up a composite picture of a country's technological capabilities, and to plead for great caution in the interpretation of any one set of empirical estimates.

One final definitional issue needs to be emphasised. The term 'R&D' often carries with it grand connotations of research at the frontiers of knowledge achieving major scientific breakthroughs. In reality, much of it involves not the 'R' of fundamental research, in the sense of inventing products and processes, but rather the 'D' of acquiring, adapting and modifying frontier technology. Even in the US, it is estimated that just 8% of the country's R&D expenditure is devoted to 'basic research', far less than is directed toward 'applied research' (25%) or 'development' activities (67%) (Rosenberg 1994, p. 13). In a developing country context, the proportion devoted to basic or applied research would certainly be lower still. In the 'R&D&D' formulation, 'development' and 'diffusion' are the major activities, and the ones to which public policy needs to be directed.

Analytics and Country Experience: A Role for Government?

The principal economic rationale for government intervention to develop a country's technological capability has to do with the latter's 'public good' characteristics (see, for example, Lall, Chapter 7; Pack and Westphal 1986). Left to the market, there is likely to be underinvestment in activities for which private agents are unable to appropriate adequately the returns from their investments. Such investments tend to enter the public domain quickly, a process facilitated in part by classic 'freerider' problems. The returns are diluted, especially in developing countries, by a weak legal system that is unable to protect intellectual property rights. High interfirm worker mobility, especially for those whose skills are in short supply, exacerbates the problem, since firms are reluctant to invest in R&D programs if there is a high probability that key staff embodying these investments will quit before returns can be appropriated.

These problems are all the more serious in an economic and political environment characterised by great uncertainty. In the best of environments, investments in R&D are highly uncertain and generally slow yielding. Where private agents heavily discount future earnings, whether owing to high real interest rates, lack of credible macroeconomic management or political uncertainty, the incentives to invest will be weaker yet. Thus, while reported rates of return on industrial R&D in the few studies to investigate the subject in developing countries (some of which are summarised in Evenson and Westphal 1994, p. 89) are high, these factors probably explain the limited scale of investment. The inability of firms to appropriate fully the returns on these investments results in social rates exceeding the private rates, and hence provides a justification for government intervention.

Other forms of market failure provide a further rationale for government intervention (see, for example, Stiglitz 1996). Where markets are underdeveloped—incomplete or missing—price signals do not function adequately. This may arise because of poor information flows, limited entrepreneurial capacity, inadequate property rights or physical infrastructure bottlenecks. Institutions (such as producer cooperatives or industry associations) that foster the cooperative behaviour sometimes needed to achieve improved outcomes may not exist. Coordination functions, especially in dealing with international markets, may be poorly developed. The cooperative and coordination functions do not necessarily need to be supplied by government, but public policy can hasten their efficient evolution.

The new growth theories, with their emphasis on the determinants of long-term growth rates, have revived interest in technological change and technology policy. 'The suspicion that [growth differentials among countries] may have something to do with technology has been around for a long time', as Fagerberg (1994, p. 1,147) observes. The new theories have shifted attention away from the approach of the earlier neoclassical modelling. However, they have yet to develop powerful policy relevance. But empirical refinement is pushing in this direction, examining, for example, the growth stimulating effects of investments in R&D and schooling, and attempting to measure the R&D spillovers from international trade and investment.[2]

The literature in this field offers some, though limited, guidance on intervention strategies (Lipsey 1997): industry support should be in the form of subsidies rather than tariffs; externalities should be targeted as directly as possible; governments ought to be involved most extensively in the 'pre-competitive research' phase. However, in spite of all the work in this field, there are still major areas of ignorance, especially in terms of explicit policy-oriented advice. If tertiary and vocational education is to be subsidised, what should the rate of the subsidy be, and should it vary across courses? Why does innovation occur more rapidly in some industries than others? How long, typically, is the 'infant industry' phase in various industries? How exactly can bureaucrats be induced to behave in a market-friendly, competitive manner?

The environment in which government intervention occurs may also influence the magnitude of the potential benefits. Here the distinction between 'forward-looking' and 'backward-looking' protection—that is, between protection that attempts to anticipate (and promote) new winners versus protection that simply props up declining industries—is relevant. In most of high-growth East Asia, Indonesia included (albeit with some egregious exceptions to be discussed below), import protection has more commonly been of the former variety, with the prospect—not always realised—of eventual liberalisation. As Lipsey (1997, pp. 105–106) observes, although economists are in general sceptical of the efficacy of selective industrial policies, such a strategy may have more chance of success in East Asia, where '... governments are mainly concerned with encouraging growth rather than brokering conflicting special and regional interests as in the USA and Canada. Also, the export orientation gives the policies a cut-off point that substitutes for the

---

[2]For a recent example of these approaches, with references to the earlier literature, see Helpman (1997).

bottom line in the private sector and prevents failed initiatives from persisting indefinitely ...'.[3]

In addition to this analytical literature, as an industrial late-comer Indonesia can benefit from the experience of the more advanced East Asian economies. The policy environments are so diverse and the case study material so vast that one hesitates to attempt to distil just a few major conclusions from them.[4] In any case, as Hadi Soesastro (Chapter 16) points out, Indonesia should not feel compelled to follow a particular model of technology development. But three general conclusions are worth emphasising.

Lessons from Korea

Arguably no country's industrialisation experience is more closely studied and admired in Indonesia than that of Korea (at least pre-crisis): its high growth appeals to developmentalists; its outward orientation strikes a chord with internationalists; its restrictive foreign investment and import regimes is attractive to nationalists; and its good equity and education record accords with deeply cherished Indonesian aspirations. Thanks to the seminal text of Kim (1997), on which the following paragraphs draw, we now have a good understanding of Korea's technology development strategies and achievements. In narrow technological terms, its accomplishments have been remarkable: expenditure on R&D as a percentage of GDP rose from 0.32% in 1971 (that is, about 50% higher than the current Indonesian figure) to 2.61% in 1994. Most other input and output technology indicators also display a steep upward trajectory.

Kim emphasises five key ingredients in the Korean experience, some of which are obviously transferable. These include export orientation as a means of ensuring that assisted firms have to meet quickly some sort of market test, a strong commitment to education, and reasonably sound macroeconomic management. It is not obvious,

---

[3]Similarly, economists have long recognised two types of infant industry: (a) where a firm shifts out to the best practice frontier over time, and appropriates the technological improvements itself, in which case there is a commercial argument for a loan but not a public subsidy; and (b) where the technological improvement diffuses beyond the firm or industry, in which case there may be an argument for a public subsidy on externality grounds.

[4]In addition to the literature referred to elsewhere in this chapter, see in particular the comprehensive literature surveys of technology transfer by Enos (1989) and Enos, Lall and Yun (1997).

however, that some of the other lessons are equally transferable. The strategy of promoting the *chaebol* produced mixed blessings; they were both an 'asset and burden', in Kim's words (p. 196). While effective in pooling resources and marshalling inputs, it had great costs in terms of political corruption and neglect of SMEs. The restrictive FDI regime denied Korean firms access to international knowhow. It would be even more costly in a country like Indonesia, which does not have such a strong commitment to scientific education. In any case, it is not obvious that, in the international commercial policy environment of the 1990s, such a restrictive regime would even be possible.

More generally, as Kim emphasises, Korea's political and historical development was unique. The harsh Japanese colonial rule, the Korean civil war and the ever-present threat from the North bequeathed a regime with an unparalleled development commitment, an intrusive, authoritarian political system and a government bureaucracy that actively cajoled firms and individuals to meet highly ambitious targets. It is doubtful whether these elements would (let alone should) be replicable in Indonesia, and whether, given the country's deep ethnic fractures, a strategy of developing conglomerates *à la* Korea would prove resistant to rampant rent seeking behaviour.

Technology Development Is an Evolutionary Process

A key strand in virtually every major study of technology is that the process of acquiring and mastering technical knowhow is *evolutionary, gradual and long term*. Governments can expedite this process in the ways discussed below, but there are no simple short cuts. This may seem so obvious as to hardly require emphasis. But since these tenets are at odds with some official thinking and popular discussion in Indonesia, and with its ambitious high-tech projects, they deserve restatement. The literature on Japan, for example, emphasises that it took that country over a century, from the beginning of the Meiji Restoration in 1868 until the mid 1980s, to reach the frontiers of global technology across many fronts (Hayami 1997).

The recent work of Hobday (1995, forthcoming) on the Malaysian electronics industry is a convenient and powerful illustration of these propositions in the Southeast Asian context. Like all studies in this area, his work confirms the importance of good fundamentals, which he defines as sound macroeconomic management, an international orientation and a good educational base (supplemented by specialised technical and management training institutes). The Malaysian industry is sometimes criticised in Indonesian (and of course Malaysian) circles as being a case of shallow, enclave-based, foreign-dominated industrialisation. However, Hobday points out that there is a good deal of innovation already occurring within this MNC-dominated industry, in response to rising real wages and increased domestic competence. This innovation, he emphasises (forthcoming, p. 24), is '... not radical or R&D-based, but [is] incremental ...'. It is also inextricably linked to domestic capabilities. The industry is sometimes admonished for the slow pace of backward integration, but as Hobday shows, the state of Malaysia's technical vocational education system has in part dictated how fast this process can occur.

Hobday's arguments are consistent with the conclusions of Lipsey (1997, p. 106), whose comments on the perils of ambitious high-tech initiatives are highly relevant to Indonesia:

> ... expectations [of successful intervention] are poor for big technology pushes that require massive changes in the existing facilitating structure or even the development of wholly new structures; the successes have tended to be those that accept the path dependency of technological change, going for significant advances that build on existing strengths and not trying for great leaps in the technological dark that unfortunately seem to attract politicians.

Clever Intervention Can Produce Results

A widely cited success story of clever and nimble intervention is Taiwan's Industrial Technology Research Institute (ITRI), well described by Lin (1998), on which this paragraph draws. Founded in 1973, the institute grew slowly in its first decade and then expanded rapidly; by 1994 it had 6,000 employees, 560 of whom possessed doctorates. ITRI is now the largest industry-oriented research institution in Taiwan, and has working relations with some 20,000 companies. It has received significant public sector funding, absorbing about 25% of the government's non-defence technology projects between 1983 and 1994. Central to its success, according to Lin, were six factors: a strong national human resource base on which to draw; access to international technology markets; a competent domestic R&D base; an emphasis on diffusion and commercialisation of domestic and international knowhow; strong ties to the private sector; and supportive intellectual property rights. As we will see shortly,

Indonesia scores reasonably well on the first, second and sixth of these general lessons. But in the diffusion of domestically generated R&D programs, and their links to the private sector, its record is much weaker.

The Indonesian Approach to Technology

This section summarises Indonesia's approach to technology issues, as enunciated in its long-term development plans, the Repelita (Five-Year Plan) documents and the writings of Minister Habibie, who, as a uniquely powerful and long-serving cabinet member, has been responsible for technology policy for over two decades. The focus here is on the official approach. As emphasised above, in reality 'technology policy' effectively encompasses a much broader array of policy variables than the official definition would suggest. Later in the chapter we offer an evaluation of government policies set against this broader canvass. All chapters in this volume include some reference to policy issues. In this section we concentrate on those concerned directly with official technology policy and its implementation. These include Samaun Samadikun (Chapter 5) and Rony Bishry and Murman Hidayat (Chapter 8), together with the overviews provided by Thee Kian Wie (Chapter 6) and Sanjaya Lall (Chapter 7), and the critique of the Habibie approach by Robert Rice (Chapter 9).

Samaun Samadikun, the former Chair of the Indonesian Institute of Sciences (LIPI), provides an overview of the government's approach to science and technology policy and its philosophical underpinnings. He emphasises the central role of the Ministry of State for Research and Technology (known as Menristek in Indonesia), in addition to the contribution of other government departments, the universities and the private sector. During the term of the Second Long-Term Development Plan (PJP II, 1994–2019), obviously drawn up prior to the current crisis, expenditure on R&D is projected to rise sharply, from around 0.2% of GDP to 2%, a figure well above that of, for example, Singapore presently. It is unclear whether this target would have been achieved even in the absence of the economic crisis. It also needs to be emphasised that the (pre-crisis) government expenditure on R&D, of about $400 million, is dwarfed by the huge investments in the nation's 'showcase' aircraft factory, IPTN, which are thought to be approximately $3 billion.

The official approach combines a commitment to the development of a strong technological base with a recognition that eventually projects must meet some sort of market test, and that FDI will play an important role in the development of indigenous capabilities. However, and reflecting in part the background of the key personalities involved, program objectives also emphasise 'technological self-sufficiency', through a four-stage evolution from basic manufacturing capacity (stage 1) to the mastery of leading edge technologies as embodied in advanced manufacturing (stage 4). (IPTN has obviously progressed farthest along this schema, and is currently regarded as being located in stage 3.) There is quite explicit reference to the need for import protection during this development phase, although the level and duration of this protection are not specified. It is also recognised that Indonesia needs to develop high-level expertise in a range of scientific fields, including biotechnology, energy and environmental sciences, in addition to the advanced engineering knowhow to sustain major industrial investments in the manufacture of aircraft, shipbuilding, telecommunications equipment and other technology-intensive activities.

It is no exaggeration to state that technology policy in Indonesia over the past 20 years has been synonymous with Minister Habibie, and it is therefore appropriate to examine his programs and his thinking carefully. In Chapter 8, two staff members of his powerful Agency for the Assessment and Application of Technology (hereafter referred to by its Indonesian acronym, BPPT), Rony Bishry and Murman Hidayat, provide an overview of the institution's evolution and mission. Established in 1978, BPPT is the major bureaucratic vehicle for implementation of the Minister's objectives. In principle it ranks equally with several other government agencies, including LIPI (see Figure 8.1), but in practice it is one of the most influential non-military agencies in the country. As Bishry and Hidayat demonstrate, practically every major project of scientific and research endeavour is at least partially under its purview, and it possesses much of the most advanced scientific facilities in the country. Nevertheless, although well funded by Indonesian standards, the authors are quick to point out that by international standards its resources are woefully inadequate. Given its highly politicised history, the authors also raise broader questions about BPPT's survival in a post-Habibie era. However, with Habibie as Vice-President, BPPT would continue to thrive and play an even larger role in years to come.

Rice (Chapter 9) provides a careful summary and critique of the Habibie approach to the development of technological capability. Although widely criticised by economists, Habibie is the only senior

Indonesian official to have expounded a systematic intellectual view of the challenges in his portfolio. He rejects orthodox economics approaches with their emphasis on cost–benefit analysis (including some measure of the effects of externalities) and comparative advantage. Instead, he proposes a measure defined as national performance productivity (PPN in Indonesian parlance), which is apparently a ratio of value added to the value of intermediate inputs used in the production process. As Rice shows, Habibie's approach rests on several guiding principles, including a strong educational base, the specification of clear scientific objectives and the development of national R&D capacities so as to minimise reliance on imported technologies.[5] A crucial element in his thinking is the importance of establishing 'vehicle industries' as a means of developing 'hands-on' advanced scientific competence. It is for this reason that the major industrial projects under his Strategic Industries Board (known as BPIS) are central to his vision. Clearly he would not countenance the practice in most countries of a separation between a government department responsible for general science and technology programs, and major industrial investments with a strong technological orientation.

One final observation on Minister Habibie and his programs is pertinent here. He derives his extraordinary influence and power from unparalleled access, and closeness, to President Soeharto, who has known the Minister since his childhood, and who personally invited him back to Indonesia from Germany to a ministerial position that involved establishment of the major projects. It is for this reason that the projects have been largely immune to pressure—from inside and outside the country—that they be subject to the usual checks and balances applied to government expenditure. Moreover, despite growing apprehension at the nature and magnitude of his funding requirements, the Minister is consistently able to draw on a sympathetic current of public support from those attracted to the notion that Indonesia, for much of its history politically downtrodden and technologically backward, will only ever catch up

---

[5]In passing, it might be noted that there are similarities between the Habibie approach and that of the prominent Japanese scholar, Yoshihara Kunio (1988), who has provided a critique of Southeast Asia's 'ersatz' capitalism. One of the bases of Yoshihara's unhappiness with the region's development strategy is its lack of attention to technological foundations, resulting in what he terms its 'technology-less industrialisation'.

through some sort of 'great leap forward' involving massive and ambitious industrial projects.

In addition to this focus on broad guiding principles and major initiatives, several chapters (especially those by Lall and Thee) include a more focused discussion of other government agencies and programs with a significant R&D content. The most important in this context are LIPI, whose Chair reports directly to the President, and the National Centre for Science and Technology Development, Puspiptek, which operates under Menristek. There are also smaller scale activities, in the form of government institutes operating outside the formal reach of Menristek/BPPT. For example, each technical ministry has an R&D division (Balitbang), and most have specialised agencies linked to extension and other services. Large government departments, such as Agriculture, and Industry and Trade, have extensive nationwide networks, some of which, as Colin Barlow points out in Chapter 13, have a long history of quite effective grass-roots interventions. More generally, however, as will be argued below, most of these agencies are poorly resourced, they have weak incentive structures, and their links to demand-driven, private sector activities are very limited.

Indonesia in Comparative International Perspective

The major indicators of Indonesia's technological base, as summarised by Lall (Chapter 7) and Thee (Chapter 6), are well known.[6] Indonesia is very much an industrial latecomer. It is only in the last 30 years that it has been open to international technology markets and has had a government commitment to the development of any sort of scientific base. The broad input and output indicators, as depicted in Tables 6.4 and 7.6, suggest it allocates a little under 0.2% of GDP to formal R&D activities. Such a figure is comparable to its lower income ASEAN neighbours, the Philippines and Thailand, less than half of the Malaysian (and the Chinese) level, and about one-fifth of India's. It is of course well below the figure for Singapore, which is similar to the lower OECD norm of around 1.5%. On a per capita basis, Indonesia's expenditure works out (using pre-crisis exchange rates) at about $1.5.

The government is the dominant actor in formal R&D activity, both as a funder and as an 'implementer'. It provides about 80% of

---

[6]See also the compilation of science and technology indicators for the five original ASEAN member countries presented in ASEAN Secretariat (1997).

the resources, and directly carries out over 60% of the activities. R&D expenditure has constituted about 2% of the government budget over the past decade—somewhat below this figure during the difficult adjustment period of the mid to late 1980s, and a little higher in the mid 1990s (Table 7.5). For various reasons, actual R&D expenditure is higher than that officially reported. During the current economic crisis, the share of this sector in the budget will almost certainly decline.

A large government role in the early stage of industrialisation is not unusual, but international comparisons suggest that the figures for Indonesia are exceptionally high. The comparable figures for the widely admired Korean model are, for example, 17% (of funds) and 4% (of implementation), although indirectly government influence would certainly be much higher. In Singapore the shares are also much lower, at 39% and 23% respectively.[7] These figures presumably reflect the 'passive' nature of most Indonesian firms on R&D issues, the heavy reliance on FDI for leading edge technology, and the activist government approach to technology policy spearheaded by Minister Habibie. A major challenge will be for the government to maintain its program of R&D support while inducing private firms to play a more active role as providers and funders, as well as implementers.

Other indicators also portray Indonesia's latecomer status. Owing to its size, it has the largest stock of scientific personnel in ASEAN, but when adjusted for quality of training, hardware support and active scientific activity, its R&D effort is again much closer to that of the Philippines and Thailand. The labour force data indicate that the proportion of scientific personnel employed in the private sector is higher than the sector's share of R&D expenditure, which suggests that many of these workers are not actively engaged in R&D activity.

Much the same picture holds for 'output' indicators as for the above 'input' indicators of R&D. Indonesia's ranking on various indicators of scientific publications is very low, although its effort here may be understated owing to the fact that only English-

---

[7] As would be expected, Singapore and Korea are much closer to OECD norms in this regard. For the OECD as a whole, in 1993 governments funded 36.2% of R&D activity, but performed only 12.7%. Moreover, there has been a steady decline in public sector involvement; in 1981 the comparable figures were 45% and 15.0% respectively ('Supplement on Universities', The Economist, 4–10 October 1997, p. 12).

language publications are generally recorded. In recent years the amount of patent activity has begun to increase. However, such a figure may be misleading, as in most cases it probably represents the response of firms to the government's introduction in the late 1980s of a patent law, rather than constituting an indication of indigenous technological effort. Over 85% of the patent applications in Indonesia have come from non-residents (ASEAN Secretariat 1997, Table 7.1).

Factor intensity analyses of the structure of Indonesia's exports reveal a similar picture. They show that, as would be expected, the proportion of 'technologically complex' or 'high-tech' exports is very low by comparative East Asian standards. Lall (Chapter 7, Table 7.3) and Soesastro (Chapter 16, Figure 16.2) show how these export composition data may be used in international comparisons. Several other studies have recently employed a similar, albeit less nuanced, approach.[8] Both Lall and Soesastro are careful to attach caveats to the use of such data, which do need to be treated with caution, in particular because much depends on how the region's very large electronics exports are classified.[9]

Finally, mention should be made of various estimates of TFP growth. Most studies suggest that Indonesia's long-term TFP growth has been comparable to East Asian norms (Chen 1997). The most detailed work has been undertaken for the manufacturing sector, and

---

[8]See, for example, Porter et al. (1996) and Ray (1996). Porter et al. construct an index of 'high-tech emphasis', with scores and rankings ranging from 0–100, lowest to highest. Their indices do not appear to have much utility apart from underlining Indonesia's low technology base. Some examples of the scores: Singapore 100, Malaysia 77.4, Japan 67.8, the US 53.2, Thailand 44.2, Indonesia 6.5.

Ray employs a broadly similar analytical schema; his time series estimates for Indonesia lead him to conclude (p. 88) that 'beginning in 1982/83 Indonesia's drive towards higher technology export production lost momentum ...'.

[9]The complication concerning electronics, which in some countries generates over 40% of merchandise exports, is whether to classify the industry as 'labour intensive' or 'technology intensive'. It is one of the few major industries whose factor endowment ranking appears to shift over the course of economic development, from labour intensive in a country like Indonesia to technology intensive in high-income economies. In intermediate cases such as Malaysia (which, owing to the dominance of electronics in its exports, displays a 'technology intensity' ranking far above that of its per capita GDP), the industry is in transition and most likely exhibits both characteristics simultaneously.

here the better database generates more robust conclusions. In particular, liberalisation appears to have resulted in stronger TFP growth, while across industries, foreign and state ownership are, respectively, associated with strong and slow growth (Aswica-hyono, Bird and Hill 1996). These results are at best illustrative: the database is weak; interpretation of residuals is hazardous at the best of times; and there are business cycle effects. But the data are at least consistent with a picture of quite rapid technological progress in Indonesia.

While broadly accurate in the picture they portray, all these quantitative indicators need to be interpreted with great caution. Formal R&D activity is only part of the story, and at Indonesia's stage of development, arguably a small part. The data take no account of the rapid introduction and dissemination of new equipment and ideas over the past 30 years, or of the myriad innovations in production processes and products. Something of the flavour of these changes is provided below. Even the formal database is most incomplete in its enumeration of the type, magnitude and sources of R&D activity. A high priority for Indonesia will be the preparation of a more comprehensive technological balance sheet, of which the STAID (1993) report is a small but important first step.[10]

International Dimensions

International connections critically shape a country's pattern and rate of technological development. This is a vast topic with many dimensions, including the following.

- R&D activity is heavily concentrated in a handful of rich OECD economies (the G7 account for over 95% of global R&D expenditure), and so a capacity to tap into global technology markets quickly and smoothly is the first requirement for a net technology importer such as Indonesia.

- International technology flows are increasing rapidly and, like trade and investment, almost certainly growing faster than global output.

- Owing to this increasing globalisation of technology flows, national technology policies need to be harmonised as never before with trade and investment policy regimes. A strategy

---

[10]An updated version of this report in Indonesian was published as BPPT et al. (1995).

which emphasises autarchy in one of these areas will almost certainly have counterproductive effects in the other areas.

- There is a steadily increasing array of suppliers in the international technology market place, although in selected technology-intensive activities markets are still oligopolistic in structure, and information flows are inherently imperfect.

- Much of the discussion focuses on the international transfer of technology, whereas diffusion within the host country is equally if not more important.

- More so than any other form of major international commercial transaction, technology flows are poorly documented, not only because of conceptual difficulties in defining just what constitutes a 'flow of technological services', but also because of empirical limitations. In most cases, we have practically no means of quantifying these flows in any sort of detail in aggregate, much less by source country, channel or the recipient sector.

The chapters in this volume shed considerable light on various aspects of these issues as they relate to Indonesia. Since 1966 Indonesia has been broadly open to FDI, and this, along with the rapid expansion in education and a stable macroeconomic environment, has arguably been the most important element of the government's implicit technology policy. Nevertheless, as both Lall (Table 7.4) and Soesastro (Table 16.1) point out, Indonesia has not been an especially large recipient of FDI by East Asian standards. There have, moreover, been quite restrictive FDI policy episodes (notably 1974–84), and it was not until the June 1994 investment deregulation package—which was prompted in part by a fear of competition from China—that a number of the major irritants facing foreign investors were removed.

Hadi Soesastro (Chapter 16) provides an excellent overview of the international dimensions of technology flows, focusing on the Asia–Pacific region and drawing out some of the implications for Indonesia. Among his major arguments, he draws attention to the various channels for technology transfer, concluding that FDI is generally the most significant source, and noting also the shift from the public to the private sector as a generator of these flows. He emphasises the rapid globalisation of flows (including not just transfers but also the international relocation of R&D activity), albeit in the context of the OECD economies continuing to be the major suppliers. An underlying theme in his analysis is the difficulty of measuring technology flows. For example, most developing countries do not estimate with any reliability a 'technology balance of payments'. Even if they did, a measure based on royalty payments would not capture much of the story (which includes such things as human capital flows, FDI spin-offs and direct licensing agreements). We are therefore left in the main with a very imperfect measure of these flows in the form of FDI. In the final section of his chapter, Soesastro surveys various East Asian approaches to technology policy, concluding that none is exactly appropriate to Indonesia's needs and context. In particular, he cautions that the earlier Korean approach of a highly restrictive policy towards FDI would not be suitable in the Indonesian environment (or indeed anywhere) in the 1990s.

An important lesson from the Asian experience stressed by Soesastro is the need to link technology strategy to international commercial policy.[11] Open economies with a good human resource base may achieve high pay-offs from carefully targeted science and technology policies—Taiwan, Hong Kong and Singapore all come to mind as nimble economies with 'below average' (for their per capita incomes) R&D/GDP expenditures but highly effective innovation strategies. By contrast, India and Russia have devoted large budgets to R&D activities, but (until recently) in the context of closed economies which thereby minimised the opportunities for a quite well-developed domestic research industry to interact with international best practices. This issue has been emphasised in a recent report by OECD (1997), which, in view of the rapid globalisation of technology flows, proposes that a nation's technological effort should explicitly include acquired technology together with the national R&D effort. As the report points out, firms purchase 'R&D effort' in the international marketplace; the OECD estimates suggest that the magnitude of such acquired technologies may be at least as large as that of domestically generated R&D activities. In developing countries, with their limited local R&D expenditure, acquired (imported) technology would obviously be a larger proportion of the total. Ideally, international comparisons of R&D effort should make more explicit allowance for the mix of these two components. Such an approach is clearly relevant to Indonesia, with

---

[11]This observation is also emphasised by Ostry and Nelson (1995), who argue that, given the speed with which technology now diffuses worldwide, for firms to obtain advantages from innovation they must have ready access to world markets, not just through trade but also through FDI and human capital.

its reasonably open policies towards trade, investment and skilled labour flows, but in the context of a technology policy which in some respects heavily emphasises self-reliance.

The globalisation of technology production and flows is also emphasised by Ostry and Nelson (1995), who point to the paradox that an increasingly open trade and investment environment is accompanied in some cases by a growing 'techno-nationalism', as firms and governments aggressively pursue competitive advantage based on technological superiority. The alleged decline in US technological leadership may be fuelling the resultant disputes over intellectual property rights. Indonesia has yet to be seriously embroiled in these international technology-based frictions, but as it shifts into more R&D-intensive activities it will begin to feel the ripples. The faster it pursues its ambitious industrial projects, the sooner its commercial interests will be exposed to these challenges.

Complementing Soesastro's general survey of the issues are a number of illuminating case studies. Two, discussed in more detail below, involve vastly different forms of transfers, and yet both affirm the key role of imported technology in driving Indonesian industrialisation. Yuri Sato (Chapter 17) examines transfers from Japan—by far the largest supplier of industrial technology to Indonesia since 1966—to one of Indonesia's largest conglomerates. Her findings point to significant transfers, and the active participation of the local subsidiary in diffusing the knowhow so acquired. In the process she challenges the common sentiment that Japanese companies are reluctant to transfer technology to Southeast Asia. William Cole (Chapter 14) delves into the experience of Bali's successful small-scale garment exporters, highlighting the key role that 'recreational tourists' played in connecting these producers to retail outlets abroad and in transmitting international consumer preferences (designs especially) back to them.

Two additional case studies involve more formal government involvement, with the public sector establishing a framework and playing a catalytic role in inducing stronger international private sector links. Shannon Smith (Chapter 18) documents Indonesia's involvement in Southeast Asia's most successful growth triangle, Sijori (Singapore, Johore, Riau), through the establishment of export-oriented industrialisation activities in Batam, the island adjacent to Singapore. With the signing of an intergovernmental agreement in 1989, Batam has quickly developed into Indonesia's major internationally oriented manufacturing zone outside Java, building on the country's strong complementarities with high-wage, land-scarce Singapore. This is a strong, market-driven relationship, but it did require initial government intervention to provide a policy framework in which Indonesia was able to capture some of the benefits of Singapore's sudden loss of comparative advantage in labour-intensive industry.

Smith's study of Batam is also important for the lessons from its misdirected development strategy prior to 1989. In the late 1970s, Minister Habibie assumed control of the island. Envisaging a grand plan of it becoming a high-tech rival to Singapore, he actively discouraged labour-intensive FDI. Not surprisingly, development on the island languished—there was practically no private sector interest other than a few politically motivated projects. It was not until the two countries' interest in the development of Batam converged, and Indonesia's technocrats were able to introduce a conducive business environment, that the island began to grew strongly. Thus the principal lessons from Batam for policies on FDI (and technology transfer) are that 'natural economic zones' can flourish across international borders provided the appropriate policy framework is in place, and that misdirected 'high-tech' strategies can seriously delay the process of labour-intensive industrialisation.

Chapter 19, by Don Scott-Kemmis, Leslie O'Brien and Remy Rohadian, provides another illustration of the role of governments in inducing stronger technical cooperation, in a special kind of 'South–South' relationship between Indonesian and Australian firms. Notwithstanding proximity and great complementarities, until the 1990s such cooperation was limited and spasmodic. Both countries have historically been net technology importers, in the context of an inward-looking industrialisation strategy. When Indonesian firms sought technology from abroad, they tended to look towards the 'OECD North', especially as Australia often lacks the manufacturing capacity to support its active innovation industry. Until recently, most Australian firms shunned Indonesia because of its perceived difficult business environment.

Strong growth and structural change in Indonesia, major deregulation reforms in both countries during the 1980s, and a warming of the bilateral relationship at the official level led to a reassessment of business opportunities. Trade and investment ties began to grow quickly, and in such an environment there was a case for governments to be involved as catalytic agents, seeking to make businesses in both countries aware of opportunities in the bilateral relationship. As Scott-Kemmis, O'Brien and Rohadian illustrate in Table 19.1, government initiatives led to an enormous array of cooperative joint ventures, some of which required modest amounts of seed money for their establishment. Australian expertise in niche markets—such as dryland agriculture, non-conventional energy, medical technology, environmental management and many others—was attractive to Indonesia, the more so because it originated from a relatively small 'non-threatening' neighbour. In the delivery of these personnel-intensive services, proximity also became a major advantage in the relationship. Other technology flows also flourished outside these formal initiatives; one (not mentioned by the authors as it is outside their portfolio) is that by the mid 1990s Australia had become the largest supplier of tertiary education services to Indonesia.

The authors also underline the challenges for governments seeking to forge closer private sector technical ties. Both countries, inevitably, approach the cooperation with different agendas, and it is not surprising that some proposals lack reality. Australian firms have been impatient for quick commercial success, and some have expected the government to underwrite commercial risk. On the Indonesian side, there have been unrealistic expectations about the scale of Australian financial assistance available. Both governments need to ensure that their role is limited to overcoming information bottlenecks and other sources of market failure, that public subsidy of commercial activities does not become entrenched (by either private or bureaucratic interests), and that projects with little chance of reasonably quick commercial pay-off are discarded. But thus far at least, the lessons from these initiatives are largely positive ones.

Micro Case Studies

Micro case research is an essential component of the study of technological development, both to supplement the macro-level indicators and to provide a deeper understanding of the factors driving the process and its outcomes. Although there is great diversity, such that the conclusions from case studies cannot necessarily be generalised, several contributions to this volume add significantly to our understanding of technology issues. We discuss these contributions, in the context of the existing literature on Indonesia, under four subheadings: industry case studies; small and medium industry; rural development; and the diffusion of imported technologies.

Industry Case Studies

Beginning in the late 1960s, a virtual technological revolution began to sweep through Indonesia. The speed of change was particularly rapid because of the economic stagnation and disruption of the previous three decades, and the consequent lack of investment in new machinery and equipment, followed by the sudden adoption of liberal economic policies and the restoration of strong economic growth. Thus changes which might have evolved over decades were compressed into a much shorter time period.

Indonesia is fortunate in that it possesses a rich history of micro-level case studies examining this rapid technological change, particularly—but not only—in the industrial sector. For example, Timmer (1973) studied the switch from hand-pounding to the mechanised milling of rice. Collier, Wiradi and Soentoro (1973) drew attention to the replacement of the traditional knife (*ani-ani*) by sickles in rice harvesting. Dick (1981a, 1981b) pointed to the major changes occurring in urban public transport, particularly the introduction of mini-buses in place of the traditional *becak*. Hill (1983) documented the shift from handlooms to mechanised weaving, while Tarmidi (1996) analysed the demise of hand-rolled *kretek* cigarettes as mechanisation spread quickly in the industry. These studies of course represent just the tip of the iceberg: the green revolution in agriculture involved the planting of high-yielding varieties of rice and other seed crops, while the range of services offered in the financial sector and the geographic coverage of the formal sector have both grown extremely rapidly.

It is important to emphasise that these changes on balance generated major benefits for Indonesia—for consumers who were able to purchase cheaper and better products, for workers who were freed of often arduous and poorly paid work, and for firms whose international competitiveness was enhanced. There was of course some displacement of workers, but in a high-growth economy these 'backwash' employment effects were generally not serious, except perhaps for older, unskilled and spatially immobile workers. It is also important to emphasise that these changes have been driven fundamentally by the availability of new technologies (new at least to Indonesia) that are privately and socially profitable. The regulatory framework has shaped the processes at the margin in some cases—the government on occasion has sought to deliberately retard the speed of technological change (for example, in the introduction of mechanised *kretek* production), while distortions introduced during the era of subsidised credit (from the early 1970s to about the mid 1980s) probably hastened the speed of mechanisation. But the principal engines driving the process were the initial stock of antiquated equipment, high rates of economic growth and investment, and open access to international technology markets.

Chapter 15 in this volume, by van der Kamp, Szirmai and Timmer, continues this tradition of detailed micro research, and confirms the findings of earlier research on rapid technological change in the two major components of the textile industry. In the case of weaving, whereas during the 1970s the major transformation involved the phasing out of handlooms and their replacement by power looms, by the 1990s shuttleless looms had been widely adopted. In spinning, too, new technologies have been introduced, mainly in the form of new investments in refined technologies rather than as discrete movements from one technology to another. The authors quantify these changes carefully, by looking at the improved human resources in the industry, by developing a technology index for each industry, and by estimating TFP growth. These quantitative indicators confirm the casual impression of rapid technological change.

Technology and Small–Medium Enterprises

Small and medium enterprises (SMEs) tend to be neglected in official thinking about technology, although they now rate a mention in the current long-term plan for 1994–2019. Such neglect results from Minister Habibie's primary concern with the major strategic industries, all of which are large in scale, and his well-known scepticism about labour-intensive industries such as garments and textiles. Such a view also finds support in the government's technical departments, such as Industry and Trade, whose members at the senior echelons have a grand technological vision heavily influenced by their predominant engineering training. Rather, much of the government's thinking about SMEs is welfare driven, as manifested in the (somewhat muddled) approaches and (allegedly corruption-prone) strategies of Depkop (Department of Cooperatives and Development of Small Entrepreneurs), which include an emphasis on reservation schemes for SMEs and *pribumi* enterprises. Admittedly, some technical departments do offer limited technical and advisory services, but as noted elsewhere these are rarely customer friendly and are generally supply driven (Lall, Chapter 7).

The Indonesian government's approach to SMEs is also at odds with East Asian experience. Korea learnt to its cost that an excessive emphasis on large firms, through its *chaebol* strategy, resulted in a top-heavy industrial structure lacking in flexibility and overly dependent on government largesse. More recently, SMEs have become a central component of the government's technology program (Kim 1997). Taiwan's industrial dynamism, resilience and flexibility, meanwhile, is widely attributed to its strong and technologically progressive base of SMEs (Lin 1998).

In fact, SMEs in Indonesia have been doing a good deal better than is often recognised by the thriving 'despair school' which exists both inside and outside government. It is difficult to obtain a macro picture of SMEs' performance, and in any case apparent trends in size distribution need to be interpreted with great care.[12] But numerous case studies have investigated the dynamics of SMEs by way of firm surveys, finding as would be expected cases of both success and failure.[13]

Coles' analysis of the Bali garment export industry (Chapter 14), based on long-term field observation, provides an illuminating account of SME dynamics in Indonesia. The industry's spectacular growth in the 1980s, almost exclusively based on small firms, was practically an 'accidental' case of industrialisation. Foreign tourists, mainly surfers wishing to support a recreational lifestyle, saw commercial opportunities in Balinese garments and the indigenous design capacity. They were able to act as marketing intermediaries, connecting local producers with retail outlets abroad, in the process dispensing important information on designs and production techniques. Later, as the island's fame spread, these links developed quickly, and the industry mushroomed from its seasonal, cottage origins to larger production units and some local design capacity.

Coles' study, and others like it,[14] suggest a model of successful and innovative SME development in which the following ingredients appear to be important:

---

[12]As Aswicahyono, Bird and Hill (1996) demonstrate, the reported decline in the SME industrial share is just as consistent with these firms graduating to larger scale units as it is with them going out of business.

[13]Recent studies of SMEs in Indonesia include Pangestu (1996), Sandee (1995), Thee (1993), van Diermen (1997) and papers presented to an international conference held at the Satya Wacana Christian University, Salatiga, in November 1996.

[14]See, for example, important studies of export-oriented SME furniture manufacturers in the town of Jepara, northern Central Java, by Berry and Levy (1994) and Schiller and Martin-Schiller (1997).

- some basic industrial competence in a particular field of activity (such as garment manufacture in the case of Bali or furniture manufacture in the case of Jepara);

- a conducive macroeconomic environment, including especially a competitive exchange rate;

- reasonable physical infrastructure, extending in these cases to proximity to import and export facilities which function without too much inconvenience; and

- injections of technical, design and marketing expertise to link small producers with new ideas and major markets.

With the possible exception of the first ingredient, all four elements are directly amenable to public policy. They may also be present in different institutional arrangements, as, for example, in the emerging subcontracting networks found in the automotive and machine goods industries (Harianto 1993). The general model developed here is equally applicable in agriculture, to be discussed shortly, and in larger scale industry, where barriers to the development of technology transfer channels are arguably lower than in the case of SMEs. It might be argued that these examples are special cases, and that the model is not easily transferred to the bulk of small firms, especially those operating in remote locations and catering to low-income markets. But neither garments nor furniture could be regarded as 'niche markets'; on the contrary they are mass consumption goods. Admittedly Bali has intense exposure to international markets through tourism, but Jepara is some distance from a major port (Semarang) and is not a tourist destination.

These case studies also have important implications for government policy. Neither resulted from any deliberate government promotional measures. The government did play an important role in providing a supportive macroeconomic environment and in the provision of a rapidly improving infrastructure. In Bali, the local government generally adopted a fairly open policy towards the presence of foreign entrepreneurs, and export procedures were not unduly burdensome most of the time. The June 1994 reform of FDI regulations, lowering the minimum capital requirement from $1 million to $250,000, made it easier for small foreign investors to operate in the country without harassment. Of course, these hardly constitute 'contributions' from government, except in the negative sense of avoiding a harshly restrictive regulatory regime.[15] By contrast, reports from Jepara in 1997 reveal that foreign workers, on whom the industry depends, were being harassed and deported.

It might be argued that an efficiently functioning government industrial extension service could play a role in lifting these sorts of SMEs into activities based on more advanced technologies, where informal technology transfer channels do not operate as effectively. Coles' case study is suggestive in this regard, as the Balinese garment industry has apparently stalled recently in its efforts to achieve further upgrading. Thee and Lall (Chapters 6 and 7 respectively) are attracted to such a model, if it can be demonstrated that the government has the imagination and bureaucratic capacity to deliver it. This is clearly an area where more research is needed. Meanwhile, it needs to be stressed that, at the very least, governments should avoid the heavy implicit penalties exacted on SMEs through the operation of a complex, corruption-prone regulatory regime. As many researchers have pointed out (Sjaifudian 1997 and others), such a regime implicitly discriminates against SMEs, which lack the political connections such a system breeds, and which, in the face of 'fixed cost' licensing requirements, pay higher per unit costs for regulatory compliance than larger firms.

Coles' study is also important because the dynamics of the process of SME technological adoption do not appear to be of interest to, or understood by, the relevant government agencies. Minister Habibie has, for example, frequently dismissed the garments industry as irrelevant for Indonesia's technological future, even though, as the author (and others such as Sandee 1995) clearly shows, a good deal of dynamic innovation is evident. Moreover, the intellectual framework of Depkop is almost completely irrelevant to the needs of the Balinese garment producers. Rather than Depkop's focus on partnership schemes (kemitraan), subsidies, regulation and protection from competition, these firms are more interested in efficiently functioning credit markets, good infrastructure, freedom from bureaucratic harassment, and perhaps some carefully targeted industrial extension support.

---

[15]As Cole puts it (p. 277), '[b]eyond these points, the role the government played seems more positive in its absence than in its actions'.

Rural Development

A similar set of issues arises in the case of Colin Barlow's four contrasting case studies of the diffusion of technology in rural areas (Chapter 13). Focusing on the generation of a range of new technologies, both internationally and domestically, he examines how effectively they are disseminated in rural areas. In some instances these processes work well, but outcomes are uneven, especially in rural areas featuring incomplete markets. Barlow pays particular attention to 'interveners', the actors—government agencies, private businesses and non-government organisations (NGOs)—from outside targeted communities who facilitate innovatory behaviour in various ways.

Barlow shows clearly how processes and outcomes vary considerably among the case studies. In the politically important rice industry, for example, yields have improved quickly, especially in the Javanese heartland. International R&D in the form of the green revolution had recently become available in the late 1960s, there was a domestic capacity to absorb the new technologies, and diffusion was facilitated by an enthusiastic (if sometimes heavy-handed) government apparatus. Input subsidies (especially for fertiliser) have been generous, and physical infrastructure such as irrigation has been rehabilitated rapidly. There was a case for some government inducement to innovation, although arguably the intervention has involved excessive 'micro-management' and persisted for too long, and a heavily centralised model has been employed in vastly different ecological contexts.

The record in tree crops is less impressive. Although some state-owned rubber plantations operate near the technological frontier, smallholder performance has lagged, especially compared to the effective, market-sensitive extension programs in neighbouring Malaysia and Thailand. In this context, the lessons from Indonesia's cocoa boom, concentrated mainly on smallholders in Sulawesi, are instructive (Akiyama and Nishio 1997): output grew by an annual average rate of 26% in 1980–94, owing principally to the provision of reasonably good physical infrastructure (especially roads), a competitive exchange rate and the absence of government distortions in marketing. (Indonesian farm-gate prices are a comparatively high proportion of export prices, and quality-based price differentials are needed to encourage farmers to upgrade quality.) In such a scenario of 'hands-off policy success', the key role for government, as in the case of the Bali garments industry, is to continue to provide more of the same, together with a range of agricultural extension services (to combat disease, for example) to facilitate continuous productivity growth.[16]

In all four case studies, Barlow shows that there is a role for carefully targeted government programs, especially in encouraging the adoption of new technologies. The case for intervention is generally strongest for roads, the provision of public education, basic research support, and inputs such as irrigation (although pricing issues—the mix of subsidies and user-pays provisions—are relevant here). He cautions that the case for market intervention is generally weak where private trading networks are well established. The provision of research support and input subsidies needs to be carefully time-bound, so that governments withdraw when private markets are operating efficiently. As with all R&D programs, it is important to distinguish between the government as funder and as provider. As the market for the private (including NGO) provision of these services expands, it is generally preferable for the government to progressively withdraw from direct service provision.

One final, if obvious, conclusion from Barlow's study is that Indonesia's enormous ecological and economic diversity dictates an equally diverse strategy of intervention. 'Markets' work better in Java, with its good transport network and information flows, well-established international connections and stronger human resource base than they do in, say, Irian Jaya. Thus there is no one single model of intervention which works without modification in all regions.

Diffusion

The general presumption in the literature is that large inflows of FDI, such as have occurred for most of the post-1966 period in Indonesia, must have raised levels of technological competence. Indeed, several contributors (Lall, Chapter 7; Soesastro, Chapter 16) argue that such flows are the best overall indicator of the magnitude of international technology transfers to Indonesia. Nevertheless, the

---

[16]Although not mentioned in these studies, it is almost certainly the case that, as in Thailand (Feder et al. 1988), improved land titling arrangements would facilitate the operation of credit markets, thereby enabling small farmers and firms to invest in new technologies. The connection is that, with clearer land titles, small landholders are able to access the cheaper credit available from formal financial institutions by offering their land as collateral.

weaknesses of FDI data in this context are widely recognised: foreign investment in extractive industries—of which Indonesia was a large recipient, especially through to the mid 1980s—can be 'enclave' in nature, with few local spin-offs (other than a fiscal contribution); foreign investors may be unwilling or unable to transfer technology, the former perhaps owing to a vast technological gulf between them and their local partners, the latter the result of deliberate commercial strategy; and even if there is a transfer to the local partner, the technology may not necessarily diffuse widely beyond it.

These issues can only be addressed with authority through micro-level case studies. There is a literature on this subject relating to Indonesia, which at a general level confirms the presumption that FDI inflows have been accompanied by significant technology transfer and diffusion. But most of the studies thus far have been somewhat anecdotal and lacking in depth.[17] For this reason, Chapter 17 by Yuri Sato makes an important contribution. Based on a detailed case study of Astra, and continuing her tradition of illuminating investigations of Indonesian conglomerates, Sato provides an all too rare glimpse of the process of technology transfer to Indonesia from Japan, which has been by far the most important foreign investor in Indonesian manufacturing. Focusing on management technology, she shows how quality control (QC) techniques were adopted and diffused swiftly within the Astra Group and its affiliates, and how QC has become as much an Indonesian as a Japanese practice. In particular, she documents how Indonesian staff have been transformed from 'passive recipients to active transmitters' of the concept of total quality control (TQC), and how local staff increasingly took the initiative in the practice. An interesting finding is that the practice has diffused quickly both within the group and beyond it to non-affiliated domestic companies. Ironically, this is the industry in which foreign investors have been widely criticised for their alleged *failure* to transfer technology. Such criticisms have provided a rationale for intervention in the form of the highly controversial (and very recently terminated) 'national car' program.

Sato provides no estimates of the productivity-enhancing effects of TQC. Also, the study focuses on a company that is widely regarded as the most 'professional' of Indonesia's conglomerates, and which may therefore be regarded as something of a special case. Nevertheless, her work is critical to an understanding of the

---

[17]Thee (1990) and Thee and Pangestu (1993) are among the few exceptions.

dynamics of technology transfer and diffusion. She emphasises the importance of having the local partner fully involved in decision-making processes; by implication, therefore, rent seeking 'crony' local partners—quite common in Indonesia—are unlikely to contribute much by way of diffusion. Sato also emphasises the sometimes neglected distinction between the transfer of production and management technology. The former can be diffused through machinery in the first instance, but the latter entails close personal contact between suppliers and recipients. Implicitly, she is sceptical of the wholesale importation of capital equipment without a corresponding effort on the part of the host country to upgrade its indigenous human resource capacities.

Policy Issues

Some General Considerations

Having outlined Indonesia's approach to technology policy issues, placed the country's technological indicators in comparative international perspective and analysed some of the salient dimensions—ranging from the international to the microeconomic—this section examines some of the major policy issues. These include education and training, policies to promote international technology flows, the question of 'megaprojects' and the operation of government R&D institutes. Our general conclusion is that the Indonesian government has had some notable successes in this area, but that most of these have arisen from the general policy environment; in the realm of specific, discretionary interventions it is more difficult to find evidence of notable achievement. Indeed, one of the major themes to emerge from this book is that, in the area of technology policy, the government is not infrequently 'doing what it should not do, and not doing what it should do'. Before looking at specifics, however, a number of observations on the general policy environment need to be highlighted.

First, as was emphasised above, a government's record of economic management will have a much more important bearing on the nation's technological development than sector-specific technology policies. This proposition has been illustrated dramatically since mid 1997, but it has always been the case. The most important contribution governments can make to a nation's technological development is to deliver rapid economic growth, which in turn is based on stable macroeconomic management, openness to the international economy, an efficiently functioning financial system, an effective legal system, including the observance of property rights, and improved physical infrastructure and human resources. There is a case to be made for more micro-level interventions, based on standard market failure considerations, but it must always be recognised that this is very much an argument of second order importance. Strong growth of itself will result in rapid technological progress, even if there may be a case for the proposition that effective intervention could hasten the process. But the converse obviously does not apply. A strong technological base in the context of a poorly managed economy does not provide the basis for a sustained improvement in technological capabilities.[18]

A second, related, general observation is that governments indirectly have a significant impact on technological development through programs that may not necessarily have an explicit R&D objective.[19] Examples include education, health and defence expenditure. A large state enterprise sector may also play a 'technology catalyst' role by design or default, as may import protection. Indirectly, competition policy may shape a nation's R&D effort to the extent that large firms are usually more R&D intensive than smaller ones; such a consideration was relevant in the Korean decision to promote the development of its *chaebol* (Kim 1997).

These arguments have some, albeit limited, relevance to Indonesia: public education has expanded rapidly over the past 30 years, although it has mainly emphasised quantitative growth at primary and secondary levels; the Armed Forces has been a comparatively small *direct* claimant on government resources, and it lacks much technological sophistication; there is a sizeable state-owned enterprise sector that is expected to fulfil broader socio-economic development objectives, but apart from enterprises within

---

[18]Curiously, this seemingly obvious proposition is often overlooked in Indonesia's most 'techno-enthusiast' circles. (Indeed, attempts by this camp to explain macroeconomic phenomena have sometimes led to greater confusion, as in the case of Minister Habibie's now infamous theory of 'zigzag' inflation and interest rates, first propounded in late 1996; see McLeod 1997.) But the point is recognised by sophisticated proponents of an activist technology strategy. As Lall (p. 147) observes: 'The history of postwar development is littered with interventions that have distorted and delayed development rather than promoting it—the risk of "government failure" is very high, and its costs often exceed that of market failure'.

[19]See, for example, the contributions in Nelson (1993) for a discussion of this issue.

the strategic industries group (BPIS), it is not generally noted for its technological dynamism; despite much discussion of the subject, the Indonesian government has never pursued an active competition policy; and import protection, which has been extensive for some periods since 1966, has more commonly been the result of lobbying by vested interest groups than introduced in a systematic fashion designed to develop infant industries (Basri and Hill 1996). Thus, in sum, some Indonesian government interventions do have implications for the pattern and pace of technological innovation, but the impacts are generally not large, certainly as compared to those of the policies which promoted rapid economic growth in 1967–97.

Third, arguments for government intervention need to be evaluated not only on their analytical merits, but also quite explicitly in the political and institutional context of the country in which they are to be implemented. The Indonesian government over the period 1966–97 generally achieved high-quality macroeconomic management, but its microeconomic interventions have often led to gross distortions.

- Programs have often lacked coordination across ministries.

- Corruption has been a continuing and serious problem (resulting, for example, in the country ranking very poorly in international comparisons such as that conducted annually by Transparency International).

- State enterprises have a poor financial record.

- Serious efforts to improve civil service efficiency—to increase accountability and to link performance and rewards more closely—have never been implemented in a determined and effective fashion, although the educational standards of civil servants are now much improved (Rohdewohld 1995).

- Important services which only government can provide (such as an incorruptible police force, an efficient and clean judiciary, and a well-regulated financial sector) have not been supplied.

- Especially since the late 1980s, a virulent, 'palace-based' cronyism has emerged and developed rapidly.

In such an environment, in which the evidence suggests that selective intervention has played a marginal role in the country's industrial success since the late 1960s (Hill 1996a), one has to be especially cautious in advocating a strategy that inevitably entails a significant increase in micro-level government intervention, especially one that involves government bureaucrats attempting to pick winners.[20] The observation of Mark Dodgson at the Update Conference that 'governments may not be good at picking winners, but losers are often good at picking governments' would seem to be particularly apposite to Indonesia's industrial policy environment.

Such an assessment is quite at odds with some recent literature emphasising high-quality, relatively corruption-free bureaucratic capacity and intervention, at both macro and micro levels, as a factor underpinning East Asian success.[21] It is probable that these authors have overstated the phenomenon: Korea and Taiwan have major problems of corruption, while the 'Korean model' of heavy intervention has not escaped the region's financial ravages. But outside the macroeconomic realm, and perhaps some agricultural and educational interventions, it is almost certainly the case that—with the possible exception of Thailand—Indonesia's micro-level bureaucratic capacity is the weakest among the World Bank's seven East Asian 'miracle' economies.

These comments are particularly pertinent to a fourth general observation, concerning the operation of, and the intellectual justification for, the high-tech projects under Minister Habibie's purview. Indonesia's technology policy for the last 20 years has been completely dominated by Habibie, an outcome that will almost certainly persist as long as Soeharto remains President. In other words, proponents of an enhanced government role in Indonesia's technological development have to confront the inescapable reality that they are in effect advocating an even larger role for the Minister. In this context, two features of Habibie's approach stand

---

[20]One is reminded of H.W. Arndt's (1978, p. 28) gloomy but highly relevant observation discussing these political economy parameters 20 years ago:

> There is hardly an economic policy—whether for the levying of income tax or an urban real estate tax, or for tariff protection of domestic industry, or for subsidies to depressed industries, or for minimal regulation of foreign investment or of road traffic, or for conservation of forests or for provision of rural credit to farmers or for priorities in investment credit by state banks, or for social welfare services or development projects of every kind—which, whatever its economic or technical merits, does not need to be weighed—and often ruled out—almost wholly on grounds of its administrative impracticability in the face of corruption.

[21]See Campos and Root (1996), the contributions in Rowen (1998), and Stiglitz (1996). Stiglitz (p. 157), for example, characterises government intervention in East Asia as '... relatively free from corruption ...'.

out. One is the extremely personalised manner in which projects have been undertaken. With the exception of the N-2130 project, still in its infancy, all the initiatives have been directly under his control: no major decision can be taken without his approval; no credible financial performance statements have ever been released; none of the usual checks and balances (such as scrutiny by the Department of Finance) is present; and not even the most powerful 'technocrat' in the cabinet has been able to challenge Habibie's direct access to the President.

The second feature of the Habibie approach relates to its intellectual foundations. As demonstrated by Rice (Chapter 9), Habibie is one of the few members of Indonesian cabinets, past and present, with a coherent and internally consistent approach to his portfolio. Moreover, some of his basic tenets are intellectually respectable, including the promotion of human resource development, the role of S&T institutions, market failure in the operation of technology markets, and a 'stages approach' to technological development. Indeed, as Rice points out, Habibie shares some of the views espoused by the widely quoted Harvard business strategist, Michael Porter.

However, the Habibie approach lacks intellectual credibility in important respects, even in the framework of Lall (Chapter 7), Porter, Lipsey and others who are prepared to countenance government intervention in certain circumstances. The Habibie strategy involves firm, not industry, based assistance, and to a firm which is state owned. The assistance is not time-bound, nor is it transparent; both features significantly increase the risk of misjudgment and corruption. His model implicitly operates within a closed economy construct, in the sense that it aims for self-sufficiency rather than an interactive capacity in increasingly competitive technology markets. The notion of opportunity cost, including at least some attempt to calculate social rates of return (factoring in some allowance for the positive externalities associated with high-tech investments), is completely absent in the Habibie approach, reflecting the dominance of engineers at the senior echelons of all of his agencies. For example, there has been no attempt to justify investment in IPTN as superior to investment in improved primary and secondary education. Since the Habibie camp has controlled both the Education and Technology Ministries in the current (1993–98) cabinet, these sorts of questions are not politically unrealistic exercises.

Moreover, the approach does not appreciate the evolutionary nature of changing comparative advantage in dynamic East Asian economies, which involves an early stage reliance on labour-intensive activities before rising real wages and an enhanced human capital base result in industrial upgrading. For the now high-wage NIEs, being a 'tailor to the world' was a crucial part of this process, and is arguably even more important for a populous labour-surplus economy such as Indonesia. Habibie's philosophy and his policies (as revealed over the period when he was in control of Batam Island) explicitly dismiss the importance of this labour-intensive phase of industrialisation. The Habibie group argues that competition from low-wage economies such as China, India and Vietnam is an additional reason for the promotion of high-tech industries, while ignoring the fact that the issue has as much to do with efficiency (infrastructure, human resources, bureaucratic red tape) as it does with comparative advantage. Moreover, part of the explanation for any loss of Indonesian competitiveness in recent years lies in the sharp real wage increases mandated by the government.

International Policies

As noted above, Indonesia scores well in most areas here, albeit mainly through inaction (Soesastro, Chapter 16). There are virtually no intentional impediments to technology inflows, for example in the form of restrictions on technology licensing and royalty payments. Although there have been calls for Indonesia to adopt a tightly regulated approach to technology flows, as Korea did (Kim 1997), these pressures have (wisely in our view) been resisted—the dangers of deterring foreign investors and of bureaucratic capture are both very great. Restrictions on FDI and the use of expatriate labour are still present, but in most cases these are now more in the form of minor (but still occasionally costly) irritants rather than significant obstacles to commerce. Probably the biggest source of apprehension—and therefore reluctance to share technology fully with local partners—among foreign investors since the early 1990s has been the extraordinarily rapid growth of crony conglomerates, and in consequence the distinct possibility that the commercial environment will be arbitrarily, and without warning, tipped in favour of these rapacious interests. Some foreign investors quite intentionally restrict the scale of their operations to avoid too high a profile, which increases the risk of arbitrary interference by these interests.

It was also noted above that technology policies and the government's broader international economic orientation lack any sort of harmonisation. While the major high-tech projects have obviously borrowed technology from abroad, the philosophical approach underpinning them—with its emphasis on self-sufficiency—veers towards autarchy. This is increasingly at odds with the rapid *globalisation* of R&D activities.

Education

Along with macroeconomic management and openness to international commerce, education is one of the key building blocks in a country's capacity to absorb, diffuse and master imported technology, and to develop an indigenous technological base. This is an area where, as Thee (Chapter 6), Lall (Chapter 7) and other contributors point out, Indonesia has achieved much over the past three decades, at least in terms of quantitative indicators such as rapidly rising literacy levels and enrolment ratios (Tables 6.1 and 7.5). Although not yet exhibiting the extremely high standards of some of its more advanced East Asian neighbours, Indonesia compares favourably in most international comparisons of educational achievement. Indicators of quality over time and across countries are not so readily available, but limited evidence suggests that the record would not be so impressive, especially at post-primary levels. There is also concern that, at the tertiary level, enrolments are skewed towards low-cost fields of study, to the neglect of programs with a significant scientific component (Tables 6.2 and 7.5). This misdirection, it is argued, will hamper the country's ability to upgrade its technological capabilities quickly in the future.

There is a well-established case, on both equity and efficiency grounds, for governments to subsidise the provision of education, especially at primary and junior secondary levels. This, broadly, has been the strategy of the Indonesian government since the late 1960s, culminating in the current drive for universal education up to mid-secondary (SMP) levels, which has in turn involved concentrating public subsidies at these levels of education. At post-primary levels the government plays a much less important role, although it does maintain a nation-wide network of public universities, some of which are among the best in the country.

How much should the government be involved in vocational secondary and technical education, some very specialist in nature? Here the answer is less clear. The general presumption is that the case for intervention (and subsidy) is less persuasive, since well-functioning markets will signal their demand for a particular skill mix, and it will be privately profitable for individuals to undertake the requisite training. Where there may be a role for governments here is as a catalyst, to stimulate the market for private education, perhaps to provide some initial subsidies to demonstrate the profitability of such training, and to provide a regulatory environment which covers such matters as skills certification. A further issue concerns institutional development and the problem of appropriability. A substantial amount of specialist training is industry and even firm specific, and is undertaken within firms. It is sometimes argued that, left to the market, firms are likely to underinvest in such training, owing to their short-term planning horizons and the frequent hijacking of trained staff. If industry associations functioned effectively, such training could be undertaken at the industry level, thus ameliorating 'freerider' concerns. But industry associations in Indonesia are, with a few exceptions, notoriously weak and ineffective.

In an important paper, Chris Manning (Chapter 10) questions the case for a significant government involvement in vocational education and training, a case advanced elsewhere by two other contributors to this volume. In one of the first systematic analyses of this issue in Indonesia, he maintains that it is preferable to concentrate scarce educational resources in the general education system. This sector is starved of resources, as indicated by its poorly paid and often underqualified teachers, very high pupil–teacher ratios and extremely limited classroom resources. He is not convinced that technical education currently constitutes an obstacle to industrialisation, since the limited firm surveys thus far undertaken are ambiguous on this issue—some report skilled labour shortages, but these often rank below other problems (including physical infrastructure and corruption). He also maintains that private education markets generally work in responding to skill shortages and in delivering market requirements. Finally, he cautions against a headlong rush into interventionist strategies on political economy grounds. There is a danger, already evident, of bureaucratic capture (sometimes in association with a fragmented foreign donor community): of supply rather than demand-driven programs, of excessive credentialism in the setting of standards, and of straight-out corruption in the delivery of services.

Thus while there is arguably a case for carefully targeted vocational education and training initiatives, in which the government plays a catalytic role designed to develop institutions and make markets work more efficiently, a heavy-handed presence, as has been emerging in recent years, does not appear to be an efficient allocation of resources, particularly if it is at the cost of the general education stream. More research is clearly required on how governments can play a creative role in lifting educational standards in this area.

Megaprojects

Indonesia has a tradition of megaprojects bordering on the grandiose. These have often enjoyed considerable political and community support, in part because they are seen as a convincing illustration of the country's capacity to overcome its technological backwardness and shortcircuit the long and arduous process of technological development. President Sukarno initiated a number of ambitious industrial projects in the early 1960s with support from the Comecon block, although few subsequently became viable industrial operations. At the height of the oil boom, the then President Director of Pertamina, Ibnu Sutowo, embarked on a dazzling array of major projects, before they too were shelved owing to the financial stringencies which arose in the wake of Pertamina's insolvency in 1975.

The projects of Minister Habibie therefore have a distin-guished nationalist pedigree which, although of great concern to economists, is rich in symbolism. As noted above, the state-owned aircraft manufacturer, IPTN, has never been subject to detailed public scrutiny, although every investigation suggests that the $3 billion or so invested in it is a poor investment.[22] The first of the major IPTN products, the CN-235, is now flying domestically, but international sales have been disappointing, in part because of problems encountered obtaining airworthiness certification from the US Federal Aviation Administration. The next major project, the N-250, has likewise been controversial, particularly since June 1994 when Minister Habibie successfully persuaded the President to grant a 'loan' to the project of $185 million from the government's reforest-ation fund. This loan was then equivalent to almost half of the funds spent on the project (Borsuk 1995).

---

[22]Although dated, McKendrick (1992) remains the most detailed assessment of the IPTN project.

The most recent Habibie initiative, to design and manufacture the N-2130 jet aircraft (known by its Indonesian acronym DSTP), at an estimated cost of $2 billion, has also been greeted with some apprehension. The project, it should be noted, is of such magnitude that it alone (and excluding other IPTN activities) accounts for more than five years of the government's entire science and technology budget. However, as emphasised by Minister Saadillah Mursjid (Chapter 11 of this volume), in his capacity as President Director of the company PT DSTP, it has one important innovation: it is intended that the N-2130 will receive no direct government funding. Under an ambitious and imaginative funding scheme, the company hopes to attract (domestic) private sector investment by selling shares to the public. It is seen as a 'people's plane', with shares denominated down to the value of Rp 5,000. The Minister frankly acknowledges the 'lofty goals' of the project, and as McLeod argues (Chapter 12) in a comment on the Minister's paper, DSTP has the great merit that its funding mechanisms and performance criteria are thus far infinitely more transparent than earlier IPTN investments.

Obviously, given the current crisis, the N-2130 project will have to be shelved, and only time will tell whether it has any prospect of being resurrected. In spite of its attractive features, the principal concern remains the element of compulsion that intrudes into the so-called 'voluntary' shareholdings acquisition. As Table 11.1 illustrates, the shareholders thus far are a very special collection of individuals and institutions, all of whom are either in, or have a special relationship to, the public sector. Indeed, it is not clear who, other than these groups and a few wealthy patriots wishing to curry favour with the government, would invest in a project with a highly uncertain pay-off, and no dividends payable until at least 2003. But, to repeat, in a world of incremental reform, the case put by Minister Mursjid represents a major improvement in Indonesia's megaproject funding, in terms of both accountability and transparency.

R&D Institutes

While the case for focused extension efforts to overcome market failures is advocated in a number of chapters, there is a general consensus that most of Indonesia's current programs are largely ineffective (see in particular Thee, Chapter 6; Lall, Chapter 7). There are two interrelated problems with most of the government

R&D institutes. First, their funding base is generally inadequate, which results in (a) salaries too low to attract high-calibre staff, (b) facilities (including equipment, sometimes from myriad, unconnected foreign aid programs) that do not meet best practice requirements, and (c) especially in those agencies working in remote rural communities, a limited capacity to engage in meaningful outreach activities for the targeted client groups. Obviously, given a more or less finite R&D budget, there is a direct and causal relationship between the resource poverty of these institutes and the extremely expensive megaprojects discussed above.

The second problem relates to the objectives and functioning of these institutes. Their mission statements and philosophies are supply rather than demand driven. Their ties with the private sector are generally weak: there is little staff mobility, and even their physical location is not always based on the needs of their client groups. Not infrequently, the institutes have a welfare rather than an efficiency orientation, and their staff regard foreign and non-*pribumi*-owned enterprises with suspicion. Such an approach compounds their funding problems, since they do not have an influential private sector constituency advocating a stronger resource base. Few attempts have been made to augment institute resources in exchange for more relevant service delivery to the private sector.

This assessment may seem unduly harsh to the many staff in these institutes who perform admirably in difficult situations. The key point, however, is the importance of structural reforms to ensure greater relevance and efficiency. In this respect, Indonesia can learn much from Singapore, Taiwan and other countries in which creative interventions have overcome market failures. Moreover, as with education and training, it is almost certainly the case that these institutes would operate more efficiently if there were dynamic, effective and far-sighted industry associations with whom initiatives could be jointly developed. This would also ensure the emergence of a demand-driven program of activities of which the private sector had significant 'ownership'. One of the key lessons from international experience in innovation is that R&D programs and institutional development need to go hand in hand (Dodgson and Bessant 1996; Stiglitz 1996). Where industry associations are weak, politicised and narrowly focused—largely the case of New Order Indonesia—technology programs that do not explicitly aim for thorough institutional reform will be less effective.

One final observation on the development of government R&D institutes is relevant. While the case for picking winners through import protection is weak, and industrial promotion strategies should be directed at overcoming broad, economy-wide market failures, to the extent that some of these failures are sector specific, an element of selectivity is unavoidable (see Lall, Chapter 7; Stiglitz 1996). It is reasonable to assume, for example, that Indonesia will want to develop and maintain competence in activities such as mining, tropical cash and food crops, forestry, the marine environment, interisland shipping, food processing, electronics, machine goods, and textiles and garments. There is no reason in principle why governments should not allocate at least a portion of R&D support resources to broad sectors, providing there are bottlenecks which such intervention can remove, and provided also that there is absorptive capacity and genuine demand in the private sector.

SUMMING UP

The 30 years of rapid economic growth through to mid 1997 resulted in a remarkable transformation in the technologies employed in virtually every facet of the Indonesian economy and society. Some of these changes would have occurred almost regardless of domestic circumstances, and some were hastened by deliberate government policies, especially in selected high-tech sectors. But most of the transformation was the direct result of the growth process, and the accompanying high levels of investment, much improved standards of education and increased internationalisation. Conversely, the pace of technological progress has come to an abrupt halt during the current crisis, emphasising again the powerful association between the two phenomena.

While providing, we hope, a reasonably comprehensive picture of all major dimensions of the issue, it is appropriate to conclude by stressing the limited state of our current knowledge and to therefore enter a plea for more detailed research. Notwithstanding the important contributions in this volume, Indonesia lacks the really detailed case study work to illuminate the pattern of technological development at the industry or even firm level. One thinks, for example, of the work of Kim (1997) on Korean technological development, or of Hobday (1995) on East Asia's electronics industry. There is also the question of defining carefully and explicitly the rationale for government intervention in this area, tightly tailoring any recommendations to existing bureaucratic capacities. The case for intervention based on market failures and the need to develop institutions is widely recognised. There is also great scepticism about some of the government's current approaches, as illustrated most of all by the megaprojects. But, as noted in the previous section, there is a yawning gap between these two extremes, in the detailed documentation of cases—from the economy-wide to the micro-level—of incomplete markets where imaginative public policy could generate high pay-offs. Finally, this study underlines the urgent need to develop a more comprehensive database covering all major aspects of technological development in Indonesia.

PART I

RECENT DEVELOPMENTS

2

TOWARDS MARCH 1998, WITH DETERMINATION

Geoff Forrester

Since 1996 President Soeharto appears to have been setting the scene for a resounding reaffirmation of his New Order in the March 1998 session of the People's Consultative Assembly (MPR). His grasp on the levers of political and military power is as strong as ever. He seems determined to leave nothing to chance in the March session. Is he being prudently cautious? Has he been contemplating revealing part or all of his plans for his successor? Or both?

The President has some cause to be more cautious than usual in approaching the March 1998 MPR session. From mid 1996 to mid 1997, Indonesia experienced unusual political and economic upheaval.

THE JAKARTA RIOTS OF 27 JULY 1996

The expulsion of the supporters of Megawati Sukarnoputri—daughter of Indonesia's first President, Sukarno—from the Jakarta headquarters of the Indonesian Democracy Party (PDI) on 27 July 1996 led to the worst rioting in Jakarta for many years:

> Darkness engulfed the long stretch of Jl. Kramat Raya, Jl. Senen Raya and Jl. Matraman Raya on the evening of 27 July. The nearby [streets] were equally quiet and dark, save for similar explosions from burning structures and the cheering and hand clapping from the

crowd. The nearly full moon ... was hidden by thick smoke rising from the burning buildings along the closed streets. Spectators, who earlier in the day were engaged in a battle with the police and military, watched quietly as the red, angry flames licked the buildings (*Jakarta Post*, 4 August 1997).

As this report from the *Jakarta Post* suggests, the rioting in Jakarta surprised and shocked the city and temporarily unnerved business. Shops and offices along several main roads in Central Jakarta were looted and burned, and three people were reported to have died. The Jakarta Stock Exchange and the rupiah fell briefly, but recovered quickly. Foreign investor confidence also recovered.

The government responded savagely to the unrest. Several hundred people were arrested, and several hundred 'disappeared'. Megawati was questioned, and the government stepped up its efforts, largely successful over the following year, to silence and sideline her. A heavy military presence protected banks and other key business sites for the first three weeks of August.

The violence was triggered by an attack on the morning of 27 July on the Jakarta headquarters of the PDI by supporters of the party's new, government-sanctioned leadership. The headquarters had been under the control of PDI members loyal to the ousted leader, Megawati. For several weeks she had led a free speech forum at PDI headquarters protesting her removal from the leadership. The forum had become a broad protest against the Soeharto regime generally.

The attack was the prelude to a year of unusually intense political, ethnic and religious unrest, culminating in the most violent general election campaign of the New Order.

A YEAR OF UNREST

Muslim Unrest

From October 1996 through to the election campaign in April and May 1997, Muslim rioting occurred sporadically in West, Eastern and Central Java. Targets of the violence included government offices, police stations, churches, factories and shops. The towns and cities affected included Situbondo, Tasikmalaya, Rengasdengklok, Pekalongan and Surabaya.

The spark for the violence in Situbondo was the trial of a young Muslim accused of defaming the Prophet Muhammad. On 10 October the prosecution announced in court its demand for the maximum sentence of five years in gaol. Local Muslims in the court were angered at the light sentence sought by the prosecution and demanded the death sentence. Crowds began to gather around the courthouse; soon they attacked the building and burned it to the ground. When a rumour spread that the accused was being sheltered in a church, the crowd began to attack and burn churches. In one church five people were burned to death. The violence spread to other towns—to Asembagus, Panarukan and Besuki, West Java. It lasted five hours, during which time 25 churches, a court house, restaurants, cars and motorcycles were destroyed.

On 26 December there was rioting in Tasikmalaya, West Java, following the torture in police detention of a Muslim religious teacher. In over 12 hours of violence in the city and surrounding districts, 12 police stations, 89 shops, eight factories, four schools, three hotels, six banks and 13 churches were destroyed and burned. Two people died and four police were injured. On 27 December three shops were burned down in the nearby village of Ciawi.

The rioting in Rengasdengklok on 30 January started with a verbal clash between a Chinese resident and some Muslims beating a mosque drum at night as part of the fasting month ritual. It resulted in the destruction of cars, homes, four factories, one school, one theatre, four churches and two Chinese temples. There was, however, limited loss of life.

Ethnic Violence

From December 1996, savage ethnic violence between indigenous Dayaks and immigrant Madurese unsettled West Kalimantan and eventually led the Malaysian government to close the border between Sarawak and West Kalimantan. On 28 December, ethnic unrest broke out in the Sambas area of West Kalimantan. Many houses were destroyed and five people were reported to have died in this initial unrest, which lasted through to 2 January. Several thousand Madurese, some of whom had lived in the area for two generations, fled Sambas for the provincial capital of Pontianak and a local airforce base.

The violence, which had quietened in January 1997, worsened in February, leading to several hundred deaths in what was undoubtedly Indonesia's most bloody episode since the civil war in East Timor in 1975 and the subsequent Indonesian invasion. The spark for the renewal of fighting in early February was an attack on a

Roman Catholic school in Pontianak. There were suggestions that a visit in late January by four Madurese Muslim leaders could have reinflamed Muslim passions. Pictures of Madurese Muslim teachers were also circulating in West Kalimantan, allegedly as a stimulus to Madurese there to take revenge on the Dayaks. Violence of this kind had erupted periodically before, in 1977, 1979 and 1983, resulting in about 40 deaths in total. On this occasion the tensions were more intense, and the violence much more costly in terms of human life.

Social and Labour Unrest

There were several examples of labour and social unrest in the same period. The capital, Jakarta, did not escape. In the Tanah Abang district of Jakarta, on 27 January small traders vented their anger against local *preman* [thugs] and police. The latter had been seeking to clear small itinerant traders from the main streets of the area. A local government office was destroyed in the resulting rioting, which took place only a kilometre or so from the presidential palace.

On 31 January, 7,000 workers at West Java's biggest textile factory in Sumedang rioted, destroying eight factory buildings, five office buildings, four warehouses, the mess occupied by Taiwanese workers on the site and many company trucks and cars. The workers rioted after failing to get a favourable response from management to wage and other demands.

Thousands of workers went on strike and demonstrated in Jakarta and Semarang, Central Java, on 22 April. The factories involved were principally in the garment and footwear sectors. The workers were complaining about delays in paying the new regional minimum wage and were demanding increased meal, transport and overtime allowances. Another factor was an earlier attempt by the ruling political party, Golkar, to create support among workers at the Nike factory in the Jakarta area, which succeeded only in emboldening workers to push their demands more forcefully.

Regional Unrest

Low-level unrest was evident in East Timor, Aceh and Irian Jaya from mid 1996 to late 1997. In February 1997 the press carried reports of a seizure of weapons in the province of Aceh. By March, 63 sophisticated weapons and a substantial amount of ammunition had been uncovered.

The seizures may have resulted in part from security preparations for the visit to the province in February by President Soeharto. Even so, the numbers of weapons found surprised observers, and raised the possibility that Aceh might be facing a return to Muslim separatist violence. (Aceh had experienced Muslim insurrections from 1954 to 1962 and from 1976 to the early 1990s.) The Muslim weekly, *Ummat*, observed that '... at a time when the tone of national politics no longer discredits Islamic movements and even advantages them, the separatist spirit has not been extinguished among some people in the Verandah of Mecca [Aceh]'.

Regional unrest continued after the general elections. Unrest in Irian Jaya at the Freeport mine in August followed the discovery of the bodies of two local Ekari people. Freeport Malaria Control personnel were rumoured to have been involved in their deaths, and Freeport became the target of Ekari demonstrations and sabotage. In one incident, two Ekari men were shot dead by military personnel. Order was soon restored, but not before non-government organisations (NGOs) had linked the fighting to local concerns about neglect of the Amungme people in the distribution of the Freeport Trust Fund.

East Timor continued to present problems for the government. Unrest persisted in the province during July and August 1997, but the main focus was the international scene. During the visit to Indonesia in July of the President of South Africa, Nelson Mandela, President Soeharto granted Mandela's request to meet the gaoled East Timor guerrilla leader, Xanana Gusmao. Soeharto may have felt confident that he could secure greater understanding of his country's position, playing on Indonesia's strong support for Mandela during his struggle against apartheid. He may also have hoped to counter the potential for Mandela's strong pro-democracy values to lead him to support the independence movement in East Timor when he succeeds Soeharto as head of the Non-aligned Movement (NAM) next year.

If this was the strategy, it backfired. On his return to South Africa, Mandela wrote to the President proposing that Xanana be released. Mandela's letter was leaked to the press by the Portuguese after it was sent by 'mistake' to the Portuguese Embassy in Pretoria, rather than to the Indonesian Embassy. The resulting publicity in August focused unwelcome international attention on East Timor and Xanana. Indonesia may still find itself under challenge in the Non-Aligned Movement (NAM) when it is led by South Africa.

THE MOST TURBULENT GENERAL ELECTION CAMPAIGN EVER

The election campaign began officially on 27 April 1997 amid very tight security. The government party, Golkar, had begun campaigning well before the official start on 27 April. It had set itself the task of winning 70.2% of the valid votes cast, up from 68.7% at the 1992 elections. Initially the campaign was quiet. Ultimately, however, the 1997 general election campaign proved to be the most violent in the New Order period.

From late March, Central Java was the focus of much of the election-related violence. It started in Pekalongan in late March, in connection with a Golkar religious event attended by the President's eldest daughter and Rhoma Irama, a prominent entertainer, Golkar candidate and former United Development Party (PPP) supporter.

Pekalongan remained a focus of violence throughout the general election campaign. On 6 April three shops and several vehicles were destroyed in the Buaran district, the same area where violence had erupted in late March. On 10 April an elite housing area in the centre of Pekalongan was attacked by a mob, inflamed by a story that two PPP members had been attacked two days earlier on their way home from a religious event. On 20 April a PPP mob attacked the office of the Deputy Governor. On 23 April about 200 villagers in the Buaran district ransacked the house of an official who had not registered them as voters in the upcoming elections.

In Central Java as well, there was fighting between Golkar and PPP supporters in the *kabupaten* [subprovincial administrative region] of Temanggung in April. On 4 April, fighting between the two groups broke out in a Muslim section of Yogyakarta. On 9 April violence erupted in the Wonosobo area, spreading during the day to the neighbouring *kabupaten* of Banjarnegara. Fighting between Golkar and PPP members occurred in Solo on 20 April.

As the campaign continued into May, PPP meetings and parades became increasingly large and boisterous. The surprise of the campaign was the brief emergence, for the first time since the 1960s, of a 'people's movement' in the form of an anti-government coalition of Muslims and Megawati supporters. Instead of silencing her, the government's actions had succeeded only in creating an embryonic national heroine.

PDI members loyal to Megawati after her removal from the PDI leadership had few choices in this election until Mudrick Setiawan, a leading PPP figure in Solo, proposed that, on this occasion, disaffected PDI members should vote for the PPP. In this way was born the 'Mega–Bintang' phenomenon: the alliance between the radical, non-sectarian nationalists who were Megawati's supporters and the Muslims represented by the *bintang* [star] emblem of the PPP.

The alliance brought disaffected youth and young devout Muslims out onto the streets of Jakarta, Yogyakarta and other Javanese cities in combined mass demonstrations. These became increasingly violent as supporters of the alliance clashed with Golkar supporters and security forces. Golkar was eclipsed in the campaign by this phenomenon. Its flags were torn down, and its street demonstrations were relatively poorly attended because its supporters began to fear physical violence. Throughout the campaign Megawati herself behaved in a restrained manner, preserving her image as a symbol of purity, courage and honesty in the eyes of politically frustrated Indonesians.

The campaign provided a brief opportunity for the expression of popular concerns, spelled out in the illegal banners and pamphlets which proliferated in Jakarta and elsewhere: 'Intimidation—NO'; 'Don't Pocket the People's Money'; 'The PPP Is Purer'; Only the PPP Struggles for Justice'; 'We Are Ready to Vote PPP without Lubricating Money'; 'My Mega My Nation'; 'Mega My Idol'; 'Liberate Yourself from Injustice'; 'Why Have Money when Our Votes Are Pawned?'; 'Corruption and Collusion Are Killing the Little People.'

In addition to the demonstrations in Java, there were attacks on polling stations by separatist rebels in East Timor and Aceh. Seventeen people were killed in election-related attacks in East Timor. Then, on 31 May, a further four military personnel were killed and 24 wounded in another Fretilin attack. One person was killed in the Pidie region of Aceh in a rebel attack which took place on election day.

The worst violence of the campaign occurred in Banjarmasin in South Kalimantan on 23 May, the last day of the campaign there. Golkar supporters disturbed people in a mosque attending Friday midday prayers, and from there ferocious rioting ensued. Shops, the Roman Catholic cathedral, homes, the main hotel and shopping plazas were burned. In one plaza, 133 people—many of them probably looters—lost their lives when they were trapped by fire. A government minister and political and religious leaders from Jakarta were trapped in a hotel and had to be rescued by the security forces. This incident and the growing fervour of the street demonstrations against the government may have encouraged a last minute decision by some Indonesians to vote for Golkar and against perceived Muslim mob violence.

FINAL ELECTION RESULTS

On 23 June, the government announced the final count in the 1997 parliamentary elections. Golkar won 74.5%, PPP 22.43% and PDI 3.07% of the vote. Golkar's final percentage had thus risen slightly compared to the early count. PDI's disastrous showing had not changed. The PPP fared less well than in the early days of the count and eventually won 89 seats rather than the 90 originally predicted, mainly because of the large number of late votes counted in Jakarta from overseas. The final figure for the *golput* [protest non-participation] vote—9.42%—was much lower than in the early days of the count.

For the first time, reflecting the broad dissatisfaction with the government evident during the campaign, PPP scrutineers in particular were assiduous in attending polling stations and reporting and protesting voting irregularities. In this way, the party collected hundreds of accounts of irregularities.

There were various concerns about malpractice in the running of the elections and in the counting of votes. For example, between 6 and 23 June, 1.3 million new votes were counted in Jakarta. They were said to be largely from overseas voters and had a significant effect on the final outcome. The embarrassing non-participation rate fell drastically to 17% in Jakarta. The consequences were more serious for the PPP, whose vote fell progressively further behind that of Golkar, ultimately costing it one projected seat. Its share of the vote in the final official count in Jakarta was only 32.8%, against Golkar's 65.3%. This contrasted with early PPP expectations of a much better result, based on informal scrutineering at some polling stations.

In the final count in North Sumatra, a last minute addition of 64,269 votes for the PDI—amazingly, twice the number of Golkar votes in a 'final' parcel of 103,897 votes—resulted in the PDI winning a second seat at the expense of Golkar. The government had been anxious to help the party following its collapse, and Golkar had even suggested that its own and PPP votes be given to the PDI to boost its representation in Parliament. When the PPP strongly resisted any such assistance, the government resorted to this manoeuvre in North Sumatra to give the PDI at least one extra seat in Parliament. However, the government was unable to save the head of the PDI, Suryadi. He was the leading candidate in the capital, where the PDI slumped so badly that it did not win a single seat. He is now out of Parliament.

In many other cases, the count bore no apparent relation to the votes cast. In one *kecamatan* [subdistrict] of Magetan, East Java, the official count gave the PPP only 700 votes for the entire subdistrict of 64 polling stations, even though PPP scrutineers reported 800 votes for the PPP from just *two* of these polling stations.

The PPP nevertheless succeeded in securing an historic first in Indonesian elections: on 4 June the poll was repeated at 65 polling stations in the Sampang district of Madura following the destruction of ballot papers and of the official forms declaring the poll when PPP supporters rioted on election day. The rioters were enraged that votes had not been counted and the results announced at polling stations, as should have occurred; instead, the ballot papers had been taken to a local government office to be counted. The rioters burned down a church, a Chinese temple, government offices and the homes of government officials. Several were wounded as the Armed Forces resorted to the use of weapons. The unrest continued in Madura up until 6 June.

The PPP had wanted the poll repeated in all of Sampang's 1,033 polling stations, where informal scrutineering indicated that the party had won about 65% of the vote. (Interim official results gave Golkar 60% of the vote.) Initially the government agreed to repeat the poll in 86 booths. Then ballot papers and official declarations of the poll from 21 polling stations were 'found', and the poll was repeated only in the 65 polling stations where ballot papers were still not available. Many registered voters in Sampang protested the limited reballot by not voting: turnout was only 58%. Golkar won 59.1% of the votes cast, but this represented only 33.9% of registered voters.

Muslim anger at the manipulation of the poll continued throughout June. On 13 June rioting broke out in Bangkalan, also in Madura. The most serious rioting was in the East Java town of Jember, which has a strong Madurese population. Local religious teachers (*ulama*) had been whipping up local anger about the election results. One was quoted in *Tempo Interactive* as saying that 'we [*ulama*] only want to give the people the will to oppose'. On 13 June, after the Friday prayer, Muslims burned down the Jember district head's office, banks and shops. Ten people were shot by the security forces, one fatally.

Within the PPP, there was strong pressure not to sign national, provincial and district results on the grounds that the party had evidence of manipulation of the poll and the counting. The PPP faced a dilemma. To sign would disappoint the supporters who had voted for it in great numbers, no matter what the final official results indicated. Not to sign would pit the party against the regime in a political contest and an act of defiance unheard of in the New Order.

Eventually the whole party leadership capitulated. The last provincial-level PPP leader to sign was Darmadi, the party's leader in West Sumatra. The Governor came to his home to tell him that if he did not sign, neither he nor the Armed Forces would deal with him or his party in the future. Darmadi told the media, 'The situation was very tense. I was intimidated'. PPP anger reflected the fact that, according to the official count, the party's vote in West Sumatra fell from 14.5% to 7.7%, reducing PPP representation in Parliament from West Sumatra from two members to one.

In signing, the PPP acknowledged the deep disappointment of its supporters at the official election results. The leadership took the view, however, that it was better to stay in the political process than to boycott it. Also important in the decision was a deeply felt reluctance to engage in confrontational politics. Although many PPP supporters would have wanted a stronger stand, the leadership opted for national stability, thus demonstrating a key factor moderating the impact of popular anger on national politics and buttressing the New Order.

The acceptance by the PPP of the official election results helped to signal an end to the election unrest. Another important factor was clear evidence of government resolve to deal firmly with unrest now that the elections were over. On 13 June, the Armed Forces had already begun to fire on rioters in Sampang and Jember in a departure from the restraint shown during the campaign itself. One person was killed in Jember. On 17 June, the Governor of East Java, where unrest after the elections had been worst, called on his military colleagues to deal firmly with rioters. The Governor and the East Java military commander both expressed the view that the unrest was the work of non-believers and crypto-communists. The Governor added that the 'people' were waiting for the command to act against the rioters. If the Armed Forces did not give the command, he would do so himself.

The Governor's comments followed immediately on the statement by the Minister for Religion, Tarmizi Taher, before senior *ulama* of East Java that it was permitted to take the lives of rioters (*halal darah*). Like the Governor and East Java army commander, he identified crypto-communists as being behind the unrest.

The violence in Jember on 13 June was the last major unrest associated with the general elections.

THE PRESIDENT AND ISLAM

The prominence of the PPP in the general election campaign and the Mega–Bintang movement were not the only developments of significance for Islam in Indonesia during the year. Of equal if not greater significance were the moves by President Soeharto to enhance his personal standing in the Muslim community by demonstrating his own personal commitment to Islam and cultivating improved relations with Abdurrachman Wahid (popularly known as Gus Dur), the leader of the Muslim organisation, Nahdatul Ulama (NU).

On 2 November 1996, when the President opened an NU meeting in East Java, Soeharto staged a very public reconciliation with Gus Dur. He warmly greeted Gus Dur on his arrival at the meeting, and the pictures of them shaking hands received wide coverage. Gus Dur in turn expressed his support for the re-election of Soeharto. The reconciliation between the two is reported to have been arranged by two of the President's children, his second son Bambang Trihatmodjo, and his eldest daughter Siti Hardijanti Hastuti Rukmana (Tutut).

Gus Dur has had a shaky relationship with the President, who did not formally receive him when he was re-elected Chair of the NU in 1994 against Soeharto's wishes. Gus Dur has been outspoken in his criticism of excessive political control and the lack of democracy in Indonesia. He is also, however, among the strongest supporters of Pancasila, the state philosophy of religious tolerance.

The two continued to enjoy good relations into 1997, enhancing the President's standing among rural Javanese Muslims. The new accord may also stem from Gus Dur's concern about threats to Pancasila and religious tolerance as reflected in the Islamic violence which took place from October 1996 until June 1997. As a strong supporter of religious tolerance, Gus Dur may be signalling his view that, whatever differences he and the President may have, the New Order is the best guarantee of a tolerant, orderly society. Gus Dur's endorsement is among the most significant Soeharto could receive.

Gus Dur's public support for Soeharto coincided with a shift in the President's own approach to the Indonesian Association of Muslim Intellectuals (ICMI). Under the patronage of the Minister for Research and Technology, Dr Habibie, ICMI had provided an avenue for access to political influence for educated Muslims—a Muslim constituency different from and less numerous than the NU's. Gus Dur has been vociferous in his criticism of ICMI, which he sees as wanting to use Islam to gain political power, and as discriminating against other religions in seeking to advance the interests of Muslims.

By mid 1997 ICMI had fallen from presidential favour. Outspoken ICMI members were purged from the list of candidates for the MPR. The Editor of the ICMI newspaper *Republika* was moved aside after the general elections. Hartono, who became Minister for Information in late June, joined ICMI in August. He could become its head, replacing Habibie. Hartono's job in ICMI would be to muzzle the dissident ICMI members whom Habibie had failed to control.

The end of the fasting month presented the President with an opportunity to demonstrate his personal commitment to Islam. Indonesians sense that they witnessed a historical watershed on the night of 8 February when the President led the prayers ending the fasting month at the National Monument in Jakarta:

> It was the first time in more than half a century that a President of the Republic of Indonesia beat the drum repeatedly and said the *takbir* prayer in front of his people. Many were disbelieving. But many more were moved to tears of joy. That deep emotion was felt not just by all the officials present, such as Vice-President Try Sutrisno and the head of the Council of Indonesian Ulama, Hasan Basri, who went on to say the *takbir* prayer after Pak Harto but it also relieved the hearts of all believers in Indonesia. [It] struck them as a sign ... that a Muslim President has become one with his people, the majority of whom are Muslims' (*Forum*, 10 March 1997).

Tens of thousands were present at the ceremony and millions more watched on television. The President's standing in the eyes of Muslims can only have risen as a result. At the same time, the regime is creating expectations of a much stronger government commitment to Islamic values and government facilitation of a sustained trend to a more Islamic society.

For much of his life Soeharto has been reputed to be a follower of traditional Javanese spiritualism and mysticism. Inevitably, therefore, there was speculation about his apparent reorientation to

Islam. Had the President undergone a conversion, and if so, when and why? Was his present commitment to Islam a stratagem to garner Muslim support in the elections? Had he always been a committed Muslim, but had concealed the depth of his faith? Or did he recognise that Indonesian society was overwhelmingly Islamic, and that he and his family must ride the Muslim wave if they were to control events?

Dr Habibie answered some of these questions. He told the press in an interview at the end of February that the President had always been a committed Muslim, but had hidden it because Indonesia needed Western aid and investment (*Forum*, 10 March 1997). In the interview Habibie quoted at length an Egyptian scholar, who had written that Muslims in Indonesia stagnated as a result of this turning to the West. Soeharto was said to have now realised that the West exploited its position by seeking to 'change the political map' of Indonesia and impose Western cultural values.

Habibie quoted Soeharto as having said to him in 1991 that he could now begin to express his faith openly because 'the time has come to stand on our own feet; we can carry out our development on our own'. Habibie is quoted in the interview as saying that he thinks Soeharto believes that if the West now wants to pressure Indonesia, 'Pak Harto can kick them in the pants'.

THE NEW PARLIAMENT AND THE NEW MPR

On the basis of the May 1997 general election results, Golkar has 325 members in the new Parliament, PPP has 89 and the PDI 11. The Armed Forces appoints another 75 members. The results assure Soeharto of the support of a minimum of 80% of the MPR, which elects the President in March 1998.

On 17 September 1997, the President announced the 1,000 members of the 1997–2002 MPR. They include ministers, senior Armed Forces officers, senior politicians, bureaucrats, all provincial governors, academics, artists, business people and religious leaders. Among those representing business are Bob Hasan and Anthony Salim. The Sultan of Yogyakarta is a member. All the top echelon of the Armed Forces leadership from both headquarters and the regions are members and, as a group, appear to dominate the body.

The President inaugurated the new MPR on 1 October. Its make-up is as follows: 149 regional delegates, including regional military commanders; 100 representatives of professional groups in society;

113 Armed Forces representatives, including 75 Members of Parliament; 488 Golkar members, including 325 Members of Parliament; 134 PPP members, including 89 Members of Parliament; and a tiny 16 PDI members, including 11 Members of Parliament.

THE PRESIDENT PREPARES FOR MARCH 1998

As discussed, the widespread unrest experienced in Indonesia in 1996–97 came to an abrupt halt soon after the elections. Since June 1997 there has been sporadic violence in East Timor and in Irian Jaya. Two police stations have been burned down in rural West Java. Anti-Chinese rioting broke out briefly in Ujung Pandang in September. But the level of unrest across Indonesia has been remarkably low compared to the 12 months to June 1997.

The suspension of unrest underlines the strength of the President's position as he moves towards the crucial five-yearly reaffirmation of his power in the 1998 MPR session. Golkar dominates the new Parliament and the MPR, thanks to its having won 74% of the vote in the general elections. After fighting a vigorous campaign, the entire leadership of the PPP signalled its acceptance of the political status quo for another five years at least. The regime has effectively neutralised Megawati Sukarnoputri, who might have presented a popular challenge to the President and his succession plans.

Although now 76 years old, the President has lost none of his strategic acuity, and still masters the political and policy detail of the presidency. No-one matches his 32 years' exercise of power in Indonesia. Military and civilian friends, allies and potential rivals have died, retired or fallen out with him. He alone has remained on centre stage since 1 October 1965. He has an unparalleled command of the job. In the eyes of Indonesians, he stands out as the person to bring development, stability and international stature to Indonesia.

The President therefore has every reason to expect renomination for a sixth term, and already there have been calls for this. In September, Achmad Moestahid Astari of Golkar called on the organisation to renominate the President quickly; as we have seen, Gus Dur of the NU also called for his renomination. Others will follow their lead.

Although he should have every reason to be confident about the March MPR session, the President is taking every possible step to ensure that he has total control and that nothing disrupts the session.

The measures he has taken in 1996 and 1997 are comprehensive, wide ranging and, in some cases, a departure from past practice.

Ensuring Armed Forces Loyalty

The President has placed in strategic military positions officers who have worked with him closely as adjutants or in the Presidential Guard. General Wiranto, a former adjutant, is the new Chief of Staff of the Army. Lieutenant-General Subagyo, the new Deputy Chief of Staff of the Army, is a former group commander in the Presidential Guard. The new head of the Army Strategic Command (Kostrad), Lieutenant-General Sugiono, is a former presidential adjutant and head of the Presidential Guard. The President's son-in-law, Major-General Prabowo Subianto, is head of the Special Forces Command (Kopassus).

General Wiranto and his deputy control the army. Lieutenant-General Sugiono and Major-General Prabowo control powerful combat units invaluable in guaranteeing control of the streets in any challenge to political stability. The placement in such positions of those whose loyalty can be trusted absolutely bolsters the position of the President enormously.

Warning the Media

General Hartono, Army Chief of Staff until being replaced by General Wiranto, was unexpectedly made Minister for Information in June 1997. Hartono can be expected to take a tough stand on any reporting inimical to the interests of the government in the lead-up to the MPR session and beyond.

Aligning New MPs

The President has acted to ensure that all Members of Parliament share a basic commitment to New Order values by holding a 'crash course' (*pembekalan*) in government and New Order ideology for all military and civilian members of the 1997–2002 Parliament. The former Minister for Information and current head of Golkar, Harmoko, was to take responsibility for this 'crash course'.

Restoring Sweeping Security Powers

It was in the first session of the 'crash course' that the President called for the restitution by the 1998 MPR of sweeping internal security powers, which had been extended each five years by successive MPRs until they were allowed to lapse in 1993. The 1988 MPR decision (Article 2, Decision VI), which is certain to be restored in 1998, gave to the President '... authority to take all necessary steps for the safeguarding and maintenance of the unity of the nation, and the prevention and elimination of social upheaval, the danger of a resurgence of the 30 September Movement/Indonesian Communist Party and other subversive movements'.

Forestalling Armed Forces Adventurism in the MPR

The Armed Forces appointments to the 1997–2002 Parliament again reflect a concern to ensure loyalty. In addition to five Regional Military Command (Kodam) commanders, the President has appointed the Armed Forces Chief of Staff for Sociopolitical Affairs, Lieutenant-General Syarwan Hamid, head of the 75-member faction. According to *Jawa Pos* of 10 July 1997, Syarwan Hamid guarantees that there will be 'no repetition' of the incident at the 1993 MPR session when Lieutenant-General Harsudiono Hartas, allegedly acting on behalf of Armed Forces interests, nominated Try Sutrisno as Vice-President before the President had indicated his own preference.

Silencing Critics

Apart from neutralising Megawati Sukarnoputri, the government has also moved to silence dissent from within ICMI. The President's dissatisfaction with ICMI critics was reflected in the dropping of some key ICMI members from the list of candidates for membership of the 1997–2002 MPR.

The government has also moved with surprising firmness to silence through the courts even minor opposition figures. As a result, young political figures from the People's Democratic Party (PRD), including its leader, Bambang Sujatmiko, have been given long gaol sentences. So also have minor parliamentary figures such as Aberson Sihaloho from the Megawati wing of the PDI and Sri Bintang Pamungkas from the PPP. The secretary of Soebadio Sastrosatomo, a leading political veteran from the 1950s, was put on trial in August 1997 for publishing, earlier in the year, a strongly anti-Soeharto pamphlet written by Soebadio. Major-General Theo Syafie was removed from Parliament in May 1997 for indicating that he accepted the legitimacy of invalid voting (*golput*) in the general elections. All these actions are a clear warning to dissidents that, with the general elections over, criticism will be crushed.

Are all these preparations merely the natural caution of a very careful and successful political leader? Or is there another plan in train for which these elaborate preparations and the tightest possible control of the political scene are necessary?

THE VICE-PRESIDENCY AND THE PRESIDENTIAL SUCCESSION

The presidency is Soeharto's for as long as he lives—or for as long as he wishes. As he grows older, his choice of Vice-President and his plans for a successor assume increasing importance, because the 1945 Constitution provides that the Vice-President automatically becomes president in the event of the death or incapacity of the incumbent.

There are already several possible candidates for Vice-President in 1998. They include the Minister for Research and Technology, Habibie; the new Minister for Information, Hartono; the head of Golkar, Harmoko; the Coordinating Minister for Trade and Distribution, Hartarto; and the Minister for Planning, Ginanjar Kartasasmita. The present Vice-President, Try Sutrisno, and a former Vice-President, Sudharmono, have also been suggested as candidates, although Vice-Presidents have tended to serve only one term under Soeharto. Another possibility for the vice-presidency is the President's eldest daughter, Siti Hardijanti Hastuti Rukmana, popularly known as Tutut.

It is important to acknowledge that probably only the President himself knows of his plans for the vice-presidency in 1998, and that in the past he has not revealed his preference until just before the presidential election. He may well follow the same course on this occasion. In all likelihood, the current kite-flying about his choice of Vice-President in March 1998 is no more than speculation, and will continue to be uninformed by any firm hint or unequivocal communication from the President himself.

The vice-presidency will nevertheless spark increasing controversy and analysis as March 1998 draws closer. Already, for example, the Deputy Head of the Armed Forces National Resilience Institute (Lemhannas), Dr Juwono Sudarsono, has drawn criticism for restating in September 1997 his view that no civilian merits the presidency *or* vice-presidency before 2005 because current civilian politics are still too divisive. His intervention would nevertheless have pleased those in the Armed Forces who want to maintain its present dominant role in Indonesian politics.

Tutut as Vice-President ... and President?

Tutut has to be considered a possible vice-presidential candidate, and presidential successor, because of the increasingly prominent and successful role she has begun to play in national politics. She was an important campaigner for Golkar in the 1997 general elections. With the apparent demotion of Harmoko following his removal from the Information Ministry in June 1997, Tutut has assumed an even more important role inside Golkar in preparation for the 1998 MPR session. She is spending an increasing amount of time beside her father at state functions. She guarantees continuity with the policies of her father.

It may be that the President believes that his health permits him to postpone the appointment of the Vice-President who will be his successor until 2003. Appointing Tutut to the position in 1998 would, however, clearly signal a commitment to a dynastic succession, and to consolidating her place in Indonesian politics while Soeharto is alive.

Such a move *would* require the careful preparations and the intense degree of control with which the President has been arming himself in 1997. The current tightening of security and the measures Soeharto has taken to ensure loyalty and compliance in the MPR could all be designed to counter any adverse reactions to a Tutut succession. The President may be resubordinating and realigning all the forces of society—Islam, the Armed Forces, the media, business—to ensure that there are *no* discordant voices from the grandstands when the baton change occurs.

The Impact of the 1997 Financial Crisis and Drought

In June 1997, just after the annual Consultative Group on Indonesia meeting in Tokyo, the then health of the Indonesian economy enhanced the President's standing as the March 1998 MPR session approached. Two developments early in the second half of 1997 have somewhat altered the outlook.

The financial crisis since July 1997 has substantially dented confidence in the economy. The value of the rupiah remains unstable, investor confidence is low, and the impact of the crisis into 1998 is still unclear. Exports should be stimulated, and government revenues buoyed, by the depreciation of the rupiah. The government has sought to counter the slump in investor confidence by cutting a substantial number of public and private sector projects, and it has invited the International Monetary Fund (IMF) to assist in restoring confidence. But the deleterious effects of the crisis will not be gone by the March 1998 MPR session.

Another emerging problem for the President is the drought, which will have a negative effect on rural incomes and which has also exacerbated the serious forest fires in Kalimantan and Sumatra. By September 1997, the fires were having serious effects on air and sea transport and on public health in Indonesia and the wider region. The President himself has had to take responsibility for the resulting dangerous levels of air pollution in Singapore and Malaysia.

Economic development has delivered steady growth in incomes and welfare to ordinary Indonesians. In late 1997 and early 1998, however, the preoccupation will not solely be with the benefits of development, but also with the impact of the economic crisis, and the negative consequences for incomes, health, safety and lifestyle of the current drought and the poor management of Indonesia's forests.

In these circumstances the judgement of the President in March 1998 may now be that he needs to be seen to stay firmly at the helm of government, and that any plans he may have had for the succession must await a more propitious time.

24 October 1997

3

THE ECONOMY:
MACRO AND MICRO REFORM
FOR GROWTH

Sri Mulyani Indrawati

INTRODUCTION

The high-performing economies of Indonesia, and East Asia more generally, have experienced turmoil since mid 1997 due to the financial and foreign exchange crises. The problem began with the Thai baht, then spread to the Filipino peso, and on to the currencies of Malaysia, Indonesia and, more recently, Korea. Initially, the speculators' attack resulted in a significant movement in exchange rates in all these countries, and others, including Singapore. Even though Indonesia is facing different problems from those of Thailand, because their exchange rate regimes have not been the same, the two countries' fundamental macroeconomic indicators— such as current account deficits, large short-run capital movements and sizeable external debt—have been quite similar.

The external shock in the second quarter of 1997/98 (that is, July–September 1997) is expected to affect Indonesian economic growth significantly. Before the crisis, economic growth in 1997 was projected to be 7.6%, slightly lower than the 7.8% recorded in 1996 and 8.1% in 1989–95. The crisis since mid year has triggered a tight money policy, as indicated by soaring interest rates for Bank Indonesia Certificates (SBI), to around 30%, and sharply reduced liquidity. Consequently, economic growth has declined significantly.

In the early period of the shock, growth was forecast to decline to around 5–6%, but by the end of the year this figure was looking much too optimistic. It is probable that the New Order is about to experience its first year of sharply negative growth. The financial sector was, of course, the hardest hit at first, but the crisis is now spreading to virtually all other sectors. Moreover, the major depreciation of the rupiah will feed into increased inflation in 1997 and 1998.

How Indonesia's economic fundamentals adjust to short-term market sentiment over the next few months will be critical. The weak point in the macroeconomic equation is the external sector, as indicated by the foreign debt and current account deficit. Large current account deficits have been a feature of rapid East Asian growth, but their sustainability depends on several factors (see, for example, Ostry 1996). One is how the capital inflow is used. In Indonesia, the external borrowing that caused the widening of the current account deficit was generally used neither for consumption nor for covering an 'irresponsible' government budget deficit. Rather, it was used mainly to expand investment in order to create growth. Nevertheless, the tremendous investment in long-gestation infra-structure and property sectors has put great pressure on the country's debt service ratio.

A second factor influencing the economy's capacity to sustain high current account deficits is policy flexibility and, related to this, the quantity and composition of external liabilities. Within ASEAN, Indonesia has the largest external debt. From 1990 to 1994, Indonesia shifted from its traditional reliance on long-term, concessional public sector debt to a much greater reliance on the private sector, including large portfolio capital inflows. This has rendered the country vulnerable to short-term shocks. Since 1994/95, increasing foreign direct investment (FDI) has helped to provide a more stable external source of capital.

A third factor has been the degree of openness and the importance of export-oriented activities. Indonesia's high export growth in the early 1990s induced strong capital inflow. Nevertheless, the sluggish export growth in all ASEAN countries since 1995 may have been an element underlying the speculative attacks on the region's currencies. Finally, there is the health of financial institutions. As the main intermediary between saving and investment activities, the banking sector plays a crucial role in the development of the real sector. Notwithstanding the 1980s reforms, Indonesia's banking sector faces many problems. These include the quality of human resources and professional staff, which have affected the quality of credit assessment; the independence of owners' interests; a lack of business ethics; and distortions in the real sector that have adversely affected the allocation of credit for efficient productive services.[23] The above problems have been aggravated by the weak supervision of the central bank, Bank Indonesia (BI), as indicated by some scandals and inconsistent policies that have tarnished its reputation.[24]

AN OVERVIEW OF RECENT ECONOMIC PERFORMANCE

Economic Growth

In 1995 economic growth was dominated by strong domestic demand, in particular household consumption and investment (Table 3.1). Nevertheless, in 1996 a sharp increase of 3.8% in government consumption occurred, up from 1.3% in 1995. This phenomenon was closely related to the increase in civil servants' salaries, and also to spending for the 1997 general election. Investment continued to grow strongly, though at a slightly lower rate than in the previous year. Inventory was run down in 1996, a reflection both of the strong growth in 1995 and diminishing business optimism in the run-up to the 1997 election. The uncertain business climate was also revealed in the very sharp decline in approved FDI, although it needs to be pointed out that part of the decline in 1996 resulted from the approval of a huge petrochemical project in 1995; in fact, the number of projects approved actually rose in 1996.

---

[23]It is sometimes argued that the problems have arisen in part because Indonesia has undertaken liberalisation in reverse sequence, in the sense that financial sector liberalisation began in 1983, before the trade and investment reforms that got under way in the mid and late 1980s. This, it is maintained, created a situation in which the high mobilisation of savings did not correspond to the opening up of the real sector, thus creating a 'bubble economy' in which most credit was allocated to a limited number of sectors (especially real estate) and investors. On the other hand, it should be noted that much of the sequencing literature is really concerned with the opening of the international capital account (see, for example, Williamson 1994), and that Indonesia has been open in this respect since 1970.

[24]The cases of Dwipa Bank, Bank Perniagaan and Bank Artha, and BI's handling of 'problematic' banks such as Bank Pacific, are illustrations of this proposition.

TABLE 3.1 Economic Growth, 1995–96
(%)

Variable	1995	1996	Value (Rp trillion)
GDP	8.2	7.8	413.8
By expenditure			
Domestic	10.1	8.7	416.7
1 Consumption	8.6	8.5	
Household	9.7	9.2	
Government	1.3	3.8	
2 Gross domestic capital formation	14.0	12.2	
3 Change of stock	7.3	−9.9	
Foreign			
Exports	8.6	6.3	111.1
Imports	15.8	9.6	113.9
By sector			
1 Agriculture	4.2	1.9	62.9
2 Mining and quarries	6.7	7.1	38.0
3 Manufacturing	10.8	11.0	101.7

Source: BPS.

Among the sectors, there was a sharp decrease in agricultural growth, down to 1.9% in 1996. This decrease may adversely affect the government's food (especially rice) self-sufficiency objectives and necessitate rice imports, especially in view of the extended dry season that occurred in late 1997. In contrast, the manufacturing sector continued to experience high growth of 11% in 1996. Cooking oil, food products and beverages all experienced rapid expansion, and there was a recovery in wood and paper products. The petroleum industry also enjoyed strong growth because of increased domestic consumption of gasoline, especially LNG and LPG as alternative fuel sources.

Other economic sectors that grew strongly in 1996 included utilities (12.6%), construction (12.4%) and most service sectors. The sources of growth within the construction sector are believed to have changed recently, away from the focus on office buildings, apartments and middle to upper-income housing, and towards smaller scale housing construction. The construction deregulation package of June 1997 is expected to accelerate this trend. Among services, the financial sector had the highest growth rate (10.5%), followed by communications and transportation (8.6%) and trade, hotels and restaurants (7.6%).

In 1996, there was a significant change in the sectoral composition of Indonesia's GDP. The continuing decline in agriculture's share relegated it from the second to the third largest sector in the economy, behind not only manufacturing but also trade, hotels and restaurants. A poor agricultural season accelerated the shift, but the fundamental longer term cause was the rapid structural transformation of the economy. There is widespread concern that, if this structural change in output shares is not accompanied by a corresponding transformation in the labour force, there could be widening sectoral, and spatial, gaps emerging, with serious distributional implications. As shown in Table 3.2, there continue to be wide disparities in labour productivity among major sectors. Although this is to be expected, and is indeed one of the key factors driving structural change, the distributional implications cannot be ignored.

Monetary Conditions

Inflation reached 6.6% in 1996, the lowest rate since 1985/86, when it was 5.2%, and well below the 1995 rate of 9%. The decline was caused by a combination of tight monetary and fiscal policies and an improved distribution of basic wage goods by Bulog, the national food logistics agency. The latter is illustrated by the trends in food prices, which actually declined in the first quarter of 1997. A similar decline occurred in the housing component, reflecting the weakened state of the property market in 1996.

It will be difficult to maintain this low rate of inflation in view of the recent exchange rate shock and its effect on imported inflation, a long dry season that may contribute to increased food prices, and the possibility that the petroleum subsidy will be abolished, leading to increased administered prices and distribution costs. Depending on the response of the monetary authorities, these various supply-side factors were expected to result in increased inflation in 1997 of around 8%, even before the crisis began to take hold. Into 1998, serious inflationary problems are likely to emerge.

TABLE 3.2 Structural Change and Productivity by Sector, 1985–95

Sector	GDP (% of total)		Labour Force (% of total)		Productivity[a] (Rp million)	
	1985	1995	1985	1995	1985	1995
Agriculture	24.1	16.1	54.7	44.0	1.28	1.75
Industry	35.6	41.8	13.4	18.4	7.71	10.88
Mining	11.8	9.3	0.7	0.8	51.25	55.18
Manufacturing	17.9	23.9	9.3	12.6	5.60	9.04
Utilities	0.6	1.1	0.1	0.3	16.65	19.79
Construction	5.3	7.6	3.4	4.7	4.62	7.75
Services	40.2	42.1	31.9	37.6	3.66	5.36
Total	100.0	100.0	100.0	100.0	2.90	4.79

[a]In constant 1985 prices.
Source: As for Table 3.1.

The money supply, broadly defined (M2), grew by 29.6% in 1996, slightly higher than in the previous period (27.6%). The rapid growth continued until the first quarter of 1997, when it began to slow. Meanwhile, the money supply, narrowly defined (M1), grew at 22.2%, broadly similar to the 1995 rate of 23.7%. Most of the M1 growth was due to the rapid increase in demand deposits. Banking credits expanded by 26.3% in 1996/97.

M1, M2 and credit all grew much faster than the targets set by the government, of 16%, 18% and 17% respectively. Domestic factors influencing M2 growth originated mainly from the fast expansion of banking credits, while the foreign factors mainly took the form of increased private loans, realised investment and, more generally, capital inflows induced by the difference between domestic and foreign interest rates. The difficulty of the monetary authorities in controlling M2 and interest rates was due to the government's own sterilisation policy (McLeod 1997). Even though the differentials eased in early 1997, the recent exchange rate shock, and accompanying tight monetary policy, has again widened the gap.

The value of outstanding credit until March 1997, including both foreign currency and rupiah, reached Rp 306.1 trillion, with rupiah

TABLE 3.3 Loan Performance, 1995–96
(%)

	Change		Share	
	1995	1996	1995	1996
Satisfactory	25.7	25.9	89.6	91.2
Non-performing	6.6	4.4	10.4	8.8
Weak	0.9	16.2	2.7	2.6
Doubtful	14.9	5.6	4.4	3.3
Bad debt	1.6	8.0	3.3	2.9

Source: Bank Indonesia.

credits accounting for Rp 245 trillion of this. The biggest credit expansion occurred in services (excluding trade) and manufacturing, which explains the high growth in these sectors. Other sectors that enjoyed rapid credit growth were commerce, reflecting the dynamism in the retail and distribution sectors, and real estate, which prior to the crisis was expected to benefit from recent deregulation measures.

Credit dominates the banking sector, as would be expected, and it accounts for about 77% of its assets. Therefore, the quality of credit portfolios affects the entire banking sector and, through it, the economy. In this context, the problem of non-performing loans continues to be serious (Table 3.3). In 1996, it is estimated that bad debt grew by about 8%, a sharp rise from the growth of 1.6% in 1995. This increase resulted in part from the reclassification of some debts within the non-performing group. However, for what the data are worth, it appears that the proportion of loans that are non-performing actually declined in 1996. The tight money policy introduced in the wake of the current crisis will, of course, exacerbate these problems enormously. Earlier in the year, BI introduced a requirement that at least 80% of foreign loans borrowed by banks be channelled into export credit. The principle here is that this would ensure a sustained capital inflow by generating an increased foreign exchange earning capacity in the export sector. It remains debatable whether, in an undistorted economy, there ought to be special concessions in favour of exports. In any case, the strategy had little impact on the exchange rate once the speculative attack gathered momentum.

TABLE 3.4 Deposit Interest Rates, 1996–97
(%)

Maturity	1996 Q1	1996 Q2	1996 Q3	1996 Q4	1997 Q1
1 month	17.15	16.95	16.88	16.43	15.92
3 months	17.29	17.35	17.25	17.03	16.47
6 months	16.88	16.90	16.93	16.78	16.37
12 months	16.68	16.42	16.93	16.70	16.39
24 months	15.39	15.78	15.87	15.14	15.95

Source: As for Table 3.3.

The trend towards declining interest rates in late 1996 continued into early 1997 (Table 3.4). The decrease was the result of falling inflation, sufficient banking liquidity, and declining interest rates in industrial countries. In the last quarter of 1996 the SBI discount rate was 8.5%, a huge fall from the same period in the preceding year, when the rate was 13.4%. The Money Market Securities (SBPU) discount rate has also been decreasing, but more slowly, to 14.3%, compared with 15.9% at the same point in the previous year. These lower discount rates have resulted in a decrease in interest rates for both domestic and foreign currency savings accounts. However, BI's efforts in reducing interest rates are unlikely to have a long-term impact, and they lack credibility. More recently, the volatility of the rupiah in response to speculative pressure has forced BI to tighten liquidity by, among other things, placing limitations on SBI purchases and temporarily terminating SBPU transactions. These changes resulted in a sharp increase in the interbank interest rate; the overnight rate exceeded 40% in early August 1997, although it later declined to 17–18%.

Balance of Payments

Current Account

Before the crisis, the balance of payments for 1996/97 registered a surplus of $3.9 billion. The current account deficit increased $1.1

billion to $8.1 billion in total, equivalent to 3.5% of GDP. The increase in the current account deficit was due to slower growth in non-oil exports, and higher growth—albeit at a declining rate—in non-oil imports. Exports in 1996/97 amounted to $52.2 billion, an increase of 9.3% on 1995/96, a slower growth than the 13.3% recorded in the previous year. Export growth in 1996/97 was fuelled by the oil sector, which expanded by 18.6% to $12.6 billion, and represented a sharp increase on the previous year's growth of 1.6%. The increase in 1996/97 is explained by a stronger oil price (averaging $20.50 per barrel) and good LNG performance. In turn, the latter has been the result of higher prices, and increased production for the Japanese and South Korean markets.

Non-oil exports in 1996/97 grew much more slowly than those of oil, by 6.6% to $39.6 billion. This compares with the 17.1% growth recorded in the previous year. The growth deceleration occurred in all sectors except agriculture. Non-oil exports are now dominated by manufactures, which account for 76.6% of the total. Considerable diversification within this sector has been occurring in recent years, with products such as electrical appliances, computers and parts, machinery and plastic materials becoming significant export items. However, manufactured exports experienced a similarly sharp fall in growth, from 20.4% in 1995/96 to just 9.3% in 1996/97. Textiles, footwear, paper and pharmaceutical products experienced much slower growth in 1996/97.

Apart from cyclical shocks, which have led to an oversupply of the above products similar to that experienced by several other East Asian countries, notably South Korean, Taiwan, Thailand and Malaysia, the sharp decline in manufactured exports has also been due to domestic structural problems. These include low levels of productive efficiency, an export structure highly dependent on natural resources and unskilled labour, and slow technological development (see Lall, Chapter 7 in this volume). Moreover, the high-cost economy—as manifested in various informal levies—and the high cost of infrastructure, including port facilities, continues to be a problem for Indonesian exporters.[25]

---

[25]These problems are well known in Indonesia, but a recent illustration serves as a reminder of just how serious they can be. According to a source in the Ministry of Industry and Trade, exporters typically have to bear 4,396 levies: 195 from the central government, 941 at the provincial level, 2,802 at the subprovincial level, 419 for various specific services and 39 charged by trade associations (*Bisnis Indonesia*, 7 October 1997).

TABLE 3.5 Non-Oil Imports by Major Commodity Group, 1995/96–1996/97
(%)

Commodity Group	1995/96		1996/97	
	Share	Growth	Share	Growth
Capital goods	21.6	16.3	24.9	27.4
Intermediate goods	72.0	23.3	68.7	5.3
Consumption goods	6.4	56.1	6.4	10.3
Total	100.0	23.4	100.0	10.4

Source: As for Table 3.1.

These domestic problems are compounded by external barriers such as environmental issues, labour standards and anti-dumping charges, which often function as non-tariff barriers on Indonesian exports. The Indonesian government is now beginning to grapple with some of these external challenges. It has established the Indonesian Committee for Anti-dumping (KADI), in particular to assist small-scale exporters, who would not be able to respond to anti-dumping investigations. Another initiative has been the establishment of the Export Strategy Review Team (TIPSE) in an attempt to resolve structural export matters. However, thus far it is fair to say that TIPSE has made little progress in addressing the problems of exporters. The government now also provides assistance in customs procedures, taxation and banking for exporters classified as registered export companies (PET) engaged in the production of 10 main commodities (textiles, wood products, footwear, electrical appliances, paper, processed foods, oil products, manufactured rubber, toys, frozen fish and shrimps). Although well intentioned, selective policies such as these are often unsuccessful because of their potential to create bureaucratic rent seeking.

Imports in 1996/97 amounted to $45.9 billion. They grew at 10.5% in 1996/97, more slowly than in the previous year (21.6%). This decline was mainly due to a sharp decrease in the growth of non-oil imports, particularly raw materials, which expanded by 5.3%, down from 23.3% in the previous year (Table 3.5). Overall, non-oil imports grew by 10.4% in 1996/97, down from 23.4% in 1995/96. Oil imports, by contrast, rose from 7.1% to 11.4%, due to increased domestic fuel consumption and an increase in transportation services and industrial development. Imports of capital goods also rose, from 16.3% in 1995/96 to 27.4% in 1996/97. These figures suggest that manufacturing activities slowed in 1996 and early 1997. Nevertheless, realised manufacturing investment increased, owing to the pipeline of already approved projects. Looking more closely at capital goods imports, locomotives, ships and aircraft all recorded significant increases. Such imports are consistent with government efforts to improve the nation's transportation infrastructure, although it is worth observing also that these sectors are under the purview of Professor Habibie's Agency for the Management of Strategic Industries, BPIS.[26] In view of the currency crisis, imports in 1997/98 will fall sharply.

Merchandise trade in 1996/97 registered a surplus of $6.3 billion. This surplus was generated by the oil sector, with a net balance of $8.2 billion, an increase of $1.5 billion over the previous year. However, the deficit on non-oil trade rose from $1.4 billion to $1.9 billion. Services trade incurred a deficit of $14.4 billion, which was mainly due to freight services ($4.6 billion) and interest payments on foreign loans ($6.3 billion). Among service exports, the major contributions in 1996/97 were tourism ($6.4 billion) and labour remittances ($0.8 billion). The repayment of foreign loans amounted to $2.7 billion, while interest payments on foreign loans totalled $3.6 billion. The latter indicates that an increasing portion of Indonesia's total foreign loans consists of commercial loans.

This phenomenon has attracted much public attention, since the loans are of short maturity and have higher interest rates. Another concern is based on the argument that private sector loans have recently been directed increasingly towards less productive sectors such as property, which might saddle the country with much uneconomic foreign debt. In response, in March 1997 BI introduced several regulations aimed at restricting certain types of foreign commercial borrowing. The regulations include a limitation on the maximum daily reserve of 30% of bank capital for foreign commercial loans of up to two years, and an obligation to provide export credit of 80% of a commercial loan. Non-bank commercial loans now

---

[26]See Bishry and Hidayat, Chapter 8, this volume.

require quarterly reporting to BI and the Ministry of Finance. This measure was taken in anticipation of the greater risk caused by large private foreign loans, and to improve the composition of capital inflows into Indonesia.

Capital Movements

Net capital inflows amounted to $11 billion in 1996/97, with the government sector absorbing $5.4 billion of this. Capital outflow from this sector amounted to $6.2 billion, thus resulting in a net deficit of $0.8 billion. The increased capital outflow from the government sector was the result of accelerated prepayments of loans with an interest rate above 7% ($1.7 billion), which was financed by the sale of shares in state-owned enterprises, and from the government's budget surplus. In 1996/97, government loan payments were accelerated in consideration of the relatively low loan prepayment penalties. Some multilateral institutions are considering waiving such penalties altogether, to provide Indonesia with additional debt relief (World Bank 1997a).

Over the same period, the private sector absorbed $20.9 billion of capital inflow. FDI accounted for $8.6 billion of this, and portfolio investment for $1.7 billion. Consequently, the biggest portion of capital inflows was still in the form of short-term flows, mainly SBIs, SBPUs and time deposits. The composition of these flows bears directly on Indonesia's external stability and its exchange rate volatility. At the same time, private capital outflows amounted to $9.1 billion, and mainly took the form of loan payments. Therefore, the net private capital inflow was $11.8 billion. The larger role of the private sector in international capital mobility induced BI to float bonds in the US capital market at the end of July 1996, in the form of 'Yankee Bonds' worth $400 million, with a coupon rate of 7.75% per year for 10 years. These bonds were expected to be a benchmark for private commercial loans and an indicator of the government's international reputation, besides functioning as an alternative form of long-term finance for the government. In March 1997, BI signed a stand-by loan of $500 million, bringing its total stand-by loan to $2 billion, which could be used both as a contingency for international liabilities and as a tool to withstand balance of payments pressures. In toto, the balance of payments recorded a surplus of $2.9 million and resulted in foreign exchange reserves of $19.9 billion, equivalent to 5.2 months of non-oil imports.

The Exchange Rate

Given the relatively strong balance of payments position through to mid 1997, the government and most market players were not expecting a currency crisis.[27] Hence the turbulence in foreign exchange markets—which became a serious problem for Indonesia in the second quarter of 1997/98 as the contagion effect of a similar shock in ASEAN regional currencies spread—came as a shock to government and business alike. In fact, during 1996/97 the rupiah had shown a tendency if anything to *appreciate*, caused by increased short-term capital inflows induced by interest rate differentials. Although these differentials had decreased in the first quarter of 1997, they were still considered large enough to attract short-term funds into the country. The nominal rupiah depreciation vis-à-vis the US dollar was 3.4% in 1996/97, lower than the 5.1% depreciation of 1995/96.

The currency's stability in 1996/97 is evident in the fact that there was no formal central bank 'intervention' during the year, unlike the one case in 1995/96 and the 14 that occurred in 1994/95. This was made possible also because of the widening of the intervention band set by BI, from 3% (Rp 66) to 5% (Rp 118) on 13 June 1996, and to 8% (Rp 192) on 11 September 1996. In the first quarter of 1997/98 BI again widened the intervention band, to 12%, in response to a strong speculative movement against the rupiah in the last week of July 1997. On 14 August the government decided to let the currency float, although initially it sought to influence the float through a high interest rate policy. Since then, however, the rupiah has slid alarmingly, quickly crashing through the Rp 3,000/$1 barrier, then, as rumours spread about the President's health, slumping to below Rp 4,000, and in the second week of December falling to·almost Rp 6,000. The rupiah fell initially because of the contagion effect, but later because of a growing lack of confidence by market players in the fundamentals of the Indonesian economy and the political situation.

During 1996/97 and the first half of 1997/98, it appeared that the rupiah's volatility was mainly due to speculative and political factors, as indicated by several attacks on the currency. In March

[27]The Thai baht crisis emerged in mid May 1997, but at that stage no precautions were taken in Indonesia by government or privately owned companies with significant foreign debt. It was later revealed that only 40 of the 207 publicly listed companies with foreign currency-denominated loans had taken out any hedging arrangements.

1996, for example, the rupiah depreciated against the dollar as the result of devaluation rumours triggered by a rapid increase in the current account deficit. Then, at the end of July 1996, the rupiah again depreciated, by about Rp 30, in response to the political unrest of 27 July 1996. Another indication of increased uncertainty at that time was the rise in the volume of daily transactions to $14 billion, compared to the normal daily volume of $8 billion. Over this period, BI intervened by allocating $500 million to prevent the rupiah from depreciating further, in addition to widening the intervention band. One year later, at the end of July 1997, the rupiah again depreciated by Rp 150 in just one day, following a wave of speculation about ASEAN currencies. The latter movement was large by the standards experienced thus far, although of course small in terms of the volatility that was soon to be experienced.

In analysing the speculation on the rupiah, it needs to be noted that Indonesia's macroeconomic fundamentals were almost universally judged to be sound through to the middle of 1997. It had high and stable economic growth, inflation had been consistently below 10% for over a decade, loan repayments had been punctually observed (in addition to a strategy of accelerated prepayment in the past year), and the balance of payments was stable. All these conditions were the result of deliberate government policy.

The managed floating exchange rate policy previously pursued by BI was often considered to be unsustainable, given the fact that it was in no sense a 'clean' float. BI was often forced to intervene in the market to maintain the stability of the rupiah, especially when the spread was very narrow. BI's policy was to maintain a stable exchange rate, at least in relation to the US dollar, and if anything to aim for a slightly undervalued currency to support Indonesia's export competitiveness. This goal was obviously difficult to achieve during the periods of large capital inflows, as the resultant balance of payments surpluses put pressure on the money supply.

An excessive reliance on one instrument, monetary policy, to achieve two objectives—price stability and targeted exchange rate movements—continued to bedevil the monetary authorities, and underlined the difficulty of attempting to pursue an independent monetary policy with a fixed exchange rate and an open capital account. The resulting high interest rates had continued to induce large short-term capital inflows. In this context, macroeconomic policy coordination becomes very important, especially the mix between monetary and fiscal policies. Nowadays, fiscal policy functions more as a tool for generating government revenues than as a means of improving income distribution and resource allocation. The World Bank (1997a), for example, has proposed that fiscal policy be oriented towards these distributional and resource allocation objectives. An illustration of how fiscal policy could play a role in an area considered to be the preserve of monetary (credit) policy relates to the property sector: the concern that 'excessive' credit is being allocated to this sector could be addressed by reforming the tax regime through implementation of a more effective property tax, rather than by the imposition of credit controls aimed at lending to the sector.

THE MICROECONOMIC AGENDA

Finding solutions to the microeconomic problems affecting efficiency and productivity is central to the maintenance of Indonesia's economic growth in the medium and long term. The deregulation momentum since the early 1990s has unfortunately slowed down, but there has been some modest progress over the past year.

Trade and Investment Policy

In early July 1997, a new deregulation package was announced. It consisted of the usual tariff reductions, on this occasion affecting 1,600 items: 1,461 manufacturing products, 136 agricultural products and three health products. Among the items affected were crude palm oil and frying oil, whose tariffs were cut from 10–12% to 5%. Government policy on crude palm oil was expected to be more progressive, however, in view of the controversy surrounding foreign investment earlier in the year. Since March 1997, foreign investors have been banned from making (new) investment in palm oil plantations, except in the eastern part of Indonesia. The sudden prohibition on new FDI created uncertainty among foreign (particularly Malaysian) investors interested in the sector. The reversal is ironic since it is contrary to the essence of the 1994 deregulation, as stated in Government Decree (PP) No. 20/1994, which permitted up to 95% foreign equity in such projects. In fact, inconsistencies in government policy towards FDI, and towards business in general, have been a major problem throughout 1996 and 1997. The national car policy and the reversal on palm oil have been among the more publicised examples.

In attempting to justify its decision on FDI in palm oil, the government argued that it was giving priority to domestic investors, especially small and medium-scale interests, and participants in the Transmigration Community Enterprise (known in Indonesia as PIR trans), to seize this opportunity in palm oil plantations, especially in the western part of Indonesia, while directing foreign investors to the eastern region. This expectation that local investors would be able to invest successfully in palm oil may not be well founded. Small investors face numerous problems, including high interest rates and difficulty in obtaining credit.[28] Moreover, the structure of closely related upstream industries poses a barrier to entry, since they are effectively duopolistic in structure, being dominated by two of Indonesia's largest conglomerates, Eka Tjipta Widjaja's Sinar Mas group and the Salim group. Then there are bureaucratic issues, such as obtaining legal authorisation for the investment and securing access to land, particularly the transfer of the legal right to use land from the state-owned forestry company, Inhutani, to private enterprise.

Currently, about 2.2 million acres of land is being used for palm oil plantations. These are owned by several conglomerates and state-owned enterprises, as follows: Salim group, 150,000 acres; Astra group, 200,000 acres; Raja Garuda Mas group, 45,000 acres; and state-owned enterprises, 441,000 acres. The remainder is owned by foreign (mostly Malaysian) companies, which have realised about 50% of their planned investments. Indonesia's current production of crude palm oil is some 5 million tons, still less than that of Malaysia (about 8 million tons). However, Indonesia is forecast to become the world's leading producer of crude palm oil, surpassing Malaysia by the year 2010 with an estimated production then of around 12.3 million tons. This forecast is based on Indonesia's strong comparative advantage in palm oil, centred on its land and labour cost advantages.

In the administration of trade, some changes have been introduced. The most important of these relates to sugar, which previously could be imported only by Bulog. During the early stages of the currency crisis, in July–September, the Coordinating Economics Minister, Saleh Afiff, raised the more general issue of Bulog's monopolistic powers, a topic that triggered some enthusiastic public

---

[28]The capital requirements are considerable. According to estimates prepared by the Head of Plantations, Department of Agriculture and Forestry at the Indonesian Chamber of Commerce (Kadin), the funds required to open up a 10,000 acre investment with a 10-year planting period are about Rp 100 billion.

discussion. However, the issue then faded away without any follow-up or official government response, which is perhaps an illustration of the political sensitivity of the subject.

The Fishing Industry

Regulatory changes have also occurred in the fishing industry. Government approval has now been granted for the import of trading and used fishing vessels regardless of weight or size. In an earlier deregulation (of 4 July 1996), the government allowed fishery companies to import new or used vessels with a minimum weight of 100 gross tons at a ratio of 5:1, meaning that every purchase of five imported vessels was contingent on the purchase of a domestically built one. One problem arising from this deregulation was the high cost of purchasing a domestically made vessel of 100 gross tons, which could reach Rp 5 billion. Another difficulty lay in meeting the credit requirements set by the banking system, which required the company to provide equity equivalent to 30% of the vessel price.

The Directorate-General of Fisheries has issued a regulation which allows the purchase of new/used vessels inside or outside the country. The new regulation consists of six requirements to be met when purchasing a vessel: (1) to be in partnership with a fishing cooperative or group; (2) to invest in at least two vessels with a minimum weight of 30 gross tons or which have a 90 hp inboard engine; (3) to invest in a chilling business; (4) to invest in a tin-canning factory; (5) to invest in infrastructure, such as ports, piers or docks; and (6) to purchase a new domestically manufactured vessel. Under this regulation, the requirement to purchase a domestic vessel can be substituted for at least one of the other requirements, partly alleviating the problems associated with the previous regulatory arrangements. Nevertheless, complaints about high interest rates and credit requirements are still being voiced, especially by small and medium-scale investors who do not have access to the foreign funds required to import a vessel. The limited number of vessels in the industry is one explanation for the low level of fishery production in Indonesia, which in 1995 was only $1.4 billion, or about one-quarter of the estimated sustainable yields.

Local Finance

Another important reform introduced during the year related to local government finance. A July 1997 deregulation measure reduced the number of local government taxes from 42 to only nine categories, and the number of retributions (levies) from 192 to 30. These reforms are expected to increase efficiency and simplify licensing procedures, as many of the taxes were cumbersome and uneconomic to collect. It remains to be seen how effective the deregulation will be in tackling problems related to the 'high-cost economy', as the structure of incentives for government employees has not been changed.

The issue of civil service reform remains one of the key items on the government's unfinished reform agenda, and it was a popular topic during the general election campaign. However, there is no evidence yet that the government is seriously intent on major reforms.

Small–Medium Industry

Another much discussed microeconomic issue concerns the role of small and medium-sized enterprises (SMEs). According to Law (UU) No. 9/95 of 1995, small-scale industry is officially defined as a business entity with assets (other than buildings and land) of less than Rp 200 million and whose annual sales are less than Rp 1 billion.[29] President Soeharto launched the small-scale strategic alliances (*kemitraan*) program on 15 May 1996, but thus far it has not shown any significant progress. This may not only be because of the wide gap between large enterprises and SMEs in productivity and efficiency, but also because market characteristics and industrial policies are unable to ensure fair competition among firms of different scale.

Although the President has frequently stated that vertical integration is not permitted if it harms SMEs, in practice there are no clear government policies to implement such an objective. Rather, the government has preferred to use moral suasion in this area, in addition to a range of focused SME support programs such as the Small Business Loans (KUK) scheme, financial and management assistance funded by a 1–3% (previously 1–5%) levy on the profits of state-owned enterprises, financial aid from venture capital, protection in the form of reservation of special business activities for SMEs,[30] and the reservation of specific locations for SMEs

---

[29] According to BPS, there are about 34.2 million businesses in this category.

[30] According to Presidential Decree (Kepres) No. 30/1995, certain business activities are reserved for small enterprises. This regulation has not been effective, as illustrated by the entry of big business in sectors that are

(especially those operating in the retail trade sector).[31] In addition, BI has also raised the upper limit for credit under the KUK scheme from Rp 250 million to Rp 350 million, and more recently Rp 750 million.[32] Whether this will have any positive effect remains to be seen. The problems of small firms have in the past more commonly been in meeting loan requirements, rather than in the volume of loans available to them.

Using the official 1995 definition of small industry, there are clearly large differences in productivity between large and small firms, as illustrated by data on manufacturing labour productivity (Table 3.6). Differentials of 10:1 or more are common, and there is no evidence that they are declining. Correspondingly, wages are much higher among large firms. These differences in productivity restrict the opportunities for mutually profitable partnerships. The most promising arrangements in manufacturing may be through subcontracting networks. However, these are difficult to implement in the absence of incentives (perhaps tax concessions) or sanctions for non-compliance. To develop such partnerships, the first requirement is to increase the capability and productivity of SMEs. Some such

---

nominally only for small enterprises. A striking example is the soy sauce industry, which is actually dominated by Indofood, owned by the nation's largest conglomerate, the Salim group. The government justified Indofood's entry by stating that it was required to establish partnerships with small enterprises. A similar situation occurred in the manufacture of simple agricultural implements (*pacul*), which is also reserved for small enterprises. In actual fact, the industry is dominated by the state-owned PT Boma Bisma, which is ignoring the regulatory requirements.

[31]Under Kepres No. 51/1996, a foreign company is not allowed to enter retail business other than franchising in Indonesia. In a recent decision by the Minister of the Interior, giant domestic retailers such as Matahari and Hero are prohibited from operating in subdistrict areas (Dati II) in order to protect traditional markets.

[32]The increases were based on a BI regulation in April 1997, which also stipulated that 25% of all credit (both in rupiah and in foreign currencies) had to be allocated to small enterprise. (The previous regulation concerning KUK affected 20% of the rupiah credit of domestic banks.) BI is now employing some additional incentives to increase lending to small enterprise. Those banks able to lend more than 25% will be granted additional resources from BI, while those falling below the target will be fined 2% of the value of their shortfall. To support banks to reach these targets, an institution known as Yayasan Dabanas is being established to provide banks with better information about small business commercial opportunities.

**TABLE 3.6 Productivity and Wages in Small and Large
Manufacturing Industries, 1994[a]
(Rp/month)**

ISIC		Productivity (value added/worker)		Wage/ Worker	
		Small	Large	Small	Large
31	Food	1,609	30,281	10,242	156,954
32	Textiles	1,907	13,754	14,416	119,182
33	Wood	2,306	14,849	15,019	145,140
34	Paper	2,913	31,562	16,396	205,529
35	Chemicals	2,153	30,282	13,216	229,022
36	Minerals	1,824	28,855	10,755	198,149
37	Base metals	2,316	112,034	39,554	119,163
38	Machinery	2,733	42,341	15,533	77,878
39	Miscellaneous	1,930	9,261	13,008	106,079

[a]Large firms are defined as those with assets of more than Rp 1 billion, and
small firms as those with assets of less than Rp 150 million.
Source: Computed from BPS, Statistik Industri [Industrial Statistics], Jakarta.

partnership arrangements do already exist, such as 'nucleus–plasma'
in agriculture and franchising in the trade sector. Nevertheless, the
SMEs' weak negotiating position, a result of limited capabilities
and market imperfections, has often led to poor outcomes for them,
including delayed payment and unbalanced risk allocation.

The success of the government's attempts at moral suasion has
depended largely on whether large companies are willing to
participate, and this in turn has depended on the range of incentives
(or penalties) meted out by the authorities. For example, the
requirement that public companies offer 20% of their shares to
cooperatives ultimately resulted in an offer of only 2%. Again, there
is the so-called Jimbaran initiative. This originated at a meeting of
the country's conglomerates in Jimbaran, Bali, where the owners
announced their willingness to form strategic alliances aimed at
minimising the gap between large companies and SMEs. In early
1997, a similar program was established, based on 79 large
companies who formed the Coordinating Agency for the Implementation of National Business Strategic Alliances (BKPK-Kunas).

Both these groups appear to be essentially political in their orientation, as few SMEs appear to have an opportunity to develop productive 'partnerships' with large companies.

Some of the challenges facing SMEs were identified in an LPEM (1996) research report as including the following: the limited quality of human resources, with most staff being high school graduates; the old capital equipment still in use; limited marketing networks in which, for example, 62% of output among firms surveyed was restricted to local (same province) markets; and production problems related to design, packaging, standardisation and quality.

The Labour Market

The most interesting recent labour market developments relate to wage policy and a new labour law. There has been a significant growth in nominal and real wages throughout the 1990s. Prior to the crisis, nominal wages had increased about three-fold over this period, and real wages had approximately doubled. In 1996, the Ministry of Manpower prescribed an increase in minimum wages that effectively increased nominal wages by 30%. This year the increase was more moderate, about 10%, or 3–4% in real terms. For the first time, too, certain export-oriented industries were granted an exemption from the increase.

The new labour law was presented to Parliament in draft form in January 1997, and has been widely discussed in public forums. Under the draft law, the right to strike is limited by the provision that employers are not required to pay wages during strikes. The freedom to form unions within firms is also curtailed, while the law stipulates the physical location where strike activity may take place. Owing to these and other controversial aspects, it was decided to postpone debate on the bill until after the elections.[33]

CONCLUSION

The currency crisis has damaged Indonesia's record of macroeconomic stability. To sustain economic growth, the government needs first of all to restore this stability. The volatility of the rupiah was caused by both a 'transactions demand' motive (that is, the need for foreign currency to finance imports and service the foreign debt) and a

---

[33]See Tubagus (1997) for more details.

speculative motive. The latter developed from a lack of confidence on the part of economic agents—both foreign and domestic—in the country's economic fundamentals, and in the ability of the government to manage the situation.

The measures taken thus far have been commended by the International Monetary Fund and the World Bank. However, structural problems still haunt the economy. In the present crisis, strong, consistent and credible signals from the government are badly needed. These signals should give a clear direction of the policies to be pursued by the government to ensure market efficiency. Transparency, consistency and fairness become crucial issues, especially as they relate to the weakest aspect of government economic policy, namely the microeconomic arena. Unlike macro policy, in micro policy authority is shared by many departments, often with poor communication and coordination. Policies pursued by the Ministries of Industry and Trade, Manpower, Education, Transportation, Agriculture, and Research and Technology are often not consistent and do not point in the same direction. This reflects both political pressures at the sectoral level and differences in policy and ideological approaches. This competition has if anything heightened since the government's impressive electoral victory last May.

How the government responds to these challenges, both macro and micro, will have an important bearing on the country's capacity to deal with its most serious economic crisis in 30 years.

30 October 1997

4

SOME COMMENTS ON
THE RUPIAH 'CRISIS'

Ross H. McLeod

Dr Sri Mulyani has given us a wide-ranging survey of recent economic developments. In this note, I would like to concentrate on what has undoubtedly been the major item in the list of matters she discussed: the so-called rupiah 'crisis', beginning in mid July 1997 with the outbreak of speculation against the rupiah, and culminating in the floating of the currency on 14 August.

At the outset, I want to suggest that the July–August devaluation of the rupiah was highly atypical, and possibly unique in history (although this assertion is probably more appropriately applied to the group of three roughly simultaneous devaluations that included also those of the Malaysian ringgit and the Philippine peso).[34] Indeed, it will not surprise me if entire books are written on this subject in the months to come!

WHAT CAUSED THE DEVALUATION?

One suggested cause for the sudden devaluation is concern on the part of investors about the quality of Indonesia's microeconomic policies. I find this explanation unconvincing. There have always been aspects of Indonesian microeconomic policy for economists to complain about. But I would argue that these are relatively minor in aggregate—in

---

[34]I shall restrict my comments here solely to the case of the rupiah.

the sense that they have been compatible with a remarkably high economic growth rate sustained over many years—and have not been becoming worse over time; on the contrary, there has been steady improvement in the 'big picture'. In particular, there has been no sudden deterioration in the quality of policy making sufficient to justify the sudden withdrawal of capital from Indonesia that has caused the currency to decline so abruptly.

Nor can we find any explanation in poor macroeconomic performance. Indonesia's fundamentals were very strong prior to the 'crisis'. Growth was high, having averaged roughly 8% per annum since the late 1980s. Inflation had been moderate for many years, averaging about 9% per annum since the beginning of the 1980s. It had actually fallen significantly in 1996, and remained low up to and after the devaluation. Finally, the balance of payments position was very strong. International reserves, already high to begin with, had increased by no less than 38%—$11 billion—in the 10 months through July,[35] reflecting the authorities' attempts to prevent the rupiah from appreciating. It is this fact in particular that makes the devaluation highly atypical. Most devaluations are preceded by reasonably lengthy periods of declining reserves or, in relatively few cases, are undertaken deliberately by governments rather than being brought about by market forces (as in Indonesia in November 1978, for example).

On the basis of these observations, it seems difficult to conclude other than that the sudden fall of the rupiah was set in motion by psychological factors: specifically, the fear that what had happened to Thailand shortly beforehand would happen also to Indonesia—regardless of the fundamentals, and regardless of important differences between the two economies. This psychological unease was then greatly amplified by the bewilderment of the Indonesian community resulting from the authorities' sudden decision to float the rupiah. This was a dramatic reversal of the policy of government control of the exchange rate that had been followed not only throughout the New Order period but, indeed, ever since Independence.[36] In such circumstances, if enough people

---

[35]The official reserves figure was understated; the figures given here are based on those that can be found in the central bank's balance sheet.

[36]Of course, the exchange rate had been changed and the intervention band widened several times in recent years, but no government had ever publicly discussed the possibility of allowing it to be determined by the market.

believe there is a significant risk of devaluation, it becomes a self-justifying prophecy.

The implications of this are, first, that the government's macroeconomic policies were fundamentally sound; and second, that nothing really has changed, except that people are now more aware of the exchange rate risk of bringing capital into Indonesia. The important thing now, I would argue, is to do everything possible to convince the markets that this episode is an aberration, and that policies will continue much as before. It is important to emphasise that there is little in the overall spectrum of government policy that can be blamed for the devaluation. Moreover, the government should be praised for the promptness with which it acted when it became clear that there was a massive tide of speculation best accommodated by floating the rupiah, rather than embarking upon a macho attempt to defend a level for the currency that the market deemed much too high.

WHAT WILL THE CONSEQUENCES OF THE DEVALUATION BE?

Many, perhaps most, observers appear to believe that economic growth will be reduced as a result of the devaluation—and indeed, that this is the dawning of a new era, in which countries such as Indonesia will never again experience the sustained high growth rates of the past. I contend that there is little or no basis for these concerns, and note that few of those who have expressed them have articulated any clear argument as to why this should be so. Most of Indonesia's fundamentals are unchanged. The exception is capital inflow, which is, for the time being at least, greatly reduced—if not negative. If this remains the case, the growth rate will indeed slow.[37] But if the return to capital remains high in Indonesia relative to the world outside—as can be expected, given that its

---

[37]It is interesting that the authorities had previously been concerned about the high level of capital inflow (and, therefore, the current account deficit), because of its negative impact (through the real exchange rate) on production of tradeable goods and services. Now that capital inflow has fallen, there may emerge a stronger appreciation for its growth promoting impact on the economy as a whole. As the singer Joni Mitchell once put it: 'You don't know what you've got til it's gone'!

**TABLE 4.1 Devaluations and Annual Economic
Growth Rates, 1977–88
(%)**

	GDP	Non-oil GDP	Comment
1977	8.9		
1978 (33% devaluation)	7.7		
1979	6.3		Slight fall following devaluation in November 1978
1980	9.9		Strong rebound
1981	7.9		
1982	2.2		
1983 (28% devaluation)	4.2		Rapid recovery from slow growth after devaluation in March 1983
1984	6.7		Continued acceleration
1985		5.6	
1986 (31% devaluation)		6.1	
1987		5.8	Minor slowdown following devaluation in September 1986
1988		7.4	Strong rebound

fundamental determinants are still much the same—it is hard to imagine that foreign investors and domestic borrowers will not be strongly tempted to exploit that differential to their own advantage by restoring capital inflow.

It may be helpful to view the possible impact of devaluation on growth from a historical perspective. Indonesia has experienced three major devaluations during the last two decades: in November 1978, March 1983 and September 1986. Table 4.1 shows the rates of economic growth prior to, and after, each such episode. First, note that these earlier devaluations were about half as large again as that of mid 1997, ranging from 28% to 33% by comparison with only

19%.[38] Second, note that only in 1978 was there a modest reduction in economic growth (of 1.4% per annum) in the following year. In 1983, the growth rate in the year following devaluation increased significantly (by 2%), while in 1986, there was little change.[39] On the basis of these observations, there can be no presumption that the 1997 devaluation will have a negative impact on growth. Moreover, when we look at the growth rate in the second year following the earlier devaluations, it can be seen that in every case the growth rate was significantly higher than at the time the devaluation occurred.

WHY HAS THE GOVERNMENT BEEN REACTING SO STRONGLY?

The government has felt it necessary to respond to the devaluation by pulling rather vigorously on the macroeconomic levers available to it. Very shortly after the rupiah was floated, it drained a large amount of liquidity from the system, employing methods similar to those used in two previous liquidity contractions (in July 1987 and February 1991; see Cole and Slade 1996, pp. 52, 58), resulting in the cost of overnight funds in the interbank market shooting temporarily to well over 100% per annum The rate most indicative of the stance of monetary policy—that on certificates issued by the central bank, known as SBIs (Sertifikat Bank Indonesia)—was more than doubled, to 30% per annum.[40] The government also made known its intention to reschedule or postpone indefinitely its own spending on a range of infrastructure projects, and eventually announced the details in mid September.

There appear to have been two reasons for following this course of action. First, notwithstanding its decision to float the rupiah, it seems clear that the government regards the devaluation as a setback and wants to limit any further reduction in the value of the currency. Its intervention now, however, is effected through the money market rather than the foreign exchange market. The drastic

---

[38]Devaluation is calculated in each case as the percentage change in the foreign currency value of the rupiah. The most recent case is based on a rate of Rp 2,432/$1 on 10 July 1997, rising to about Rp 3,000/$1 by late September.

[39]Where possible I have used the growth figures for non-oil GDP, since the oil sector is not affected in the same manner by changing currency values.

[40]This was the rate for the three-month maturity.

increases in interest rates were designed to make it very expensive to speculate against the rupiah (that is, to hold foreign currency assets financed with rupiah borrowings), and to encourage capital inflow by offering high returns to investment in rupiah assets. In view of the discussion above, however, which suggests that economic growth is as likely to rise as to fall as a result of the devaluation, it is not clear why the government should be so concerned about the possibility of further weakening of the rupiah.

Second, the Minister of Finance has drawn attention to what he regards as the negative consequences of the devaluation for the budget. In particular, he noted that the cost of the existing subsidy to domestic consumers of petroleum products would increase, since the rupiah cost of oil had risen significantly. He noted also that company tax revenues could be expected to fall because the many companies that had unhedged foreign borrowings would experience large increases in their borrowing costs (and therefore large reductions in profits) as a result of the devaluation.

It is not clear that the devaluation will cause a budgetary shortfall. First, it is entirely possible that the value of the rupiah will recover, at least partially. Reflecting a high degree of confusion and uncertainty in the market, the exchange rate remains highly volatile at the time of writing (late September, 1997). Second, the grounds for increasing the domestic fuel consumption subsidy by failing to increase prices are weak. The subsidy is an extremely inefficient method of redistributing income, and the argument that it is needed to help Indonesian firms compete on world markets is a nonsense—since all that is achieved is to distort the allocation of resources towards firms and industries that use fuels relatively more intensively.[41] Third, the impact on company profits (and hence company tax revenues) is not straightforward. Companies producing exports and import substitutes will receive higher rupiah prices for their output, which could increase profits significantly.[42] Companies which use imports as inputs are adversely affected, but the former effect will dominate.

There are also certain other budget items not mentioned by the Minister that are influenced by the devaluation. One is the govern-

---

[41]It was suggested initially that domestic petroleum prices would not be adjusted upward, but within a week this stance was being reconsidered.

[42]Although most borrowers are understood not to have hedged their foreign borrowings, companies in the tradeable goods and services sector enjoy this natural hedge.

ment's oil and gas sector revenues, which are still a significant proportion of the total, and which increase in rupiah terms because of the devaluation. A second item, surprisingly not mentioned by the Minister (at least, in press reports I have seen), is the cost of servicing the government's foreign debt. Principal repayments are not of concern, since new borrowings are of the same order of magnitude (so the increased rupiah value of new borrowings matches the increased rupiah amount of principal repayments). But interest payments will be about 23% higher (if the exchange rate stays around Rp 3,000/$1). A third is import duties, which tend to increase because of the higher rupiah prices of imports, but also to fall because of reduced import volumes. The net impact here is hard to predict.

I do not have enough information to estimate with any confidence the overall impact of all these items on the budget. In very rough terms, however, extra oil and gas revenues should be sufficient to cover the extra petroleum products consumption subsidy, and increased company tax from profits resulting from higher revenues should more than outweigh reductions because of higher borrowing costs. The net result may well be to generate a small surplus; a significant unplanned deficit seems an unlikely result. Whatever the case, various other budgetary items are always subject to fairly high degrees of uncertainty, and in the past have often diverged widely from their initially planned magnitudes. In this sense, fluctuations introduced as a result of changes in the value of the rupiah are nothing special.

In addition, it should be borne in mind that the government has built up substantial deposits with the banking system as a consequence of its fiscal conservatism in the past. I have reported these elsewhere at around $11 billion (McLeod 1997, p. 13). One would have thought that, if the devaluation does result in the emergence of a substantial deficit, it would be entirely appropriate to make use of these funds to sustain planned budgetary spending.

THE IMPACT OF GOVERNMENT POLICY

By now it should be clear that the danger to Indonesia is not the devaluation itself, but the government's reaction to it. There are no strong grounds for believing an economic slowdown to be an inevitable consequence, but the policy stance of the government after floating the rupiah seems capable of bringing on such a slowdown, quite unnecessarily. Many companies will have been badly hurt by the increase in their borrowing costs—especially those involved in real estate development, which does not enjoy the natural exchange rate hedge available to producers of tradeable goods and services. To push interest rates up significantly simply adds an additional burden, without good reason, and increases the probability that bankers to the companies concerned may also be severely damaged.

Cutting back spending on infrastructure projects poses an additional threat to continued economic growth by virtue of the negative impact on aggregate demand. As with higher interest rates and higher external financing costs, this impact will be especially strongly felt in the construction sector. Moreover, there will also be a negative impact at the microeconomic level, since many of the postponed projects—especially the construction of toll roads and harbour facilities—would appear to have a high social rate of return. After all, observers have been arguing for years that this kind of infrastructure is quite inadequate to the needs of the rapidly modernising Indonesian economy, and that this inadequacy is an important obstacle to growth.[43]

CONCLUSION

The authorities acted wisely, first in widening the foreign exchange market intervention band, and then by floating the rupiah. But they remain unwilling to allow the markets to determine important prices such as the exchange rate and interest rates. This has led to the introduction of policies that reflect a crisis mentality and that seem to exaggerate greatly the magnitude of the 'crisis'. These policies are more likely to harm economic performance unnecessarily, rather than help the economy to adjust to changing circumstances.

19 September 1997

---

[43]It may be argued that some of the postponements were of grandiose projects that were unlikely ever to go ahead in any case, since they were economically unfeasible (for example, the construction of bridges between Sumatra and Malaysia, Sumatra and West Java, and East Java and Madura). Although there may be some minor benefit here in persuading investors that the government does not intend to proceed with these flights of fancy, more sophisticated observers will note that many highly desirable infrastructure projects appear to have suffered the same fate.

PART II

TECHNOLOGY:
OVERVIEW OF THE ISSUES

5

INDONESIA'S SCIENCE
AND TECHNOLOGY POLICIES

Samaun Samadikun

INTRODUCTION

Indonesia is now in the fourth year of its second 25-Year Long-Term Development Plan (1994/95–2018/19), and of its sixth Five-Year Development Plan (Repelita VI). During the past 25 years, the Indonesian economy has grown at an average annual rate of 7%, and now has a GNP (measured at mid 1997 exchange rates here and throughout this chapter) of about $200 billion. The role of science and technology (S&T) during this period has been to support the development process in the fields of agriculture, health, housing, infrastructure, natural resources, the environment, industry and defence, and to contribute to the solution of socioeconomic problems. In the current 25-year plan, S&T has been designated a development priority.

The Asia–Pacific region has been growing rapidly, a process which, notwithstanding the current turbulence, can be expected to last well into the 21st century. The Deutsche Bank, for example, predicts that in the year 2004 world GDP will reach $39 trillion, with the US, Japan and Europe generating about 57% of the total, Asia (excluding Japan) 31% and other regions 12%. But by 2019 this picture will have changed dramatically: world GDP is predicted to almost double to $76 trillion; the share of the US, Japan and Europe will decline to 32%; Asia's share will rise to 51%; and that of other nations will also increase, to 17%.

Indonesia is planning to sustain its present annual economic growth rate of about 7% during the second 25-year plan. If this growth rate can be achieved, Indonesia's GDP will approximately double by the year 2005, and almost quadruple, to about $800 billion, by 2019. With an estimated population of at least 230 million people in that year, per capita income should be around $3,500. Moreover, according to recent World Bank estimates, China, Indonesia, India, Brazil and Russia will be contributing about 50% of total world trade by the turn of the century.

INDONESIA'S S&T POLICY AND ITS IMPLEMENTATION

Indonesia's S&T policy is outlined in a decree by the People's Consultative Assembly (MPR), together with guidelines covering other major areas such as the economy, industry, education, welfare and defence. Activities in the S&T sector are expected to produce basic science, applied science, technology and production techniques. The policy guidelines issued by the MPR apply to both the five-year and 25-year plans.

Policies for implementing S&T strategies in each Repelita are issued by the Minister of State for Research and Technology through a publication entitled *Punas Ristek* [National Priority Program for Research and Technology]. This guidebook is based on proposals from the National Research Council (DRN) to the Minister. A budgeting mechanism in the form of a 'one-door policy' is used as the policy instrument for implementing the program. Detailed planning for the implementation of S&T policies is conducted by the research institutes residing within the technical ministries (Health, Industry and Trade, Agriculture, Mining and Energy), as well as by national research institutes (Indonesian Institute of Sciences, Atomic Energy Agency, Space Agency) and, in a limited way, universities. Some kind of division of labour is being implemented in the Indonesian system similar to that normally practised in industry, namely between research conducted at the corporate and at the divisional level. Long-term basic research is conducted by universities and national laboratories, while applied R&D is carried out mainly by the ministries. This is in addition to engineering work done by manufacturing enterprises.

Quality control is incorporated into research activity through selection and monitoring panels, and peer review. The mixing of public and private funding for research is a new development in Indonesia, with the first jointly funded activities being implemented in the 1995 budget. Cooperation with venture capital and banking institutions has been organised to ensure continuous funding for a full range of activities, from research through to development, engineering, production and marketing.

Total expenditure on R&D amounts to about 0.26% of GDP at present, with the private sector contributing only around 25% of this. The Second 25-Year Plan has set a target for total annual research funding of 2% of GDP by the end of the period, with private sector participation making up 70% of the total. Total government funding for the S&T sector in 1996/97 was about $400 million, and this is projected to increase to about $16 billion by 2020. In view of the steep projected increase in national research spending in the years ahead, especially by the private sector, special incentives will need to be devised, and socially acceptable success criteria formulated, to achieve the necessary social and political support to obtain the required public funding.

HUMAN RESOURCES DEVELOPMENT

Programs to improve the quality and quantity of Indonesia's human resources are funded through several budgets. Tertiary education is funded by the education sector, and post-tertiary professional development by the S&T sector. The role of the private sector in tertiary education is considerable: there are about 500 private tertiary education institutions, compared with only about 40 such public institutions. Total enrolment in tertiary education is about 2.2 million students. This corresponds to an enrolment rate of 15%, which is still low. The rate is projected to increase to 25% by 2020.

In 1989, a parliamentary decree was issued covering education in general. That part of the decree related to higher education stipulated a common standard of performance evaluation for both public and private higher education institutions. To achieve this, a National Accreditation Board for Higher Education was established by ministerial decree in 1995. The Board has reported the results of its first evaluation, of the quality, efficiency and relevance of 1,500 study programs at major public and private higher education institutions, to the Minister of Education and Culture.

THE INSTITUTIONAL FRAMEWORK FOR S&T
IN INDONESIA

Indonesia is experiencing rapid structural change as it diversifies from agriculture to industry, and shifts from a reliance on oil and other natural resource-based exports to manufactures. To facilitate this transformation, an appropriate techno-structure will need to be built. S&T activity will be just one part of this, but looking ahead, the role of S&T institutions will be increasingly important in making the transition a success.

The establishment of an institutional framework for the management of S&T began very early in Indonesia's national history. Responsibility for the development of science was assigned first to the Ministry of Higher Education and Science in the 1950s, then to the Ministry of Research during the 1960s and early 1970s, and subsequently to the Office of the Minister of State for Research and Technology. The DRN, which was set up in 1984, drafts S&T policy and sets research priorities. These are then submitted to the Minister of State for Research and Technology for approval. The DRN has also been given the task of monitoring, evaluating, guiding and controlling the implementation of research programs. At present it has 164 members, from academia, national and departmental research institutes, and industry.

The DRN consists of five groups, representing national development needs rather than scientific disciplines. These groups are: (1) basic human needs, including health, agriculture, food and nutrition, housing and education; (2) natural resources, energy and the environment; (3) industrialisation, including manufacturing industry, small industries, engineering industries, and science, technology and human resources for industry; (4) defence and security; and (5) social dynamics, economy, law, culture, philosophy, politics, law and regulation.

The Indonesian Academy of Sciences (AIPI), established in 1991 by parliamentary decree, has 27 members grouped in five commissions. These commissions represent the disciplines of engineering sciences, medical sciences, basic sciences, social sciences, and art and culture. AIPI is more of an advisory body than an implementing agency. It publishes its reports and recommendations in the form of public documents, and is the highest scientific body in the country. Members of AIPI are entitled to join the DRN.

Priority S&T Areas

Several S&T areas have been identified as priorities because of their potential to make a substantial contribution to accelerating the pace of national development in the years ahead. These areas are biotechnology, medical technology, food technology, product and production engineering, materials science, chemistry and process engineering, energy technology, electronics and informatics, and environmental protection technologies.

Downstream Policies

To justify the use of public funds, most S&T development activities are expected in the long term to boost economic or social welfare. The need for industries in Indonesia that can meet present and future domestic demand for physical infrastructure, and create wealth through value added activities, has provided the basis for planning the development of national S&T capabilities. The industries that have been targeted as meeting national needs include aeronautics and aerospace, maritime and shipbuilding, land transportation, telecommunications, energy, engineering, agricultural machinery, defence and related support industries. Since the late 1970s, 10 industries have been designated 'strategic industries', including aerospace, shipbuilding, railroads, telecommunications, electronics, explosives, steel and machine goods.

Measuring Technological Capability

The methodology devised to improve the capability of the strategic industries—commonly referred to in industrial circles in Indonesia as 'begin at the end, and end at the beginning'—consists of four measurable stages (see also Bishry and Hidayat, Chapter 8, this volume). The first of these is for a firm to achieve the capability to manufacture a marketable product, usually through a licensing agreement. This should generate the cash flows needed to allow the company to develop its capabilities further, while giving it a 'feel' for the total system for eventual unbundling of the product into subsystems. The introduction of the product should also generate feedback from consumers about reliability and customer service, which can be used as an input to improve the next generation of the product. In a mature industry, this is usually the end stage of the product development cycle, coming after research, development and engineering have been completed.

Mastery of the first stage—production and marketing—is measured by the firm's ability to compete in the market based on quality, price and delivery time. It is also at this point that preparations are made for the second stage, the technology integration stage. This mainly involves training enough human resources to be able to integrate existing technologies into the design of a new product, and building facilities where the new design incorporating capability acquired during the first stage can be tested and produced.

The implementation of the second stage is measured by the success of the new product in the marketplace, again using such market criteria as price, quality and delivery time. Preparations also need to be made for the third stage, the technology development stage. This involves training skilled personnel, and providing additional testing and production facilities for the next new product to embody the newly developed technology. New technology produced through the integration of existing science is considered an economic commodity, so that economic criteria are also applied to its evaluation.

The successful implementation of the third stage is once more measured by the marketability of the new product embodying the newly developed technology, using the criteria employed in the previous stages. Preparations for the fourth stage, the scientific development stage, now commence. In this stage the firm is able to build on the valuable assets accumulated during the previous three stages, including the number of skilled people with certified professional qualifications in production, engineering and design, other knowhow developed within the company, and its supporting techno-structure. The fourth stage is necessary to maintain competitiveness in the product line and to become a centre of excellence in the technology.

The strategic industries have been applying this methodology consistently over the past 15 years, and are at various levels of development. Most notable is the aircraft manufacturer, PT IPTN, which is at present in stage 3. IPTN started out in the late 1970s assembling helicopters and fixed wing commuter aircraft through licensing agreements, selling them worldwide to pass the first development stage. In the mid 1980s the firm launched a 35-seat commuter aircraft, jointly designed with Spanish aircraft company CASA, which was later certified and marketed worldwide. IPTN now has about 20,000 employees, including some 4,000 engineers. It recently launched a newly designed, 70-seat, wide body commuter aircraft, the N-250. This aircraft uses home-grown fly-by-wire technology, making it the most advanced product in its class. IPTN has set up joint venture companies in the US and Europe for the manufacture and marketing of the N-250. It is thus set to become a technology exporter, and possibly the first Indonesian company to export high technology through royalty agreements.

Next in line is PT PAL, the shipbuilding firm, which has launched a modern 500-berth passenger ship for interisland routes, and PT INKA, which designs and manufactures high-speed railway carriages. Less successful are the telecommunications and electronics strategic industries, which are still in stage 1, and which are trying to identify major products to be used as a platform for their second stage of development.

Other Technology Development Models

The development of an S&T capability in the chemicals industry is following a different path. A systematic effort has been pursued since the 1960s to raise the industry's capabilities in plant design and engineering. At present, several public and private engineering companies are providing design, engineering and contracting services for complete fertiliser and cement factories—mostly for the domestic market, although they are slowly beginning to market their services in neighbouring countries.

With Indonesia now the world's largest exporter of LNG, efforts are being made to develop this industry's plant design and engineering services. The Natuna gas field, with a projected investment of $38 billion in the next 10 years, will be used as a platform for the development of technological capability in this high-technology area. The special nature of Natuna gas, which contains 70% CO_2, requires separation plants that use cryogenic techniques and involve the reinjection of the CO_2. Huge offshore structures are needed to anchor such installations.

The capabilities that Indonesia develops as it tackles projects like Natuna will be needed as it continues to develop its natural gas resources. Together with research institutes, industry and Pertamina (the state-owned oil company and the owner of the gas field), the AIPI and the DRN have set up committees to plot an upward path for the country's technological capability. The intention is to enable Indonesia to capture most of the projected investment, and to build up a national techno-structure to facilitate the implementation of even larger national development projects in the future.

S&T POLICY ON FOREIGN INVESTMENT

The huge flow of foreign capital into Indonesia in the last 15 years has brought with it a commensurate inflow of technologies, embodied in the associated equipment, organisation, information and human resources. Programs have been devised to increase the contribution of local companies to support these investments, with the objective of improving their competitiveness in the international market and of introducing the discipline of quality, price and delivery time to the local companies. A project supported by the World Bank, the Industrial Technology Development Project (ITDP), is being implemented to provide funds and technical assistance for these types of activities. The DRN is a member of the steering committee, to guarantee the continuation of the project after the termination of World Bank funding.

Indonesia's S&T policy on international investment can be summarised as follows: 'to be good partners in tackling the world together'. This policy is not new, but has been practised since the time when most foreign investment in Indonesia was in the oil and gas sector. Indeed, the Indonesian scientific community and foreign investors have succeeded in making the country's oil and gas industry one of the best in the world.

INTERNATIONAL COOPERATION

ASEAN undertakes S&T cooperation through its Committee on Science and Technology (COST). Member countries[44] have in common their support for economic development through S&T as part of a broader national development agenda. The S&T indicators commonly used in ASEAN are similar to those employed elsewhere, and include the ratio of R&D expenditure to GDP; industrial R&D expenditure by the public and the private sector; the number of scientists and engineers engaged in R&D; scientists and engineers as a percentage of the labour force; and patent applications by domestic and foreign inventors.

Although these indicators are routinely used to assess the general state of S&T in each member country, comparisons between countries are not easy owing to the region's diversity. Indonesia, which has the largest population of all the ASEAN countries, has

---

[44]ASEAN has nine members: Brunei Darussalam, Indonesia, Malaysia, the Philippines, Singapore, Thailand, Vietnam, Myanmar and Laos.

concentrated on developing basic industries to exploit its rich natural resources in oil, gas, minerals and timber. Singapore, on the other hand, owing to its size and location, began by trading and later shifted to manufacturing and finance. While both Indonesia and the Philippines started with import substitution, Malaysia and Thailand were quicker to emphasise export orientation.

ASEAN has agreed to implement the ASEAN Free Trade Area (AFTA) by 2003, and discussions are being held to bring it forward to the year 2000. Member countries are preparing for this date, and considering how best to mobilise S&T to reap maximum benefits from the opportunities created by the enlarged market potential of AFTA. In 1992, total value added produced by Indonesia's manufacturing sector was twice that of Singapore and larger than that of the Philippines; according to projections for 2020, value added in Indonesian manufacturing will be about twice that of South Korea currently.

Cooperative efforts in the field of S&T are now being extended to APEC. During the 1995 meeting of APEC's S&T ministers in Beijing, it was agreed that member economies should accelerate their common efforts in science, technology and research. Indonesia strongly supports this initiative.

LOOKING AHEAD

A new *Punas Ristek* is being prepared by the DRN, and will be submitted to the Minister for implementation during Repelita VII (1999–2004). Several issues needing special attention have been identified, as follows.

- *Excellence and self-reliance in several key technologies, to be achieved by the end of the Second 25-Year Long-Term Development Plan.* The strategic industries will be the driving force behind this effort. Participation from other industries and from the scientific community will also have to be mobilised. The recently launched, top-down research, development and engineering (RDE) initiative, RUSNAS, is a step in this direction.

- *International competitiveness.* This is a new program to support manufacturing industries to become more competitive in the world market. Support is most needed in production techniques and in the training of skilled operators and maintenance staff.

To achieve these goals, about 100 new polytechnics will be established in the next 20 years. Manufacturing laboratories are also being established at universities.

- *Development of technology-oriented small and medium-sized enterprises.*

- *Intellectual property rights.* A conscious effort will be made to import international technology through licensing agreements, and to monitor the trade in technology. To participate actively in this technology trade, the production of domestic patents will need to be intensified.

- *Development of the eastern part of Indonesia.* The challenge here is how the S&T sector can participate in such development while using the opportunities thus created as a springboard for the development of new technological capabilities.

Indonesia still has a long way to go to develop its technological innovation capabilities, creating wealth through technology. Its present S&T policies are directed towards improving the quality of work performed by engineers and scientists in their professional endeavours. Some shifts in priorities, from institutional building towards 'people' building, are being implemented. With the limited resources available, a good balance must be struck between bottom-up and top-down activities.

6

DETERMINANTS OF INDONESIA'S INDUSTRIAL TECHNOLOGY DEVELOPMENT

Thee Kian Wie*

In view of the relatively sluggish growth of Indonesia's manufactured exports in recent years, particularly among unskilled labour-intensive and resource-intensive products, some policy makers and economists have argued that Indonesia can no longer continue to rely on these exports to sustain rapid industrial growth. Instead, it is argued, Indonesia should henceforth concentrate on developing industries that are more technology and skill intensive, which in turn would require improved industrial technological capabilities. To understand how and why manufacturing firms would make a concerted effort to raise their technological capabilities, it is important to identify the major determinants of industrial technology development in developing countries, including Indonesia.

MAJOR DETERMINANTS OF INDUSTRIAL TECHNOLOGY DEVELOPMENT

Based on extensive empirical research on the process of industrial technology development in several developing countries, Lall (1993,

*I wish to acknowledge with thanks the valuable comments on an earlier draft of H.W. Arndt and Colin Barlow.

pp. 730–743) concluded that there are five major sets of factors that have favourably or adversely affected such developments. These are the incentive system, human skills, technological information and support services, finance, and science and technology (S&T) policies. This chapter assesses these five sets of factors in the context of industrial technology development in Indonesia.

The Incentive System

Whether manufacturing firms deem it important to make the necessary investments in upgrading their technological capabilities depends on the incentive system they face. This incentive system is largely shaped by government policies, particularly in the area of macroeconomic policies, the trade regime and domestic competition policies.

Macroeconomic Policies

The New Order government has from the outset attached great importance to sound macroeconomic policies. It was realised that maintaining macroeconomic stability was essential to encourage firms to undertake the long-term capital investments necessary for rapid and sustained economic growth. Despite the fact that during the past three decades the Indonesian economy has been buffeted by several major shocks—such as the financial crisis in the state-owned oil company, Pertamina, in early 1975, the steep decline in the price of oil in the 1980s and, more recently, the currency crisis of mid 1997—the Indonesian government has never shrunk from taking the necessary measures to contain these shocks and restore macro-economic stability. As a result, its record on controlling inflation has been fairly good, although Indonesia's inflation during the past decade has always been slightly higher than that of most of its Asian neighbours, except for the Philippines (Hill 1996a, p. 7).

The Trade Regime

Since the end of the oil boom in the early 1980s, the Indonesian government has introduced several deregulation measures, including successive trade reforms, to reduce the 'anti-trade bias' of its trade regime. These measures included a gradual but steady reduction in tariff protection and non-tariff barriers (NTBs), specifically quantitative import restrictions, and the introduction of a duty exemption and drawback scheme for export-oriented companies in

May 1986, which enabled these firms to procure their inputs at international prices. Despite these trade reforms—including the latest package of mid 1997, which lowered tariffs on a number of products—further reductions in protection are needed to correct the remaining, still significant, anti-trade bias of the trade regime.

Domestic Competition Policies

While trade reforms have led to increased import competition, domestic competition is still subject to extensive regulations and restrictions. Obviously, these regulations adversely affect the competitive business environment in Indonesia, as they unnecessarily increase the costs of doing business (giving rise to longstanding complaints about Indonesia's 'high-cost economy'). They also reduce efficiency and limit economic opportunities, often for the less privileged small businesses, which tend to lack political and administrative connections. Restraints on domestic competition include marketing controls, pricing, industrial licensing, public sector dominance in certain industries, and controls and 'taxes' (including illegal levies) on intracountry trade (World Bank 1997a, p. 118). These various regulations and restrictions should be reduced, if not abolished outright. By improving the competitive business environment for manufacturing firms, private and state-owned alike, through the removal of price distortions caused by import protection and restrictions on domestic competition, Indonesia's scarce resources could be deployed more efficiently, and firms' competitiveness would be enhanced.

Human Skills

One of the key determinants of a developing country's industrial technology development is the skills and work ethos of its human resources, specifically its technical human resources (scientists, engineers, programmers, technicians) as well as the managers who deploy and manage these technical human resources. While basic literacy and low levels of education are adequate during the early stages of industrialisation, at higher industrial levels advanced training in science and engineering and more widespread vocational education become increasingly necessary in order to cope with rapidly evolving industrial technologies. More complex industrial technologies also necessitate on-the-job training by firms, including the development of a range of specific skills (Lall 1990, p. 45).

During the past quarter of a century Indonesia has made rapid strides in education, as is clearly indicated in Table 6.1. The primary school gross enrolment rate, which was just 62% (13 million students) in 1973 (World Bank 1997a, p. 64) rose quickly to 115% and 100% in 1980 for male and female students respectively as a result of the introduction of compulsory primary education. Using revenues from the oil booms of the 1970s, the government initiated an ambitious primary school construction program extending to rural areas all over the country. This factor, together with the abolition of primary school fees in 1984 and the vastly increased output of primary school teachers through a crash training program, accounted for the widely acclaimed success of Indonesia in having achieved universal primary education within a relatively short period of time. As a result of the rapid spread of primary education, the adult illiteracy rate dropped from 43% in the early 1970s to 16% in 1995 (Long 1997, p. 17).

The data in Table 6.1 show that, while Indonesia's educational achievements in primary education are comparable to those of its Asian neighbours, in secondary education it is behind them, with the exception of Thailand for male students. These differences are especially pronounced with regard to female students, even though in 1980–93 the gender gap in secondary enrolment has narrowed appreciably. Tertiary education has also expanded during the past quarter century, but here too student enrolments are still far behind levels in neighbouring countries. Moreover, the rapid expansion of education has not yet been sufficient to produce an adequate number of skilled workers at all levels to sustain Indonesia's rapid industrial growth and upgrading. Despite the rapid spread of primary education, nearly one-third of all primary school pupils drop out before graduation. As a result, the educational standard of the Indonesian workforce remains low, with 68.6% of the total workforce having completed or dropped out of primary school, 28.6% having completed either junior or senior secondary school, and only 2.6% having completed tertiary education (Long 1997, p. 17).

While secondary (including vocational) and tertiary education in Indonesia still needs to be expanded, a much greater emphasis should also be given to raising the *quality* of education at all levels. This is evident from several studies which show that the quality of education (defined by competencies in numeracy, reading and reasoning skills) is on average quite low. For instance, findings of a comparative study revealed that the reading ability of grade 4

TABLE 6.1 Student Enrolment Ratios in Indonesia and Selected Asian Countries, 1980–93[a] (%)

Country	Primary				Secondary				Tertiary		Adult Illiteracy (% of population 15 years and older)
	Male		Female		Male		Female				
	1980	1993	1980	1993	1980	1993	1980	1993	1980	1993	
Indonesia	115	116	100	112	35	48	23	39	–	10	16
Malaysia	93	–	92	93	50	56	46	61	4	–	17
Philippines	113	–	112	–	61	–	69	–	24	26	5
Singapore	109	–	106	–	56	–	59	–	8	–	9
Thailand	100	98	97	97	30	38	28	37	13	19	6
Hong Kong	107	–	106	–	63	–	65	–	10	21	8
South Korea	109	100	111	102	82	93	74	92	15	48	<5
Japan	101	102	101	102	92	95	94	97	31	30	<5

[a]In countries with universal primary education, gross enrolment ratios may exceed 100% because some pupils are younger or older than the country's standard primary school age.

Source: World Bank (1996b), Table 7, pp. 200–201.

TABLE 6.2 Student Enrolments in State and Private
Universities in Indonesia, 1982/83–1992/93

| Year | State Universities | | | | | |
| | Science and Engineering | | Other Disciplines | | Total | |
	No. ('000)	%	No. ('000)	%	No. ('000)	%
1982/83	83.1	30.3	191.3	69.7	274.4	100.0
1983/84	90.8	30.6	205.8	69.4	296.6	100.0
·1984/85	98.1	31.2	266.7	68.8	364.8	100.0
1989/90	120.7	23.8	385.7	76.2	506.4	100.0
1990/91	123.0	27.7	321.6	72.3	444.6	100.0
1992/93	135.5	31.7	291.5	68.3	427.0	100.0

pupils in Indonesia was slightly lower than that in the Philippines, and much lower than that in Thailand, Singapore and Hong Kong. Another study found that grade 5 and 6 pupils had not mastered many basic number skills. Worse, it was found that teachers made mistakes similar to those of their students. A follow-up study of year 1 and 2 students in junior secondary schools found that errors made in primary school were still prevalent in secondary school. Hence, poor scholastic performance begins in primary school and is perpetuated through secondary school (World Bank 1997a, pp. 66–67).

Another matter of concern is the imbalance between graduates from the social sciences on the one hand, and the natural sciences and engineering on the other, in both state and private universities (Table 6.2).This imbalance has led the Minister for Research and Technology (Professor Habibie) and Minister for Education and Culture (Professor Wardiman) to vigorously promote university study abroad in the natural sciences and engineering, notably by providing scholarships under the Overseas Fellowship Program to talented senior high school and university graduates. Minister Wardiman has also issued a decree precluding the establishment of new tertiary institutions that do not provide teaching in the natural

TABLE 6.2 (continued)

| Year | Private Universities | | | | | |
| | Science and Engineering | | Other Disciplines | | Total | |
	No. ('000)	%	No. ('000)	%	No. ('000)	%
1982/83	32.7	19.3	137.0	80.7	169.7	100.0
1983/84	80.7	18.0	369.0	82.0	449.7	100.0
1984/85	89.2	18.7	386.9	81.3	476.1	100.0
1989/90	150.7	15.4	828.8	84.6	979.5	100.0
1990/91	250.3	25.2	743.7	74.8	994.0	100.0
1992/93	233.3	23.3	767.3	26.7	1,000.6	100.0

Source: Republic Indonesia (1995).

sciences, including engineering. This decision is expected to halt the proliferation of higher education institutions that provide training in the social sciences and humanities only (Australian International Education Foundation 1996, pp. 29–30).

Meeting the large demand for university graduates in the natural sciences and engineering is not only a matter of increasing the output of such graduates and improving the quality of higher education, but also of enticing more of them to look for employment in national industry rather than, as is currently the case, in the public sector or in higher education (Table 6.3). Owing to such employment patterns, there is a great shortage of high-level expertise in national industry, which is being met by foreign managers and engineers, often from the Philippines and India. The latter are cheaper than their Western counterparts (Long 1997, p. 17).

Technological Information and Support Services

For many manufacturing firms, establishing linkages with the domestic S&T infrastructure is important, as they often do not know how to undertake the technological effort themselves to solve production and other problems, reduce costs and diversify output.

TABLE 6.3 Number and Proportion of Scientists and Engineers by Sector of Employment and Level of Higher Education, 1995

Level of Education	Employed in Government		Employed in Industry		Employed in Higher Education		Total	
	No.	(%)	No.	(%)	No.	(%)	No.	(%)
D3	99,055	87	15,375	13	75	0	114,505	100
S1	80,945	71	17,620	15	15,250	13	113,815	100
S2	4,360	61	425	6	2,100	13	7,185	100
S3	3,215	76	230	5	775	18	4,220	100
Total (S1, S2, S3)	88,520	71	18,275	15	18,425	15	125,220	100
Total (D3, S1, S2, S3)	187,575	76	33,650	14	18,500	8	239,725	100

D3: Diploma; S1: Bachelor's degree; S2: Master's degree; S3: PhD degree.
Source: Papiptek–LIPI.

This process of upgrading of firms' technological capabilities often involves drawing on technical information and other technological support services provided by the country's public S&T infrastructure (Lall 1993, pp. 737–739). Such infrastructure consists of a wide range of state and private university research institutes, government R&D institutes, and technology support institutes providing metrology, standards, testing and quality assurance (MSTQ) services.

A recent study concluded that the linkages between the public R&D infrastructure and Indonesia's export-oriented textile, garment and electronics firms were very weak, if not non-existent (Thee and Pangestu 1994). Managers of some firms that had tried to establish linkages with R&D institutes generally expressed their dissatisfaction, particularly with the researchers who, in their view, had little understanding of the technological needs of the firms they were supposed to advise, and who often were not aware even of the most recent technological developments in their fields of expertise. But in addition, a large majority of the firms interviewed were either unaware of the R&D capabilities of the country's S&T institutes, or sceptical of the relevance of their activities for their own specific technological needs.

Indonesia's public S&T institutes consist of the 12 national-level and several regional R&D centres of the Agency for Industrial Research and Development (BPPI) in the Department of Industry and Trade, and the R&D centres of the non-departmental government institutes (LPND), particularly the Indonesian Institute of Sciences (LIPI) and the Agency for the Assessment and Application of Technology (BPPT). As is discussed in more detail in Chapter 5 (by Samaun) and Chapter 8 (by Bishry and Hidayat), the R&D activities of the LPND are coordinated by Professor Habibie, the Minister of State for Research and Technology.

BPPI's R&D centres are mostly engaged in product certification, training and testing activities rather than in R&D activities proper. Their research staff are generally not well trained, and are often not aware of the latest technological developments in their fields. Moreover, much of their laboratory equipment is obsolete because the centres are underfunded (Lall and Rao 1995, p. 84). Hence in general they are not in a position to provide adequate technical information or technology support services to Indonesia's manufacturing firms.

The LPND R&D centres, particularly BPPT and LIPI, are for the most part better funded, staffed and equipped than those of BPPI. Many BPPT and LIPI researchers have pursued advanced postgraduate training abroad. Hence these are the two most suitable public S&T institutes to provide technical information and technology support services to national industry. However, like the BPPI R&D centres, LIPI and BPPT have thus far not played a significant role in Indonesia's industrial technology development, because they generally have not, as alluded to earlier, been able to establish mutually profitable linkages with national industry, particularly private industry. As a result, most of their research work has been supply rather than demand driven. Moreover, because of their lack of contact with national industry, they are generally not aware of the technological needs of private industry and lag behind world technological frontiers (Lall and Rao 1995, p. 84). The limited interaction of LIPI and BPPT with national industry can be attributed to the fact that their internal management systems and administrative procedures tend to be overly bureaucratic, leaving little room for managerial autonomy or results-oriented accountability (LIPI 1994, p. 16).

To encourage the public S&T institutes to carry out research that is more relevant to the needs of national industry, the Office of the Minister of State for Research and Technology (Menristek) introduced the Priority Partnership Research Program (RUK) in 1995/96. Under this program the government, in cooperation with private and state-owned manufacturing enterprises, finances research, development and engineering (RDE) activities. This is conducted jointly by the public R&D institutes and manufacturing enterprises, and focuses on the problems facing the latter (particularly in the areas of production techniques and technological improvements and innovations) for a maximum period of three years. The main objective of RUK is to encourage national industry to improve its competitiveness by raising its investments in RDE capabilities (Menristek 1997, pp. i, 5–7).

As the RUK program has been going for only two years, it may be premature to assess its effectiveness. However, judging from the bulk of research proposals submitted thus far, it appears to be supply driven—in the sense that the public R&D institutes have been the main initiators—rather than demand driven, that is, initiated by national industry itself. Hence, it remains to be seen if much of the research output of the program will be of direct relevance to the real needs of national industry. Manufacturing firms require speedy access to all kinds of technical information in order to solve the various technical problems they encounter in the production process or in the modification of products to suit local conditions and tastes. While some of this technical information can be obtained freely from technical journals, visits to trade fairs and the like, or can be acquired on a commercial basis from the licensor as part of a foreign direct investment (FDI) package or technical licensing agreement, there is also a body of technical information that is not accessible through the market in view of its 'public good' nature. The provision of this kind of technical information has to be undertaken by government, or as a joint venture with interested firms (Lall 1993, p. 738).

At present two public technical information services are in operation: Pusdata, administered by the Department of Industry and Trade, and Ipteknet, administered by BPPT. The Ipteknet system links and provides Internet access to the six public S&T institutes (World Bank 1996c, p. 47). However, in view of the rapid growth in the number of private providers, these two public information services need to restructure their operations to raise their efficiency if they are to survive in an increasingly competitive business environment.

Manufacturing firms also need an effective system of MSTQ services. To assess the quality of a product and meet the increasingly strict demands of importing countries, physical and chemical properties need to be measured against properly specified standards, including technical and sanitary standards (Menteri Perdagangan 1995, pp. 7–8; World Bank 1996c, p. 45). MSTQ services, which can be provided by public or private technology support institutes, are therefore vital in assisting firms to improve their international competitiveness.

Thus far, however, MSTQ services in Indonesia are still inadequate. Many firms do not realise that their products need to conform to strict standards and performance requirements, both national and international, if they want to enter export markets. Unfortunately, the testing and calibration services provided by existing laboratories are rather weak and not yet internationally recognised. As a result, exporting firms often have to turn to internationally recognised laboratories abroad to provide them with the necessary—but expensive—testing and calibration services (World Bank 1997a, p. 45). Moreover, many Indonesian exporters are unaware of the need to introduce a quality assurance system that meets ISO 9000 standards (World Bank 1996c, p. 45). An internationally recognised certification institute is being planned to promote the widespread application of such standards among Indonesian firms (Moedjiono 1995, pp. 9–12). In addition, Indonesia's National Standards Council (DSN), with technical assistance provided by the Japanese and German governments, has developed a master plan to improve MSTQ services in Indonesia (World Bank 1996c, p. 45). This could conceivably be achieved by privatising the public MSTQ laboratories, or by inviting the private sector to provide MSTQ services, which they could most likely do more efficiently than the public laboratories.

Finance

To promote industrial technological development, the governments of the East Asian NIEs have used fiscal incentives, and have established specialised financing instruments and institutions. These fiscal and financial mechanisms have included the provision of credit to purchase modern capital equipment, technical assistance to firms to upgrade their technologies, and fiscal incentives to encourage firms to conduct R&D (World Bank 1996c, p. 29). Compared to these countries, Indonesia's technological effort, in terms of R&D expenditures and activities by manufacturing firms, remains inadequate. This is reflected in Indonesia's ratio of R&D expenditures as a percentage of GDP, which is the lowest among the countries included in Table 6.4. The data in Table 6.4 also show that the major source of R&D funding in Indonesia (and India) is the government. This is in marked contrast to the industrially more advanced Asian countries, where the major source is the private sector. Similarly, most R&D activities in Indonesia are conducted by the government, specifically by the public R&D institutes, again in contrast to the Asian NIEs. The fact that the bulk of R&D activities in Indonesia is being carried out by government research institutes and laboratories means that these institutes should be playing a crucial role in providing technical information and support services to private firms. This, as we have seen, is not yet the case.

Aside from their lack of managerial autonomy, the public S&T institutes are also constrained by the relatively small government budget for the sector (Table 6.5), which limits the scope for the Indonesian government to promote directly the country's technological development (Hill 1995, p. 114). This becomes even more evident when the government budget for S&T activities is examined more closely. During the period 1985/86–1989/90, total government expenditures on S&T declined in both absolute and relative (that is, as a percentage of total government expenditures) terms. In 1991/92–1993/94, R&D expenditures began to rise gradually in absolute and relative terms, although they still ranged between only 1.5 and 2.1%

TABLE 6.4 R&D Expenditures, Sources of R&D and Locus of R&D Activities in Indonesia and Selected Asian Countries, 1992

	R&D Expenditure as % of GDP	Sources of R&D Expenditures (%)			Agents of R&D Activities (%)		
		Government	Industry	Other	Government	Industry	Other
Indonesia	0.16[a]	80.0[b]	19.0[b]	1.0[b]	62.0[b]	33.0[b]	5.0[b]
Japan	2.8	17.4	76.0	6.4	8.9	33.5	19.6
South Korea	2.2	17.2	82.4	–	3.7	72.7	24.6
Singapore	1.3	39.2	60.8	–	22.7	60.8	16.4
Taiwan	1.7[b]	52.1[b]	45.5[b]	–	9.1[b]	53.6[b]	37.3[b]
India	–	74.0	26.0	–	74.0	26.0	–

[a]1994.
[b]1991.

Source: Republik Indonesia (1995), *Indikator Ilmu Pengetahuan dan Teknologi Indonesia 1994* [Science and Technology Indicators of Indonesia, 1994], 2nd edition, BPPT, Ristek and Papiptek–LIPI, Jakarta, March.

TABLE 6.5 Government Expenditure on S&T and R&D, 1985/86–1993/94

Year	Total Govt Expenditure (Rp billion)	Govt Expenditure on S&T (Rp billion)	Govt Expenditure on R&D (Rp billion)	Total S&T and R&D Expenditure (Rp billion)	Total S&T and R&D Expenditure as % of Total Govt Expenditure	Govt S&T Expenditure as % of Govt R&D Expenditure (%)
1985/86	23,046	255	181	436	1.9	71
1987/88	22,783	220	138	358	1.6	63
1989/90	36,575	360	197	559	1.5	55
1991/94	50,556	617	328	945	1.9	53
1993/94	62,322	837	469	1,306	2.1	56

Source: Republik Indonesia (1995), *Indikator Ilmu Pengetahuan dan Teknologi Indonesia 1994* [Science and Technology Indicators of Indonesia, 1994], 2nd edition, BPPT, Ristek and Papiptek–LIPI, Jakarta, March.

of total government expenditures. No discernible shift towards a greater government effort to promote S&T development is evident as yet.

The figures in Table 6.5 should be treated with some caution, as they were derived from the data contained in sector 16 of the government budget, which records total government expenditures on S&T and on training and education. Projects listed in other sectors of the government budget may also contain some S&T activities that should be added to sector 16 expenditures (Republik Indonesia 1995, pp. 43–44). Nevertheless, the rise in government expenditures in 1991/92 did not reflect a rise in spending on S&T activities proper, but instead was caused by greater expenditures on physical construction, specifically the laboratories and R&D centres of the various LPND (Republik Indonesia 1995, p. 44).

It should also be pointed out that total government expenditures refer to 'routine' plus so-called 'development' expenditures. Routine S&T expenditures consist mainly of salaries paid to researchers and personnel in S&T institutes and material expenditures, and amount to about 55% of total government R&D spending. In 1993/94, however, this percentage dropped to 41%. The data in Table 6.5 also indicate that R&D expenditures as a proportion of S&T expenditures have declined steadily since 1985/86, although this trend began to be reversed slightly in 1993/94. This relative decline also reflected the fact that a larger proportion of R&D expenditures was spent on physical construction (Republik Indonesia 1995, p. 46).

In view of the rather modest government budget for S&T, the challenge facing the Indonesian government is to induce private firms to undertake more S&T activity, which, as indicated in Table 6.4, is currently very limited. To achieve this, it would be necessary to improve the incentive system further, so as to encourage private and state-owned firms to become more efficient. As developing countries, including Indonesia, are largely dependent on imported technologies, the major technological efforts of firms in these countries will initially need to be focused on mastering and adapting imported technologies, rather than on R&D proper. It is only at a more advanced stage of export-oriented industrial and technological development that firms will be forced to shift their technological efforts to the more costly acquisition of greater innovative capabilities. Meanwhile, the limited funds of Indonesia's public institutes should be directed more towards the support of activities that are still beyond the technological competence of firms, particularly small and medium-sized enterprises (SMEs). Such activities could include the introduction of new products and processes on a trial basis, minor technical modifications to reflect local conditions and tastes, and the improvement and expansion of MSTQ services (Hill 1995, p. 114).

S&T Policies

Another major determinant of a developing country's industrial technological development is the set of explicit S&T policies being pursued by the government. Direct S&T policies with an impact on technology development include fiscal or other incentives to stimulate a greater technological effort by firms; regulation and control of the import of foreign technologies through FDI or technical licensing agreements, to prevent foreign licensors from overcharging; and the direct targeting of specific technologies for research by the public S&T institutes. The most important of these policies are those related to FDI, the control of technology imports and 'mission-oriented' R&D strategies (Lall 1993, p. 740).

Compared with several other Asian countries, Indonesia has pursued a relatively 'open door' policy to technology imports through technical licensing agreements and FDI. No effort is made to monitor or control foreign technology imports, as is done in some developing countries. While the Investment Coordinating Board (BKPM) screens the applications of prospective foreign investors, no questions are raised about technology issues, including payments. As a result, almost no data are available on the amount of technology imports flowing into the country (as reflected in the outward flow of royalty payments), except for the data on capital goods (embodied technology) imports and the number of applications for foreign patents.

While there have been calls for government intervention in this regard, the Indonesian government has thus far not indicated any interest in changing its technology import regime. This stance is to be commended, particularly as imposing restrictions on technology imports might very well slow down the inflow of new FDI, and the accompanying inflow of technology imports. This would ultimately be harmful to Indonesia's industrial technology development, and would also run counter to the deregulation momentum, which was intended to improve the investment climate.

Moreover, government officials generally do not have the necessary business experience or knowledge about industrial technologies to make informed decisions on the appropriate levels and forms of royalty payments (Hill 1995, pp. 113–114). Even if they did, restrictions on royalty payments could lead foreign licensors to circumvent them by resorting to other means—for instance, 'transfer pricing'—to obtain what they felt was the right amount of royalty payments. However, to obtain data on the net inflow of technology imports, the government could consider requiring firms to report payments for technology imports, while at the same time guaranteeing that it would not intervene in negotiations between foreign licensors and Indonesian licensees. To strengthen the bargaining power of the Indonesian licensee, the public technical information institutions could, to the extent that it would be possible, provide the Indonesian licensees with valuable information about the sources of the available, relevant technologies, and the prevailing prices of those technologies in the international technology market.

The other set of S&T policies with a direct impact on Indonesia's future industrial technology development has been the explicit policies formulated by Professor Habibie, Minister of State for Research and Technology since 1978. Habibie's views on S&T policies for the Second 25-Year Long-Term Development Plan (1994/95–2018/19) are based on his assessment that the international competitiveness of Indonesia's manufactured exports has been declining, and that S&T development is imperative to raise the competitiveness of national industry (see Bishry and Hidayat, Chapter 8, and Rice, Chapter 9, in this volume). This goal is to be achieved by establishing partnerships (*kemitraan*) between the public S&T institutes and private firms, and developing highly skilled human resources in this field. In addition, the capabilities of the 'strategic industries' will be raised to accelerate the transfer and mastery of technology, while technology transfers will also be accelerated by the mastery of 'progressive manufacturing' (Menristek 1996).

In essence, Habibie believes that Indonesia's development strategy, which in the past was based mostly on the principle of comparative advantage, will in coming years need to put an increasing emphasis on the development of highly skilled human resources so as to permit the mastery of science and sophisticated technology (McLeod 1993, pp. 4–5). This can be achieved by temporarily protecting and subsidising a range of state-owned strategic industries (10 at present, including IPTN, the state-owned aircraft company, and PT PAL).

Although the development of the 10 strategic industries, particularly IPTN, has provided valuable training opportunities and exposure to advanced technologies for a large number of young scientists and engineers, the challenge now being faced is how to deploy these valuable human resources for maximum economic gain. This should be done by drawing a clear distinction between the technology development objectives and the manufacturing objectives of the strategic industries. To achieve the manufacturing objectives of these industries, they should be put on a fully commercial basis; the high protection and large subsidies they are currently receiving should gradually be reduced (World Bank 1996c, p. 47). At present the protection, assured government procurement, monopolistic position and huge explicit and implicit subsidies enjoyed by the strategic industries—much of it non-transparent—impose high social opportunity costs on the country without a reasonable prospect that these industries will become commercially viable in the foreseeable future.

CONCLUDING REMARKS

This chapter has provided a brief assessment of the five major determinants of industrial technology development in Indonesia, namely the incentive system, human skills, technical information and technology support services, finance, and the government's S&T policies.

Our assessment indicates that, despite considerable improvements over the past decade, the incentive system needs to be reformed; the trade regime still contains several import and export restrictions, and domestic competition is being hampered by various regulations. Although the government has put great emphasis on the development of human resources, notably by expanding educational opportunities at all levels, much remains to be done in the field of education. In particular, educational opportunities in the field of natural sciences and engineering need to be expanded, and a much more determined and comprehensive effort made to raise the quality of education at all levels.

Although the government has taken several steps to upgrade the capabilities of the public S&T infrastructure, linkages with the private sector remain weak, mainly because of bureaucratic constraints. To improve this situation, the government should grant these institutes greater managerial autonomy and allow them to establish linkages with the private sector without being hampered by bureaucratic restrictions. In addition, the private sector should be encouraged to provide MSTQ services to manufacturing firms. The decreased fiscal capacity of the government has hampered its ambitious S&T projects. A greater effort therefore needs to be made to encourage private firms and state-owned enterprises to fund their own technological efforts, as is already the case in most of the East Asian NIEs. To achieve this, the incentive system needs to be improved to make the business environment more competitive and raise the competitiveness of firms.

The government has sensibly pursued an 'open door' policy for technology imports, but it should proceed with the utmost caution in regard to the 'mission-oriented' strategic industries. Despite the fact that the 'technological mission' of the strategic industries is based on an ambitious and systematic technology policy, nurturing them for an unlimited time through continued high protection and subsidies should be avoided at all costs. The strategic industries should not be allowed to absorb large amounts of public funds without a reasonable prospect of becoming economically viable in the foreseeable future.

7

TECHNOLOGY POLICIES
IN INDONESIA

Sanjaya Lall

INTRODUCTION

Indonesia is a relative newcomer to modern industry and, by any standards, a highly successful one. Its manufacturing sector has enjoyed high and sustained rates of growth, its shares of GNP and exports rising sharply. A modern electronics and transport equipment industry is taking root; the textile industry is well equipped and highly competitive; several heavy intermediate industries have been established; foreign direct investment (FDI) is flowing in strongly; national conglomerates are strong in financial terms and flexing their muscles overseas. Nevertheless, Indonesia's industrial structure has several weaknesses in terms of technology—the life blood of its longer term growth and upgrading. The technological base is shallow and backward in comparison to many countries in the region, particularly the large East Asian Tigers; domestic capabilities to absorb and improve upon complex imported technologies are narrow and weak; the capital goods sector, a vital ingredient of industrial deepening, is underdeveloped; the relatively small amount of technological effort that is conducted is concentrated and distorted. Traditional, small-scale, low-productivity activities still dominate manufacturing. Indonesia's export industries, driven by foreign investment, remain concentrated in simple labour-intensive assembly or resource processing activities.

It is widely accepted that future industrial growth in Indonesia will have to move away from this initial pattern, and that this will require a significant upgrading of technological capabilities. Such capabilities are needed, not to undertake 'innovation' at the frontiers of knowledge, but to import the most cost-effective technologies, master them efficiently and adapt and improve upon them, and take on more complex technologies and more difficult technological tasks. This does not mean 'doing without' imported technologies; quite the contrary. It means constant, even aggressive, imports of new technologies. It also means a move from a passive dependence on technology transfer to a more active role in mastering and building upon imported technologies. Without such a move, the economy cannot grow out of its shallow and stagnant technological base, out of growth driven by the inherited base of cheap labour and primary resources to that driven by higher productivity, skill and technology-intensive activities. The former is a good starting point for industrialisation, but without subsequent deepening and diversification it is likely to be eroded by rising wages and slowing export growth. The engine of growth in the Asian Tigers has been technological diversification and deepening, not a continuing passive dependence on simple foreign technologies.

Indonesia presently has practically no culture of indigenous research, design and engineering; even large enterprises tend to rely on imports of 'ready-made' technology. This raises the costs of importing new and more advanced technologies and constrains its assimilation. The skill base, especially at higher levels, is relatively poor. The science and technology (S&T) infrastructure lags behind the needs of world-class manufacturing. In a setting of rapid globalisation, dynamic technological change, and spreading trade and investment liberalisation (to which the Indonesian government is also committed), it is imperative for it to broaden and deepen this base.

This chapter considers Indonesia's technology policies in terms of the competitiveness of its industrial sector. It starts with the structure and technological content of Indonesia's manufactured exports. It goes on to analyse its human resource endowments for technology development and formal technological activity. Finally, it considers the main features of Indonesian technology policies in relation to the more advanced East Asian economies.

TABLE 7.1 Exports from Asian Countries, 1980–94

Country	Total Merchandise Exports			Manufactured Exports
	1994 Value ($ million)	Growth Rate (1980–90)	Growth Rate (1990–94)	1994 Value ($ million)
Hong Kong[a]	28,739	11.5	−0.3	27,302
Singapore[a]	57,963	12.1	10.9	56,224
Korea	96,000	13.7	7.4	89,280
Taiwan	92,847	11.6	5.9	86,348
Indonesia	40,054	5.3	21.3	21,229
Malaysia	58,756	11.5	17.8	41,129
Thailand	45,262	14.3	21.6	33,041

[a]Excluding re-exports.

Sources: World Bank, *World Development Report, 1996*; ADB (1994), *Key Indicators of Developing Asian and Pacific Countries; Hong Kong External Trade* (February 1996); *Singapore Trade Statistics* (1996).

EXPORTS AS INDICATORS OF TECHNOLOGY DEVELOPMENT

Manufactured export performance can be used as an indicator of the competitiveness and technological sophistication of a country's manufacturing sector. It is not a perfect indicator; export data do not provide a good indicator of the technological competence of large economies with significant inward-oriented activities such as Indonesia. Export data cannot distinguish between different levels of technology used in a country within a broad product group, so that a country undertaking only the assembly of a high-tech product appears as a very advanced exporter. Nor do they distinguish between exports by foreign and domestic firms; this distinction may be important if it reflects different local technological inputs. However, these deficiencies may be offset by looking at the local technological content of exports and the role of FDI in trade.

The aggregate figures are shown in Table 7.1 for seven Asian countries—the four Tigers (Hong Kong, Singapore, Korea and Taiwan) and the three 'new Tigers' (Indonesia, Malaysia and

TABLE 7.2 Technological Basis of Competitive Advantage, 1980–95

Activity Group	Major Competitive Factor	Examples	World Manufactured Trade (%)	
			1980	1995
Resource intensive	Access to natural resources	Aluminium, food processing, oil refining	18.8	15.1
Labour intensive	Cost of unskilled, semi-skilled labour	Garments, footwear, toys	17.4	17.9
Scale intensive	Length of production runs	Steel, autos, paper, chemicals	27.8	23.7
Differentiated	Products tailored to varied demands	Advanced machinery, power generation equipment	24.3	23.4
Science based	Rapid application of science to technology	Electronics biotechnology, pharmaceuticals	11.4	19.9

Source: Classification from OECD (1987) *Structural Adjustment and Economic Performance*, OECD, Paris; data from World Bank trade database.

Thailand). The largest manufactured exporters in the group in 1994 were Korea and Taiwan; the smallest was Indonesia. The fastest rates of growth in 1990–94 were for Thailand, Indonesia, Malaysia and Singapore. Hong Kong was the only country in the group to have declining exports (re-exports excluded), a dramatic deterioration on its earlier performance. On the whole, however, the general export performance was far better than in other developing regions, suggesting considerable technological competitiveness and dynamism.

These data reveal little about the sophistication or complexity of export activity. One way to do this is to look at the technological composition of manufactured exports. This is done by classifying exports according to categories used by the OECD (Table 7.2). The categorisation is far from perfect. There are overlaps between the categories (resource-based industries can be very capital intensive), and the groups are broad (for instance, many electronics exports are labour intensive), but if carefully used the classification is helpful to portray general trends. Labour-intensive products are generally at the low end of the technology spectrum, with low requirements for technological skills and activity. Products in the scale-intensive group tend to use complex, capital-intensive technologies, but are generally not at the cutting edge of technology. Within the scale-intensive group, process industries (for example, chemicals) should be distinguished from engineering industries (for example, automobiles). The latter tend to have more difficult learning requirements, be very linkage intensive, and involve a larger variety of skills. Differentiated manufactures comprise more sophisticated engineering products, with advanced design, research and operational skills, while science-based products use leading edge technologies.

In broad terms, we call the last three categories (scale intensive, differentiated, science based) 'technologically advanced', and the last two of these 'high-tech', products. Resource-based products are not considered further here because their competitive edge is very specific and does not as such tell us much about industrial maturity.

Table 7.3 shows the technological breakdown of manufactured exports over 1980–95. In 1995, the highest concentration in labour-intensive exports is in Indonesia (43.3%) and Thailand (35.8%). Moreover, Indonesia is the only country in the group with the share of such exports rising over time (by a massive 34 percentage points); this goes against the general tendency for share to decline with growing industrial complexity and technological competence. While Indonesian differentiated exports, led by recording equipment and various types of electrical power machinery, have grown fast (to $2.5 billion in 1995), they are still small by regional standards.

The most technologically advanced exporter in the group is Singapore, followed by Malaysia, Taiwan and Korea. Indonesia is the most technologically backward by both indices, by a considerable margin. Table 7.3 underlines how its performance lags behind that of the other rapidly industrialising countries in the region.

While this shows the technological structure of exports, it does not reveal the domestic technological content of the activities involved. In particular, it is important to know the level of technology involved in the export activity, and the role of domestic as compared to foreign firms in production and the technological activity performed locally. These are difficult to quantify directly for lack of relevant data, but some less direct indicators may be considered.

Level of Technology

The technological content of exports varies between countries according to the local input of components, equipment and technical knowledge. For instance, a 'high-technology' export in one country may come from locally assembled imported components, with few local inputs, physical or technological; in another, the same export can have substantial local inputs of equipment, components, design, development and engineering. These imply very different technological capabilities in the activities involved. Thus Malaysia's high-tech exports, driven primarily by electronics and electrical assembly operations owned by multinational corporations (MNCs), have relatively few linkages or technological inputs locally; practically all the equipment, technology and components are directly imported or provided by other MNCs in export processing zones. While Malaysian export activity has upgraded considerably over time in process and product technology, the level of domestic capabilities remains low—certainly in comparison with Singapore, Korea and Taiwan, which have lower shares of high-tech exports.

Singapore, although also MNC-driven, is at a considerably higher level of product sophistication, uses more advanced skills, and contributes more local design and development. High-tech exports in Korea and Taiwan come mainly from local enterprises, use a great deal of local equipment and components, and rely heavily on local technological inputs down to basic design stages. (Korea is ahead of Taiwan, with a more diverse and 'heavier' industrial structure and greater R&D effort; see below.)

In this context, Indonesia is at the bottom of the technology ladder. Much of its exports of high-tech products emanates from fairly low-level assembly activities, many relocated from Taiwan, Singapore and Malaysia. (The 'growth triangle', with Singapore, plays a significant role; see Chapter 18 below.) While these activities are bound to upgrade their processes and products over time, as they did in Malaysia, the process is bound to be slow. In any case, the Malaysian lead is unlikely to be eroded or its experience of electronics-led export growth replicated.

SANJAYA LALL

TABLE 7.3 Distribution of Manufactured Exports by Technological Categories, 1980–95[a] (%)

Category	Korea			Taiwan			Singapore		
	1980	1990	1995	1980	1990	1995	1980	1990	1995
Resource based	9.0	6.8	11.6	9.8	8.2	9.2	44.4	26.9	13.2
Labour intensive	49.2	40.8	23.2	54.3	41.2	31.0	10.6	10.3	7.6
Scale intensive	23.6	19.3	21.0	9.1	10.3	10.6	9.3	5.9	5.5
Differentiated	11.3	15.6	18.7	12.4	20.6	20.4	20.5	22.3	21.2
Science based	6.9	17.4	25.5	14.5	19.8	28.9	15.1	34.6	52.5
Technologically complex[b]	41.8	52.3	65.2	36.0	50.6	59.9	45.0	62.8	79.2
High technology[c]	18.2	33.1	44.1	26.8	40.3	49.3	35.6	56.9	73.7

[a]Data for Hong Kong are not shown because of the high proportion of re-exports in the total.
[b]'Technologically complex' products include scale-intensive, differentiated and science-based products.
[c]'High-technology' products include differentiated and science-based products.

TABLE 7.3 (continued)

Category	Indonesia			Malaysia			Thailand		
	1980	1990	1995	1980	1990	1995	1980	1990	1995
Resource based	78.5	48.2	33.6	17.6	9.9	8.6	21.7	13.8	10.7
Labour intensive	9.6	39.2	43.3	18.7	18.9	12.7	47.0	45.5	35.8
Scale intensive	5.1	9.3	8.3	5.2	7.7	5.9	7.8	6.3	7.7
Differentiated	0.8	2.2	10.3	12.3	23.3	26.8	22.2	14.1	19.5
Science based	6.1	1.1	4.5	46.3	40.3	46.0	1.2	20.2	26.4
Technologically complex[b]	11.9	12.7	23.1	63.8	71.3	78.7	31.3	40.6	53.6
High technology[c]	6.8	3.4	14.8	58.6	63.6	72.8	23.4	34.4	45.9

Source: Calculated from World Bank trade database at the three-digit SITC level.

Role of MNCs

A strong foreign presence has mixed implications for local technological development in export activities. In industrially mature countries, MNC export activity generally has significant indigenous input and design, and interacts with and contributes to the local technological base. In less industrialised countries, MNC exports are usually driven by cheap labour, and have low local technological content. As just noted, there are various combinations of MNC presence and local technological activity in Asian countries. Table 7.4 shows recent data for FDI inflows to Indonesia and its comparator countries.

Singapore is the most FDI-intensive economy in the region (and probably in the world) in terms of the share of FDI in gross domestic investment, followed by Malaysia. At the other end of the spectrum, Korea has very low levels of reliance on foreign investment, with Taiwan at a somewhat higher level. Both these countries, in particular Korea, have traditionally emphasised externalised technology transfer, in strong contrast to Singapore, which promoted internalised modes but with extensive targeting in the FDI selection and technology development process (Lall 1996). All three countries had clear strategies for technology development and used a variety of selective interventions to achieve them. Korea and Taiwan used infant industry protection, local content rules, FDI restrictions and technology promotion; Singapore used subsidies and other incentives to induce MNCs to bring in more advanced technologies and raise local technological activity.

Indonesia does not have a high overall dependence on FDI, but has relied heavily on foreign firms to lead its recent manufactured export growth. Its foreign investment strategy has become more liberal over time, but, as far as domestic market-oriented FDI is concerned, remains discretionary and cumbersome, marked by rent seeking. The export-oriented sector is more open to investors. While the government has tried to induce import substituting foreign investors to become more export oriented, in general Indonesia has not had a coherent technology strategy for using FDI to upgrade its technological development. As noted below, much of this thrust has come from the public sector.

In sum, Indonesian industry appears to have specialised in relatively low-technology segments, and not to have improved its technological status over time, in comparison with its fast growing neighbours.

TABLE 7.4 Role of Inward FDI, 1984–94

Country	Annual FDI Inflows ($ million)							FDI as % of GDI[a]	
	1984–89	1990	1991	1992	1993	1994	1995	1984–89	1990–94
Hong Kong	1,422	1,728	538	1,918	1,667	2,000	2,100	12.2	6.7
Singapore	2,239	5,575	4,879	2,351	5,016	5,588	5,302	28.3	28.4
Korea	592	788	1,180	727	588	809	1,500	1.4	0.7
Taiwan	691	1,330	1,271	879	917	1,375	1,470	3.3	3.0
Indonesia	406	1,093	1,482	1,774	2,004	2,109	4,500	1.6	3.5
Malaysia	798	2,333	3,998	5,183	5,006	4,348	5,800	8.8	22.4
Thailand	676	2,444	2,014	2,116	1,726	640	2,300	4.4	4.3

[a]Gross domestic investment. The figures are simple annual averages.
Source: UNCTAD, *World Investment Report 1996*, Geneva.

Does This Matter?

If manufacturing output and exports are growing rapidly, is there any need for concern about the underlying technological base? Does it need to upgrade at all? And if it does, will this process not occur naturally, under pressure from market forces—that is, from rising wages and as a consequence of greater openness to trade and FDI?

The argument of this chapter is based on the premise that technological upgrading does matter in a latecomer to industry like Indonesia, just as it does in more advanced industrial countries. A weak technological base does not matter in the first stages of industrial growth, when low wages and the exploitation of natural resources provide the base for investment and exports. But this cannot serve as the base for sustained growth; growing mastery of more advanced technologies and technological functions becomes necessary for long-term development and industrial maturity. This is particularly true of an economy as large as that of Indonesia, which can support the development of a diverse and deep industrial structure with domestic production of heavy intermediates and capital goods.

To do this efficiently and competitively requires a growing local base of capabilities; imported technology, necessary as it is, cannot substitute for indigenous technological effort. This is true regardless of the mode of technology import; both internalised and externalised modes call for local technological capabilities. There are differences between the two, of course—an FDI-dependent strategy may reduce the need for domestic design and development capabilities. However, as the Singapore experience shows, the attraction of higher 'quality' FDI, with more value added activities, calls for larger supplies of high-level managerial and technical skills, stronger support institutions, more capable local suppliers and a very targeted strategy to attracting FDI and guiding its upgrading. A more nationalistic, Korean-style strategy calls for more intensive efforts to develop local capabilities at all levels, including that of R&D. Both strategies can be successful, though which gives the optimal base for longer term growth and competitiveness remains debatable. What is obvious is that technological upgrading is a pressing and constant need. This is the case with Indonesia.

Can market forces by themselves drive the desired pace of upgrading? They can—if all the relevant markets function 'efficiently', with free market prices giving appropriate signals for physical, human and technological investment. Such efficiency cannot be assumed. Market failures abound whenever information is created and exchanged, technologically linked agents make investment decisions and undergo lengthy and risky learning processes, new institutions have to be set up, and the benefits of technology and skill development are not fully appropriable. This is even more true of developing countries (Lall 1996; Stiglitz 1996), where, almost by definition, markets are less efficient and support institutions lacking. Given widespread market deficiencies, well-placed interventions can play a vital role in catalysing and accelerating technological development. The experience of the Tigers shows clearly the vital role played by such interventions in technological deepening.

This experience also shows, however, that government interventions, to be effective, have to be placed within a *coherent, stable and well-implemented strategy* to remedy market failures. More importantly, the strategy must not seek to supplant markets; it should be led by the private sector; and it should be implemented with administrative skill, autonomy and flexibility. The history of postwar development is littered with interventions that have distorted and delayed development rather than promoting it—the risk of 'government failure' is very high, and its costs often exceed those of market failure. However, this is not an argument against government intervention per se: market failures have their own costs, and, given their prevalence in technology development, there is a strong case for devising effective strategies to remedy them. That this can be done effectively is shown by the three technologically dynamic Tigers.

The main issue is whether the Indonesian government can do something similar. This question can be examined by looking at the two main components of technology development: skills and technological activity; and the policies adopted by the government to promote these.

THE HUMAN CAPITAL BASE FOR TECHNOLOGY DEVELOPMENT

Table 7.5 presents data on educational enrolments in Indonesia and other neighbouring countries. These figures are imperfect indicators of skill creation: they cannot capture differences in educational quality and drop-out rates; and they do not measure the (very

TABLE 7.5 Educational Enrolments (latest available year)

Country	Enrolment Ratios			Vocational Training Enrolments	
	Primary (% of age group)	Secondary (% of age group)	Tertiary (% of age group)	(No.)	(% of population)
Korea	95	99	55	1,483,198	3.33
Taiwan	100	88	38	513,700	2.46
Singapore	107	68	19[a]	9,391	0.32
Indonesia	115	45	9	1,440,869	0.77
Malaysia	93	61	10	90,079	0.48
Thailand	87	49	21	545,791	0.92

na = not available.

[a]Singapore's tertiary enrolment figures exclude polytechnics, which enrol 27% of the age group. If these are counted as tertiary institutions, its tertiary enrolment figure would increase to 46%.

[b]'Core technical' subjects are natural science, mathematics and computing, and engineering.

TABLE 7.5 (continued)

Country	Tertiary Enrolments in Technical Fields									
	Natural Science		Maths/Computing		Engineering		Core Technical[b]		All Technical[c]	
	(No.)	(% of population)	(No.)	(% of population)	(No.)	(% of population)	(No.)	(% of population)	(No.)	(% of population)
Korea	81,222	0.18	171,147	0.38	437,537	0.98	689,906	1.55	730,346	1.64
Taiwan	16,823	0.08	32,757	0.16	179,094	0.86	228,674	1.09	303,964	1.45
Singapore	1,281	0.05	1,420	0.05	13,029	0.47	15,730	0.56	16,767	0.60
Indonesia	22,394	0.01	13,117	0.01	205,086	0.11	240,597	0.13	315,325	0.17
Malaysia	8,776	0.05	4,557	0.02	12,693	0.07	26,026	0.14	32,222	0.17
Thailand	39,045	0.07	na	na	51,949	0.09	90,994	0.15	816,256	1.37

[c]'All technical' subjects include core technical subjects plus medicine, architecture, trade and crafts, and transport and communications.

Sources: UNESCO (1996), *Statistical Yearbook, 1996*; Singapore government data.

important) element of skill formation by employee training. Never-theless, they are the only comparable data available and they do show the base of educated manpower which allows technological activity to be undertaken.

At the secondary and tertiary levels of education, Indonesian enrolments lag well behind those in the Tiger economies; they are also somewhat lower than those in Malaysia and Thailand. Malaysia is considerably more advanced in secondary enrolments. It is nearly the same as Indonesia in tertiary levels (possibly at higher levels of quality), though it has some 25% higher enrolments than shown due to the large numbers of students enrolled abroad (the comparable figure for Indonesia is much smaller, though it has been growing quite quickly). Thailand is somewhat ahead of Indonesia in terms of secondary enrolments, but much further ahead in tertiary enrolments. In terms of tertiary enrolments in technical subjects—a better indication of the skills available for technological development—Indonesia is again well behind the Tigers and marginally behind the new Tigers.

These data suggest that Indonesia is not badly placed in terms of human capital compared with Malaysia and Thailand, but this is not a cause for complacency. Malaysia is facing severe problems of skill shortages in maintaining its high-tech production base, which it is relieving by allowing the liberal use of expatriates. However, its government accepts that this is not a solution, and that it will have to boost its educational base significantly. Thailand is also facing serious skill deficiencies, and trying to upgrade the educa-tional system. All three are very far from the levels reached by the mature Tigers, but they will have to narrow the gap if they are to move up from their present status as essentially low-level imple-menters of imported technologies.

The enrolment data may be misleading once quality and relevance are taken into account. This is particularly true of Indonesia. According to various surveys, there is a mismatch at all levels between the skills produced by schools and those needed by industry. The training system is inflexible and unresponsive to industrial and technological needs. There is a shortage of skills for industry, which manifests itself in various ways. At the low end, less than 1.7% of the workforce has a post-secondary education, and there is a dearth of technicians at the middle level. At the top end, there is an acute shortage of high-level technical and managerial personnel. Industrial engineering, quality management, strategic technology planning and information technology are particular areas of weakness, and these are particularly significant for export competitiveness. Only 21% of medium and large firms have at least one employee with a science and engineering degree at the college level; such degree holders comprise about 1% of total employees in these firms.

In addition, there is an extreme shortage of qualified teachers and poor provision of laboratories and experimental equipment. Textbooks are scarce and library facilities inadequate. Technical students receive insufficient tuition and undertake little practical application of theory. Graduate programs are well below international standards. University staff are paid poorly, and relatively few hold advanced degrees. The quality of their research is poor, and less than a fourth of the faculty actively conducts research. University research is typically uncoordinated, and lacks the means to establish priorities or to disseminate its findings. Indonesia also suffers from a scarcity of managerial skills, poor management systems and the lack of a technical orientation in management. Public secondary technical schools have deficient equipment and teaching materials, weak links with industry, and an inflexible and overcentralised system (World Bank 1992).

Increasing skills at the top technical and managerial levels is not by itself sufficient to ensure the upgrading of productivity and competitiveness of industry. The level of skills of *shop-floor* employees must also be raised, so that they can be trained to handle new technologies calling for numeracy, flexibility and multiskilling. This would involve not just raising the quality and quantity of education and vocational training, but also investments in continual training and retraining of employees. However, firms invest little in training beyond the minimum required for operations. Most lack the teaching resources to provide effective in-house training, and only a handful of large firms have good quality training programs. Even in export-oriented firms training is weak, confining their operations to less demanding operations.

Technological Activity

Formal R&D in Indonesia, with its recent industrialisation and a specialisation in low-technology activities, is low (Table 7.6). Indonesia has the lowest ratio of R&D spending by productive enterprises in the group of countries listed in the table (the leading

TABLE 7.6 R&D Expenditures

Country	Year	As % of GDP		R&D per Capita
		Total	By industry	($)
Hong Kong	1995	0.1	na	19.8
Singapore	1992	1.0	0.6	153.6
Korea	1995	2.7	2.0	271.1
Taiwan	1993	1.7	0.8	179.6
Malaysia	1992	0.4	0.17	11.2
Thailand	1991	0.2	0.04	3.1
Indonesia	1993	0.2	0.04	1.5
China	1992	0.5	na	2.4
India	1992	1.0	0.22	3.1
Pakistan	1987	0.9	0.0	2.6
Japan	1995	3.0	1.9	1,225.6
France	1994	2.4	1.5	544.8
Germany	1991	2.3	1.5	674.8
UK	1994	2.2	1.4	383.6
US	1995	2.4	1.7	655.2

Sources: UNESCO, *Statistical Yearbooks;* various country data.

OECD countries are included for interest). In 1993, Indonesia invested around 0.2% of GDP in R&D, with the government providing 80% of the funds. This was at the level of Thailand and higher than Hong Kong, but lower than all the other comparator countries. In per capita terms, Indonesia spent only $1.5 on R&D, compared to $271 in Korea, $180 in Taiwan and $154 in Singapore. By this measure, Thailand spent more than twice as much as Indonesia, Malaysia some seven times more, and even Pakistan, a noted technological laggard, over twice as much.

Data on R&D by industrial firms are not readily available in Indonesia. The STAID (1993) survey attempted to measure formal R&D as well as production engineering efforts in industry. There were many gaps in the replies, but the data collected suggest that industry spent around Rp 165 billion on R&D in 1991 and about four times that on production engineering. R&D came to 0.2% of the output of the reporting establishments. The largest spenders were chemical and fabricated metal manufacturers. It is not clear, however, that the activities reported were in fact genuine 'R&D' as defined in developed countries, or whether they also included (as is likely) quality control and testing activities.

Formal R&D is not necessarily a good indicator of technological activity in private industrial firms; informal engineering and shop-floor effort may be more important for technological upgrading and productivity improvement. Even at this level, however, Indonesian industry is well below levels needed to mount sustained technological upgrading and diversification. Technological activities in Indonesian manufacturing firms have been evaluated recently in a number of detailed studies, including the contributions to this volume (see also Hill 1995). These generally reinforce the general impression that technological capabilities are weak in Indonesian industry. There is little independent design and development activity. Many operational capabilities are underdeveloped. Many advanced activities in the capital goods and electronics subsectors are missing. The country remains overwhelmingly dependent on foreign sources for its primary technology and specialised skills.

At the same time, however, there is also some dynamism in areas of manufacturing activity and exports, though with wide variations between industries and groups of firms. Some large conglomerates have accumulated good technological capabilities in complex activities, though their lack of exposure to export markets has held back the development of skills in product development and cost reduction. MNCs seem to have the best operating capabilities in Indonesia, but are reluctant to deepen their local technological activities. The public sector displays a wide range of technological capabilities, but again the lack of competitive pressures seems to hold back the intense technological effort needed to establish a sustainable export presence.

The public sector research institutions, divided between those under the Ministry of Industry and Trade (MOIT) and those under the Minister of State for Research and Technology (Menristek), have various deficiencies.[45] The national S&T institutions employ about 240,000 scientists and engineers, of whom around half are diploma holders (rather than graduates) and less than 10% hold a Master's degree or doctorate. According to STAID (1993, p. 4), 'Such

---

[45]This draws on Lall and Rao (1995).

shortages of advanced degrees make it difficult to achieve significant levels of R&D and innovation'. Moreover, 70% of the technical
staff engaged in R&D are agricultural specialists, and about 80% of
the total R&D effort comes from public research institutions. Only
about 6% of trained R&D personnel work in manufacturing industry.

The 12 national-level R&D laboratories and several regional
R&D institutes under MOIT are primarily engaged in training and
testing, and product certification. They undertake little technology
development and tend to have poor linkages with industry. Incentives to sell technology and services to industry are limited, and
their abilities to market technologies are weak. Salaries are low
and staff undertrained; the laboratories operate with outdated
equipment and are starved of resources.

The laboratories under Menristek are responsible for the bulk of
government and national expenditures on R&D. They are non-
departmental, and tend to be better funded and staffed than MOIT
institutes. However, their contribution to industrial technology has
been very limited. They have established few linkages with
private industry, are supply rather than demand driven, and often
lag behind world technological frontiers. Their internal management and procedures tend to be bureaucratic.

Other technology support services provided by the government
are also weak. MOIT has established semi-autonomous extension
services, called Technical Services Groups, to support the textiles,
engineering products and pulp/paper industries, with a 90% subsidy
provided to firms that use their services. While these have done
some valuable work, their geographical and industrial coverage, as
well as their firm-level reach, is limited. Much of the work has
been done by expatriates, perhaps reducing their own capabilities.

The metrology, standards, testing and quality assurance
(MSTQ) services system in Indonesia is hampered by low quality
awareness and the absence of a comprehensive set of industrial
standards in the country. Until 1994, many of the standards adopted
were below international levels, and were not recognised abroad.
Measuring and testing laboratories lacked international accreditation, and firms found it difficult to access international standards.
The standards system was very fragmented, with many ministries
involved in setting standards and the private sector poorly
represented, if at all. The adoption of ISO 9000 standards was not
supported by a national plan for its implementation, though the line
ministries have launched strategies for its spread.

Metrology and calibration services are not adequate to modern industrial needs in Indonesia. The national metrology institute cannot meet demand for its services, and its calibration network is weak, especially outside Java. The regional calibration network, operated by MOIT, is not coordinated with the national institute. It provides very limited services and its staff is not adequately trained. Some of the testing laboratories under MOIT are poorly equipped and out of date, and need to be substantially improved before they can be accredited internationally. By contrast, the laboratory run by the former Ministry of Trade, now part of MOIT, operates as a commercial enterprise and is efficient, though it benefits from its monopoly on trade. Menristek's testing facilities are primarily geared to the public sector (especially its strategic industries). Private testing facilities are efficient, but are concentrated in West Java.

Indonesia lacks an effective productivity centre to provide technological support and training in generic technologies relevant to large subsets of firms (for example, CAD, quality control systems, just-in-time management, flexible manufacturing). The Institute for Machine Tools, Automation and Production Technology was designed to provide such services, but has focused on the needs of strategic industries and ignored the needs of small and medium-sized enterprises (SMEs). Similarly, technology information services, offered by BPPT and MOIT, are in need of coordination and improvement to help smaller firms access international technology markets effectively.

TECHNOLOGY STRATEGIES AND INSTITUTIONS IN INDONESIA

Background

The Indonesian government does not have a *strategy*—a coherent set of policies to encourage or remedy market failures— for technology development. The incentive structure under which Indonesian industry operates is not conducive to vigorous technological activity; trade and industrial policies have traditionally favoured protected import substitution, with strong government interventions in the allocation of favours and public ownership of important industries. This strategy has evolved in an ad hoc manner in response to various political pressures, and has not been truly 'selective'. In other words, it has not been geared to selecting promising activities and inducing them to 'learn' and develop world-class capabilities behind protective barriers. Nor has it geared interventions in factor markets to the needs of technology development. As in many other developing countries, such a strategy has fostered some technological development, but at high cost and in a relatively small set of activities. It has certainly failed to achieve the dynamic technological and competitiveness gains achieved by selective interventions in countries like Korea and Taiwan.[46] Though the structure of incentives has improved over the past decade with liberalisation, the design of liberalisation is haphazard and, again, lacks a clear strategy for promoting technological development or providing supply-side support. Moreover, the process remains opaque and subject to considerable rent seeking and arbitrariness.

In Indonesia today, there is still relatively little pressure on industrial firms to invest in technological activity. Trade and ownership restrictions, backed by market power in the hands of large domestic conglomerates, hold back technological activity, not just by firms that enjoy privileges but also by those that are relatively deprived. Some policies conflict with each other. Some are geared to meeting the needs of special sections of industry, while others are deficient in addressing the needs they are supposed to meet. Responsibility for policies is spread over different agencies, with little effective coordination, and sometimes active rivalry. The desire to promote technology development in a small group of public sector enterprises has not resulted in general stimulation of industrial technology.

This section considers policies aimed directly at industrial technology development (ITD). Policy has two broad thrusts. One is to provide research, information and other support infrastructure for industry in general. The other is to achieve the technological 'mission' of the strategic industries—of entering high-tech activities and building a range of production and design capabilities there. The dividing line between the two is not very clear. Part of the infrastructure for industry is devoted to strategic industries and their mission; part is operated independently by the scientific research establishment; and part is operated by MOIT, which is concerned with the more mundane needs of industrial extension.

---

[46]For analytical background and a description of more successful interventionist policies, see Lall (1996). On Indonesia, see Hill (1996c).

The organisation and implementation of technology policy in Indonesia is shared between the line ministries and a special ministry in charge of technology. The most important in terms of technology policy is the latter, Menristek. Each line ministry has its own research laboratories dealing with problems concerning its sector. In the manufacturing sector, MOIT has set up, through its Agency for Industrial Research and Development (BPPI), several research institutes and extension services. It also handles standardisation for a number of industrial products. Apart from the ministerial laboratories, there are six non-departmental government institutes (LPNDs),[47] which conduct the bulk of R&D in the public sector (which in turn comprises the bulk of national industrial research).

The LPNDs subcontract research from the ministries (whose own R&D facilities are not well equipped or staffed), and are supposed to provide research services to private industry and promote linkages with research at universities. The LPNDs are under the aegis of Menristek, which has overall responsibility for technology policies and development in the country. The Minister of State for Research and Technology provides the coordination for all S&T organisations in the country. In his personal capacity, he also controls several other crucial institutions in the S&T area. These are reviewed below.

BPPT and the Strategic Industries

One of the LPNDs is the Agency for the Assessment and Application of Technology (BPPT), in charge of devising and implementing policies for industrial as well as other forms of technology, and whose Chairperson is the Minister for Research and Technology.[48] For industry, the BPPT has selected eight activities as 'vehicles for technological transformation'.[49] The 'transformation' is to be carried out by various means, including studies, research and direct

---

[47]These LPNDs, set up in the mid 1970s, are in the fields of industrial technology, surveying, atomic energy, statistics, aeronautics and sciences.

[48]For a fuller discussion of BPPT, see Chapter 8 in this volume, by Bishry and Hidayat.

[49]These are aeronautics and aerospace; maritime and shipbuilding; land transportation; telecommunications; energy; engineering; agricultural equipment; and defence.

participation in production. Direct ownership of industries is mainly in the 10 strategic industries under the Agency for Management of Strategic Industries (BPIS), also chaired by the Minister. These industries have ambitious technological strategies for increasing the local physical and technological content of their operations. These interventions have world competitiveness as their *ultimate* objective, but in the meantime they have been granted protected markets (mainly by government purchases), soft budgets and various subsidised inputs and access to services. Their position is privileged relative to other public sector enterprises that are under the control of the MOIT, and much of the subsidisation is hidden.

The best-known example is the aircraft manufacturer IPTN (McKendrick 1992). This is the pride of the BPIS complex, the largest and most ambitious investment made by the Indonesian government to promote ITD in the country, and an archetypal 'mission-oriented' project. It is aimed at a very demanding activity, much in advance of existing capabilities, involving high risk, expense and government support. It is supposed to substitute efficiently for imported aircraft and create local linkages, skills and a technology culture in Indonesian industry. While the firm has various technological achievements to its credit, it has been heavily subsidised and its financial performance has been very poor—on normal commercial criteria the enterprise may not start to break even for two decades. It has minimal linkages with local industry; hardly any of its skills have spilled over to Indonesian industry (indeed, very few seem to have relevance); and it has absorbed a lot of the infrastructure at Puspiptek (see below).

In 1993 some Japanese consultants conducted studies of technology in most strategic industries, under the aegis of the Japan–Indonesia Science Technology Forum. They found great variation in the efficiency and capabilities of the enterprises concerned. On the *negative* side, all suffered from managerial and organisational weaknesses. Most served protected or captive markets, exported relatively little or nothing, and lacked incentives to be cost efficient. There was very high dependence on imported technologies. The older enterprises had outdated equipment. There were widespread quality and process technology weaknesses, and design and development capabilities were extremely limited. Local linkages varied, but their promotion was not an important objective of management. The large numbers of engineers employed may have crowded other Indonesian enterprises out of the limited skill market.

On the *positive* side, there were some signs of technology development within the enterprises. Managers were keenly aware of the need to upgrade technology, improve quality and develop design and research capabilities; the holding company exerted constant pressure to increase these efforts. Plans were being formulated to broaden the range of capabilities and obtain ISO 9000 certification. A range of training programs was being launched, with the stated objective of each enterprise being to become a 'centre of excellence' in its industry. Morale and motivation were generally good. In comparison to other Indonesian enterprises, local or foreign, these firms had the most ambitious technological objectives and the most systematic programs to achieve these objectives.

It is thus too early to discount completely the technological contribution of the strategic industries. If they were able to strike closer relations to the rest of industry, operate under more competitive conditions and share their growing technological capabilities with the private sector (perhaps with the ultimate objective of becoming private), they could make very positive contributions to changing the technology 'culture' of Indonesian enterprise. The evidence at hand does not, however, suggest that these conditions are being met.

Puspiptek Research Structure

The Indonesian government has set up 11 research facilities under the coordination of the National Centre for Science and Technology Development (Puspiptek). Puspiptek was established some 17 years ago to form the nucleus of a science city at Serpong to bring together the most advanced research facilities for industrial and scientific development. In the field of industrial technology, it was supposed to complement and link up with other industry-related research done by MOIT, and to have strong relations with the private sector. However, the immediate industrial objective of these laboratories was to *serve the strategic industries*. The proposal to establish Puspiptek came initially from the Minister of State for Research and Technology, whose intention was to use the 'science city' as an integrated centre for research, education, training and production,[50]

---

[50]The Indonesian Institute of Technology has a campus next door, and the University of Taramunagara is setting up a new campus there. Production has not yet started, but there is an industrial estate under construction. This is designed to attract skill and technology-intensive foreign and local firms that

to promote the mission-oriented technology development. This led to Puspiptek facilities being internalised by the self-contained system of technology development fostered by Menristek. As far as manufacturing goes, their central role remains with the strategic industries; Puspiptek remains largely delinked from the mainstream of Indonesian industry.

There are six BPPT laboratories: LUK (material strength, components and structures), with 80 staff (12 scientists and engineers); LCDC (energy and energy resources), with 462 staff (126 scientists and engineers); LAGG (aerodynamics, gas dynamics and vibration), with 177 staff (117 scientists and engineers); LTMP (thermodynamics, engine and propulsion systems), under construction; LMBA (natural disasters), under planning; and LPT (processing technologies), also under planning. Of these, LUK, LAGG and LTMP were designed specifically to serve IPTN, though they may ultimately have wider application. The one that may have the widest relevance, the processing technologies laboratory, is still on the drawing board.

There are four laboratories under LIPI (the Indonesian Institute of Sciences): LKIM (calibration, instrumentation and metrology); LMT (applied metallurgy); LRT (applied chemistry); and LFT (applied physics). The first is a national facility for standards, and should be providing essential services to all industry. However, standardisation in Indonesia is spread over a number of different ministries, and efforts to develop a coherent system are still to bear fruit. The other laboratories are more oriented to 'pure' scientific research, and are not expected to have much direct impact on industrial technology.[51] As a fairly minor part of their operations, they do provide some testing services for industry.

It may be too early to judge the longer term benefits of the Puspiptek network for Indonesian technology development, but its overwhelming focus on the strategic industries has certainly reduced its value to the bulk of industry. The linkages of the laboratories with Indonesian industry may improve over time, but this would need a significant change in their orientation and culture. It would

---

will draw upon the technological facilities at Puspiptek and help to commercialise its technologies.

[51]The only LIPI facility outside Puspiptek to have a direct bearing on industrial technology was LEN, which entered into the design and manufacture of electronic components. This was hived off as a separate firm, and is now one of the strategic industries (see above).

also need a change in the technological capability of private firms themselves, since at present they lack an appreciation of their need for technological effort, both in-house and from outside agencies. Without a growth of internal effort, they cannot effectively seek or use research institutes.

MOIT Technology Institutions

BPPI coordinates the activities of nine sectoral research institutes, 10 regional testing laboratories and five industrial research and testing centres. Of these, the sectoral research institutes are the most significant for industrial technology. [52]

These sectoral research institutes suffer a variety of problems. They have little contact with the LPND laboratories, even though six of the MOIT institutes are located at Serpong, near Puspiptek. They are all wholly government funded, earning little from the sale of services to industry. They mainly serve medium-sized, and some large, firms, though they were set up to provide technology services for all types of enterprises. Being poorly funded, they have to work with outdated and insufficient equipment and facilities. Their salary structures are bound by government rules, and are too low to attract the best quality engineering staff.[53] They lack the resources to train their staff effectively or send employees abroad for

---

[52]The nine institutes are: MIDC, in metals and machinery (casting, heat treatment, design of machinery, consultancy and technical services, training); IRDMTP, in materials and technical products (quality assurance of steel structures and diagnosis of industrial plant); IRDCI, in chemicals (development of pesticides, packaging technology); IRDLAI, in leather and allied products (tanning and finishing of leather, footwear production technology, rubber and plastic substitutes for leather); CRDI, in ceramics (development of building materials, fine ceramics, insulation and refractory materials); IRDHBI, in handicrafts and batik (quality improvement, process improvement, development of machinery); BBT, in textiles (research, testing, training, quality certification, consultancy services); BBK, in construction materials; BBS, in pulp and paper (pilot plant testing, engineering and consultancy services, training, standards development).

[53]For instance, in 1991 the Metal Industries Development Centre could offer a new engineering graduate around $50 per month, while in the same city IPTN was offering $200 (which was itself less than the equivalent in the private sector in Jakarta, around $300–500, though probably with fewer perks). The Puspiptek network also offered substantially better remuneration than the MOIT institutes.

advanced training, unlike many of the LPND institutions, which
have government as well as donor and World Bank financing for
foreign training programs. Under past rules, they were not allowed
to charge for services to industry. This was changed in 1990, but the
institutes are still able to earn only a very small proportion of their
budgets from the sale of services. Several institutes suffer from weak
management, poor staff morale and the lack of a culture to serve
industry actively.

The combination of poor funding, obsolete facilities, weak
management, inappropriate incentive structures, low salaries and
flagging staff morale means that the institutes have generally not
been able to keep up with technological frontiers or establish
credibility in industry. Most large firms, especially those with good
overseas connections, find they have little to gain from the MOIT
institutes. Very small and traditional firms lack the motivation and
information to access their facilities. In addition, the passive stance
of management holds staff back from actively promoting the
institutes' services.

There is practically no tradition of private firms contracting
universities or research institutes to carry out research, or of collaborative efforts by groups of firms (in the same industry or comprising
supplier and user industries). In more industrially advanced
countries, strong industry–institution linkages and industry–industry
linkages are critical to undertaking the necessary technical and
human effort for technology development. The absence of such
linkages and of an efficient technology infrastructure prevents the
local solution of many technological problems in Indonesia.

Despite these constraints, some institutes do render valuable
services, especially in training and testing for medium-scale
enterprises.[54] More important, the network of institutes and testing
stations is the only technological infrastructure in Indonesia
potentially able to meet the operational technology needs of most of
industry. The institutions in it are not intended to produce innovative
technology, but that is not the prime need of Indonesian industry at
this stage.

The real need is to raise the productivity of existing (and
emerging) technologies, improve quality and design, and raise the
skills of the workforce. These are all relatively mundane (but

---

[54]The textile institute is reputed to be the best of the nine, and many large
textile firms also draw upon it.

essential and difficult) tasks for which many enterprises need assistance and information. The relative neglect of the MOIT institutional framework for the more appealing high-tech LPNDs and the strategic industries, with their 'mission' that has few evident benefits for ITD in the rest of industry, would seem to be a misallocation of the government's technological resources.

CONCLUSIONS

Indonesian technology is at a low level in relation to the Asian Tigers. It is widely accepted, by analysts and by the government, that future growth and competitiveness require that technological development be accelerated. Indonesia can draw on the experience of the mature Tigers to stimulate ITD. However, the Tigers offer, not a single model, but a *variety* of models of technology development policy with a certain core of common elements (Lall 1996). For instance, the deepening of local R&D capabilities in Korea and Taiwan involved selectivity in foreign investment, trade interventions and a range of supporting measures. These measures comprised controlled exposure to international competition, combined with infant industry protection; the creation of an appropriate human capital base; the promotion of the local capital goods sector; the provision of infrastructural technology institutions and services; the development of technology finance; and the underwriting of exceptionally risky investments. It also involved a range of specialised institutions to meet the information and assistance needs of SMEs.

At the other end of the spectrum, Hong Kong practised a laissez faire trade and industrial policy, and developed a light industrial structure specialised in the assembly and manufacture of consumer goods. However, its enterprises, mostly small and medium in size, faced various needs for technology upgrading even in this relatively undemanding structure. The Hong Kong government provided considerable support and subsidies to its SMEs for raising productivity and adopting new technologies. Singapore combined free trade with targeted strategies to attract and guide foreign investment, backed by incentives and institutional, human capital and infrastructural measures to make upgrading attractive for MNCs. Its policies allowed it to develop a much 'heavier' industrial structure than Hong Kong, and also a much greater technological capability.

Of these strategies, Korea's involved the most detailed and pervasive interventions. One of its unique features was the creation of giant private conglomerates to internalise deficient markets for finance, skills and information, and to bear the large risks inherent in entering world markets in complex technologies without a direct reliance on MNCs. This strategy was costly and prolonged, and led to high levels of concentration, but it did succeed in giving Korea a technological base unmatched in the developing world. The Taiwanese strategy was based on the development of SMEs, and so involved the government in playing the role of guide, helper and coordinator—orchestrating the development of frontier technologies and providing extensive support to firms that were too small to undertake their own R&D. In Singapore, the strategy relied exclusively on MNCs and investments in the human and institutional infrastructure. The results are impressive, though the research base remains narrow and relatively weak. Hong Kong did not have a technology strategy until recently, but it supported its SMEs to master and utilise the most modern technologies. While this allowed an impressive expansion of manufactured exports for a period, it also resulted in technological weaknesses and a massive de-industrialisation. Singapore chose to develop its technology base by a combination of MNC and local government-sponsored effort in very highly specialised niches of high-technology industry.

The choice of policies in Indonesia has to reflect its particular economic and political realities.[55] The country's size, spread and resource base point to the need for policies to promote considerable deepening; its weak human capital base and administrative capabilities suggest, on the other hand, that interventions should not be very detailed or pervasive if they are to be effective. Let us consider briefly some of the main policies that may stimulate Indonesia's technological growth.

As far as the incentive regime is concerned, while trade strategies are being reformed, protection rates are still high and variable, with little apparent relation to the encouragement of technological learning in the country (that is, their extent and duration are not related to the time and effort needed to master the technology). The pace of future import liberalisation is unclear, and the net effect of the regime may be to deter long-term technology development in all firms not geared to export markets. The best reform strategy would be to have a clear schedule of tariff

---

[55]These are succinctly analysed by Hill (1996b).

reductions, pre-announced and with sufficient time for industry to restructure and develop competitive capabilities (as in Korea in the 1980s). The program should be credible, with no deviations for special interest groups—the recent granting of special privileges to the national car project is clearly counterproductive. No special preference should be given to public enterprises, strategic or other. However, a certain degree of selectivity between broad types of activities needs to be retained to take account of their different learning (or 'relearning') needs.

Industrial policy reform also has some way to go. There remain barriers to competition in the large-scale private sector, the unclear rules of the game, and the privileged position of public enterprises, especially the strategic industries. Such barriers and privileges distort resource allocation and prevent the most efficient and technologically dynamic firms from realising their potential in the marketplace, while diverting resources to rent seeking activity. They create undesirable segmentation between size classes by imposing taxes or other penalties on growing beyond a specified size. While Indonesian conglomerates can be a potential source of technological strength, like the *chaebol* [conglomerates] in Korea, so far they have not been a source of technological dynamism. It would require very careful policy formulation, and very effective implementation, to ensure that they become more export oriented and develop a technological culture.

On the 'supply side' of technology development, Indonesia needs to enhance its base of skills, technology finance and technology support institutions.

If there is one overwhelming lesson from East Asia for Indonesia, it is the need for a larger, better and more specialised base of industrial skills. The lags are particularly large at higher levels of technical education, but they also apply to educational quality at all levels. Both the quantity and quality of Indonesian education have to be improved. So does the rigid and centralised vocational training system, producing the wrong kinds of skills of the wrong quality using the wrong equipment. The level of in-house training by industrial enterprises, currently very low, has to be raised, and SMEs need to be assisted in human resource development by specialised training institutes. Given the constraints on government capabilities and resources, the government should encourage the private provision of education and training, with appropriate measures of accreditation and certification and controls on the quality of the training provided. While there are large numbers of private establishments in existence, the monitoring of their quality is inadequate, as is the information available to potential trainees and employers.

It is particularly important to persuade SMEs of the need to recruit better trained labour and to invest in more training for their employees, both in-house and external. As in other developing countries, traditional SMEs are unaware of their skill deficiencies and invest little in training beyond what is needed for basic operations. However, the maturation of the industrial structure involves SMEs moving into highly specialised market niches where high levels of skill are often required. This process calls for a massive change in the attitudes of SME managers and in the surrounding institutional environment that provides support for training. Indonesia should start planning for such a change.

The provision of foreign training has often been regarded as an immediate way to solve the most pressing skill needs in Indonesia. However, this is very expensive and can only address some special needs; in the past the strategic industries have tended to exploit foreign training schemes to their advantage. This valuable resource must be used carefully to provide the greatest spillover benefits to industry at large.

Finance for technology investments is needed by all sectors of industry, though the content of schemes may vary considerably by groups of firms and by industry. Traditional SMEs are perennially constrained by a shortage of investment and working capital. They do not need 'technology finance' in the normal sense of the term (to set up new technology-based ventures), but special facilities linked to extension may prove beneficial in inducing them to use consultancy and other services. More proactive financing schemes need to be devised that can link special financial instruments to technological upgrading, training, design and development efforts within firms, and that subsidise them in contracting research or testing and other technical functions to S&T institutions. Many countries provide a certain proportion of the expense of using such institutions as an inducement to industry to 'develop the habit', with the subsidy element declining over time. Finally, given that the government is thinking of incubator schemes for potential technology-based entrepreneurs (say, from Puspiptek), it may be useful to plan for venture capital financing.

In the large-scale sector, there is a need to set up financial instruments for technological activity in firms not related to well-funded conglomerates. In the short term, formal R&D is unlikely to be an important activity, but financial provision (at concessional or market rates, depending on the need to induce firms to start) for quality upgrading and standardisation, subcontracting, external training and design activities can play an important role. The crucial element in all concessional schemes, including grants, should be that the enterprises themselves put in a large part of the money, so that there is a commitment on their part. Joint R&D between local and foreign firms, and projects undertaken jointly with private and public laboratories, may be supported by grants or preferential credit. There is a need for specialised financial intermediaries to undertake these tasks, but the government has to play the initiating role.

The technology institutes, both under Menristek and MOIT, need to establish closer linkages with industry, upgrade their facilities and capabilities, adopt a more aggressive approach to selling their services to the private sector, and disseminate information to industry. In addition, it is essential to have full coherence and coordination between the ministries responsible. It is clearly wasteful for each to pursue its own scheme for spending on its system, perpetuating the divisions that have existed till now. Menristek, the institution designated to lead technology development, seems to have an overambitious and risky approach, and needs to be more sensitive to widespread criticisms of its approach and tactics.

The diffusion of technologies should be accelerated mainly by means of extension programs to improve the technological capabilities of SMEs, in particular by setting up effective and customer-oriented productivity centres. These should also be able to provide comprehensive information services to firms on sources of technology, in the way that the Tigers have been able to do (saving the heavy costs involved in searching for and importing the best technologies). Several extension programs are in place in Indonesia, but their results so far have been disappointing. Part of the difficulty may lie in the relatively passive attitude of many agencies in charge of extension and financial support services, and perhaps in the bureaucratic requirements of providing support. A more proactive approach, with qualified teams visiting enterprises, offering free diagnoses and putting together packages of technology, training and finance (initially at low cost to the enterprise, but over time moving towards full cost), would be more productive. The Taiwanese experience (the China Productivity Centre and the Industrial Technology Research Institute), and that of the Hong

Kong Productivity Centre, shows that more aggressive outreach efforts, backed by government financial support, can yield striking results.

In sum, the Indonesian government needs to do a great deal to promote ITD. Part of this entails less intervention and part requires more, but of a very different sort from before. The vision that is guiding technology policies at this time is narrow and overambitious, and is neglecting the mundane needs of industry. There is a lack of coordination and coherence in policy making and implementation, as well as enormous rent seeking and lethargy. The private sector is forging ahead on its own, but there is a need for an effective framework of policies to overcome the numerous market failures that inevitably face technological deepening. It does not appear that the government is fully aware of the nature and extent of the problem.

PART III

TECHNOLOGY:
THE POLICY FRAMEWORK

8

THE ROLE OF BPPT IN INDONESIA'S TECHNOLOGICAL DEVELOPMENT

Rony M. Bishry and Murman Hidayat

BACKGROUND

Indonesia, like other developing countries, is a net importer of technology. To minimise spending on imported technology and to develop an indigenous technology, it is important to have a local capacity to assess and apply technology. This is one of the main reasons for the establishment of BPPT, the Agency for the Assessment and Application of Technology. Indonesia's national program for research and technology has involved the setting up of various kinds of infrastructure, including the national education system, technological centres of excellence and laboratories, the Indonesian Academy of Sciences (AIPI) and government research institutions (Habibie 1995, pp. 185–188).

BPPT is one of six non-departmental government research institutes (LPND) engaged in multidisciplinary research activities and under the coordination of Professor B.J. Habibie. LIPI, the Indonesian Institute of Sciences, is another of the six. The other four agencies—Bakosurtanal (the Agency for Aerial Survey and Mapping), Batan (the National Agency for Nuclear Energy), BPS (the Central Bureau of Statistics) and Lapan (the National Institute for Aeronautics)—are more focused on specific areas.

Other institutions related to science and technology (S&T) in Indonesia are the Office of the Minister of State for Research and Technology (Menristek), the National Research Council (DRN),

AIPI, the Agency for the Management of Strategic Industries
(BPIS)—including the 10 state-owned strategic industries it
manages—and the National Centre for Science and Technology Development (Puspiptek) in Serpong, West Java. The key institutions of
Indonesia's S&T infrastructure are presented in Figure 8.1.

Thus far, little has been written about the role of BPPT. Studies
which contain some reference to the agency include Raillon (1990)
and Hill (1995). This chapter, which extends an earlier paper by one
of the authors (Bishry 1992), discusses the role of BPPT, including its
tasks, programs, perceived role, unique characteristics and future
orientation. Since BPPT is a social investment, its allocation of
resources should be based on a broad social calculus. Our analysis
also investigates the structure of the agency's budget.

AN ASSESSMENT OF S&T POLICY IN INDONESIA

For the past two decades, the Indonesian approach to S&T policy
has been heavily influenced by the ideas of B.J. Habibie, Minister of
State for Research and Technology, and Chair of BPPT, BPIS and
DRN. Habibie plays a crucial role in formulating and coordinating
Indonesia's S&T policies, and reports directly to the President.
Consequently, most of the organisational S&T infrastructure
established since the late 1970s (including BPPT) has been set up to
support the Minister in his efforts to develop a national
technological capability. Whereas the former Minister, Professor
Sumitro,[56] pursued an S&T policy aimed mainly at achieving high
economic and employment growth, Habibie's approach has
emphasised much more the need to develop Indonesia's human and
technological resources for nation building.[57]

According to Habibie, Indonesia's strategy for technological
and industrial transformation has two features: a stage-wise
approach; and the selection of vehicle industries to implement the
strategy. The first has involved the identification of four
consecutive stages of industrial transformation, to be implemented in
all S&T activities. This feature is intended to improve the value

---

[56]Professor Dr Sumitro Djojohadikusumo was Minister of State for Research
from 1973 to 1978. He was replaced by Professor Dr Ing. Bacharuddin Jusuf
Habibie (known as B.J. Habibie) in 1979.

[57]For further discussion of the policy approaches adopted by Habibie and
Sumitro, see Rice (1990) and Chapter 9 by Rice, this volume.

FIGURE 8.1 Key Elements of Indonesia's S&T Infrastructure

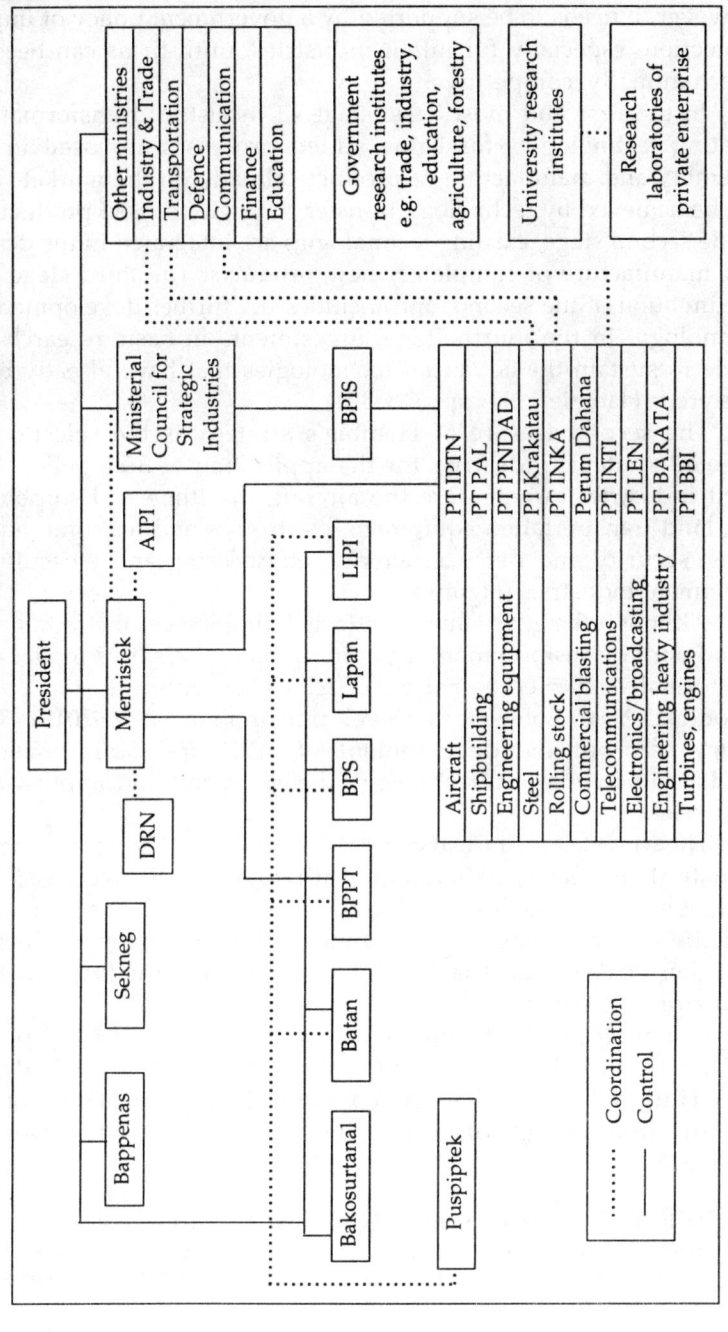

Source: Adapted from Djojonegoro (1991), Figure 3, p. 10.

added process by developing a progressive manufacturing plan. However, it needs to be supported by a government policy of import protection, especially for infant industries, until firms can become internationally competitive.

In the first and most basic stage of industrial transformation, existing technologies for value added processes are used in the assembly and manufacture of products already on the market. This can be achieved by technology transfer through licensed production. In the second stage, existing technologies are integrated in the design and manufacture of completely new products. The third stage is a continuation of the second, and includes the further development of technology. In the fourth stage, investments in basic research are made to sustain the advanced technologies that have already been mastered (Habibie 1995, pp. 348–351).

The second feature of Habibie's strategy is the selection of targeted strategic industries for the application of S&T policy. The eight industries targeted are the aircraft, maritime and shipbuilding, land transportation equipment, electronics and telecommunications, security and defence, energy, engineering and agricultural equipment industries (Figure 8.1).

Although the government's efforts to implement the four stages of industrial transformation began in the mid 1970s, this model was not formally adopted as national S&T policy until the start of the Second 25-Year Long-Term Development Plan (1994–2019). This policy sets out the development of R&D and S&T, covering production techniques, technology development, and applied and basic sciences.

The Ministry of Industry and Trade has also been pursuing an industrial development strategy with four broad objectives: the development of resource-based industries aimed at increasing value added; the improvement of technological capability; the production of higher quality industrial products; and more active participation by the private sector.

Although this S&T approach is recognised as national policy and was included in the Broad Outlines of State Policy (GBHN) in 1993, Hill (1995, p. 110) has argued that technology policy is not the central determinant of industrial development outcomes in Indonesia:

> As in all countries, technology policy—and industrial policy more generally—is the sum of a range of macro and microeconomic interventions. Some have technological objectives, while others have

incidental and often unintended effects on technological development. It must be stressed again that the general economic policy framework is the most important factor in the development of technological capability. That is, factors which influence a country's rate of economic growth, its investment rate, its degree of international orientation, and its stock of skilled labour are likely to be the major determinants of the pace of technological progress.

All of these latter policies have contributed to industrial and export diversification in Indonesia, and no ministry or organisation has supreme control over industrial or S&T policy. However, BPIS, through its strategic industries policies, is able to influence a series of key industries. In light of this, BPPT's role in the implementation of Indonesian S&T policy is analysed in the following sections.

THE ROLE ASSIGNED TO BPPT

Since its establishment in 1978, BPPT has been assigned a number of tasks and functions. Its main tasks are: (1) to prepare a general policy program for the assessment and application of technology; (2) to ensure coordination in the implementation of programs for the assessment and application of technology; (3) to provide services to government and private sector agencies in the application of technology; and (4) to carry out activities in the assessment and application of technology in support of government technology policies.

BPPT's functions are broadly defined, and include the following: (1) to control and assess the implementation of programs for the application of technology, and to foster activities for the transfer of technology; (2) to promote cooperation between government and private sector organisations in Indonesia and overseas in the assessment and application of technology; (3) to promote the application and assessment of technology in primary and secondary industry, the socioeconomic system and regional development; (4) to promote the development of basic and applied sciences in engineering, biology, engineering and marine biology; (5) to promote the assessment, application and development of technologies in such areas as habitat and the environment, industrial processing, energy, electronics and information, manufacturing, certifications, defence, security and strategic industries, the processing and engineering industries, and natural resources (the latter including energy and land resources) and the mitigation of natural disasters; (6) to promote administrative development, services and the planning of programs in support of the implementation of the primary duties of BPPT.

Research and technology activities in Indonesia can be categorised as comprising basic research, R&D, and assessment and application activities. In the implementation of technology development, Habibie has formulated a general strategy for the transformation of S&T to include integration, utilisation, development and basic research.

According to the Minister, this implies that Indonesia's technology development has to 'begin at the end, and end at the beginning', because 'integration' means that the production of goods based on existing technologies has to start first, even if it involves the integration of foreign technology. This could then be followed by the assessment and application of technology, R&D, and finally basic research. In line with this policy, BPPT has to become the spearhead for technology development in Indonesia, since the stages of integration and utilisation involve the assessment and application of technology.

To carry out its tasks and functions, BPPT has designed programs in the fields of aerospace, ocean science, energy, biotechnology, agriculture and agro-industry, materials technology, transportation, electronics, telecommunications and computation, processing technology, defence and security, regional and environmental development and management, training and education in engineering, and S&T infrastructure and institutional development. It has conducted technical cooperation with foreign counterparts in many technologically advanced nations. One recent example is the collaborative S&T program concluded with Australia's Department of Industry, Science and Tourism (DIST), termed Collaboration on Science and Technology between Australia and Indonesia (COSTAI). This program aims to promote cooperation between the two countries in the field of scientific research and technological development over the period 1995–2000.[58] Other collaborative programs include those in ocean science with the Scripps Institution of Oceanography, University of California, San Diego; with Japan's New Energy Development Organisation (NEDO); and in applied research on

---

[58]For a fuller description of the program, see Chapter 19 by Scott-Kemmis, O'Brien and Rohadian.

recombinant biopharmaceuticals with German firms Linde-KCA-Dresden GmbH and Rhein Biotech GmbH.

The role of BPPT is reflected in its development projects, which can be classified as follows:

(1) an integrated technology management system;

(2) solar energy for remote areas;

(3) improving the socioeconomic environment;

(4) environmental pollution control, including mitigation of disasters;

(5) production technology projects in the fields of engineering, natural resources and energy utilisation, and in the mineral and pharmaceutical industries;

(6) technology acquisition, affecting all major sectors of the economy;

(7) a program of applied sciences;

(8) the development of S&T facilities and infrastructure;

(9) an aeronautics vehicle utilisation project;

(10) information systems development; and

(11) education and training programs for government officials.

BPPT's scientists have also been involved in other national projects, such as the Natuna gas field, Batam Island and interisland bridge planning. To provide technical/engineering services to society, BPPT has developed a Technological Service Provision (Yantek) unit. Under this system, units in BPPT can sign contracts with and be paid for their services by government institutions or private companies. It is expected that BPPT will receive additional funds from this source, while being able to utilise government research facilities more efficiently, serve national needs and collaborate with the private sector.

BPPT'S RESOURCES

Human Resources

BPPT employs a large number of university graduates from local and foreign universities; 2,058 of its employees held university degrees

TABLE 8.1 BPPT's Employees by Degree and Field of Study (no.)

Degree	Engineering	Natural Sciences	Agriculture	Other	Total	Total Employees (%)
S1	961	198	48	310	1,517	73.7
S2	301	39	35	56	431	20.9
S3	84	12	8	6	110	5.4
Total	1,346	249	91	372	2,058	100.0

S1: Bachelor's degree; S2: Master's degree; S3: PhD degree.
Source: Bureau of Personnel, BPPT.

as of 31 March 1997 (Table 8.1). Since 1985 BPPT has provided scholarships to 3,367 students to study abroad and within Indonesia, especially in the fields of science, technology and engineering. During the period 1985–92, for example, the Overseas Fellowship Program, managed by BPPT, provided scholarships to 1,321 students, while the Science and Technology Manpower Development Program assisted another 521. Current programs include the Science and Technology for Industrial Development (STAID) programs for 1990–98 (1,074 students) and 1995–2003 (451 students).

High-Quality Laboratories

BPPT manages several high-quality laboratories utilising the latest technology. At Puspiptek there are three Technical Operating Units: the Laboratory for Testing and Construction (LUK), the Laboratory for Aero Gas Dynamic and Vibration (LAGG) and the Laboratory for Energy Resources (LSDE). Puspiptek's other laboratories include the Centre for the Assessment and Application of Industrial and Agricultural Biotechnology (PPP Biotek) and the Laboratory for Thermodynamics, Engines and Propulsion Systems (LTMP).

There are also a number of laboratories located elsewhere in the country, including the Hydrodynamics Laboratory in Surabaya, the Technical Operating Unit on Ethanol, Protein and Sugar in

TABLE 8.2 BPPT's Budget, 1995/96–1996/97
(Rp million)

Budget	1995/96	1996/97
Domestic budget		
Development budget (DIP)	109,249.5	110,161.0
Routine budget (DIK)	45,173.7	53,143.1
Foreign aid budget (BLN)	50,188.2	35,572.7
Total	204,611.3	198,876.8

Source: Planning Bureau, BPPT.

Lampung, and the Laboratory for the Study of Coastal Techniques in Yogyakarta. Other facilities are the Baruna Jaya Research Vessel, the Weather Modification Technical Operating Unit, and the Technical Operating Unit for the Development of Balinese Ceramic and Porcelain Art and Technology.

The Research Budget

To finance its activities, BPPT is provided with an annual budget consisting of development, routine and foreign aid components, which in 1996/97 totalled Rp 198.9 billion (Table 8.2). This trans-lates into an average research budget for each of the agency's 2,058 scientists of Rp 96.6 billion ($34,000).[59] However, this figure is effec-tively an overstatement. The routine budget is not used directly for research, but is allocated to salaries, office supplies, building maintenance and travel. Neither are the development or foreign aid budgets used solely for direct research. Part is spent on construction and part on miscellaneous items not directly related to research. Although there are no exact or detailed data on the allocation of the development budget, as a rule of thumb only 50% of it is used directly for research activities. The annual research budget per

---

[59]The rate used throughout this chapter is the exchange rate for August 1997 of Rp 2,839/$1.

TABLE 8.3 Budget Structure of Two BPPT Projects
(%)

Budget	Project 1	Project 2
Additional salary	10	15
Materials	5	17
Travel	12	18
Service contracts	28	12
Miscellaneous	9	14
Building/machinery	35	23
Total (Rp million)	253.35	786.66

Source: Planning Bureau, BPPT.

scientist is therefore approximately $9,500,[60]—although this figure may be an *underestimate* to the extent that not all of BPPT's graduates are directly involved in research. With such a low annual budget per scientist, it would not be easy for BPPT to aspire to international quality research, especially in engineering. Still, the opportunity cost of the research budget—which is around eight times the country's GDP per capita—also needs to be considered.

The budget for BPPT's 11 major projects is divided among 28 subprojects. The smallest amount allocated is Rp 253.4 million, for an information electronics technology application project, while a larger amount of Rp 786.7 million has been allocated to an energy utilisation project (Table 8.3). In the case of these two projects, 10% and 15% of the budget respectively is used to supplement the monthly salaries of the 25 and 125 scientists engaged on the projects, by Rp 55,000 ($19) to Rp 105,000 ($37). The annual budget per scientist for the first project is about Rp 10 million ($3,522), while for the second it is a little over Rp 6 million ($2,113). Both these figures are much lower than our earlier estimate of $9,500. The salary supplement is quite low, and most of the budget is used for non-salary purposes. These include travel (12% and 18% of the

---

[60]That is, 50% of Rp 110.2 billion, divided by 2,058 scientists, equals Rp 26.8 million, or $9,500.

**TABLE 8.4 Examples of Projects with Foreign Aid Component
(Rp thousand)**

Project	Total Budget	Rupiah Component	Foreign Aid Component
Project 1	11,150,181	3,989,722	7,160,459
Project 2	13,633,961	2,910,324	10,723,637
Project 3	18,398,507	4,922,798	13,475,709

Source: Planning Bureau, BPPT.

respective budgets), and facilities, infrastructure and service contracts (78% and 67% respectively). In both cases the budget is too low to allow high-quality research.

In some cases a project has a larger budget because a foreign aid component is attached to it. Table 8.4 shows the rupiah and foreign aid components of three such projects. Although there are several projects in this category, aid revenues are spent either on physical investment or on non-physical service contracts. The rupiah portion is also mostly spent in this way, since rupiah matching funds are required to obtain foreign aid in each project.

Even in the absence of foreign aid, a larger budget will always mean an increase in either physical investment or non-physical service contracts. Table 8.5 presents some examples of projects in which the budget for additional salary and travel remains low, even though the total budget has increased considerably. Since service contracts and physical investment are not spent directly by scientists, they do not have direct access to most of the budget. It can be concluded that managers and administrators have more say in the spending of budgets than do the scientists.[61]

---

[61]Although not directly comparable, it is instructive to view these figures in light of the allocation of the Japanese research budget of ¥7.9 trillion in the mid 1980s. Some 46% was allocated to personnel, 17% to tangible fixed assets, 16% to raw materials and 20% to other costs (Foreign Press Center of Japan, 1986). The budget allocation for tangible fixed assets investment was much lower than in Indonesia, indicating that most of the Japanese budget was used for research activities.

TABLE 8.5 Examples of Large-Budget BPPT Development Projects (Rp million)

Budget	Project 1	Project 2	Project 3	Project 4
Additional salary	3.6	8.3	3.7	4.4
Materials	2.3	7.8	7.3	3.5
Travel	3.2	26.6	5.1	16.4
Service contracts	151.2	682.5	45.0	5,238.8
Non-physical service contracts	4.5	11.0	8.0	6.0
Building/machinery	88.6	292.8	2,289.2	5,602.4
Total	253.4	1,029.0	2,358.4	10,871.5

Source: Planning Bureau, BPPT.

OTHER ISSUES

There are at least five other issues which in one way or another affect BPPT's performance. The first concerns the 'uniqueness' of BPPT. Habibie, BPPT's Chairperson, wears several 'hats'. Of the various S&T institutes in Indonesia, approximately 50% are his direct responsibility. The remainder, although led by others, are mostly under the coordination of the Minister. The number of influential positions held by Habibie has created a perception that BPPT is very influential in Menristek, in BPIS and in the 10 state-owned strategic industries. In addition, since the establishment in 1990 of the Indonesian Association of Muslim Intellectuals (ICMI), which was until recently chaired by Habibie, a perception has arisen that BPPT is influential in ICMI, and conversely that BPPT has political influence because of its ICMI ties.

The second issue concerns the post-Habibie era. As the Minister plays a crucial role in BPPT, a pessimistic view has been expressed that the agency would not be able to survive his departure. The Minister is known to be familiar with both technology and business, and possesses powerful religious and political affiliations. A more optimistic view holds that BPPT will not have to continue to depend on Habibie to carry out its mission. If BPPT's leadership is separated from its role in the strategic industries and its political connections, the agency may continue to function under a person who is concerned exclusively with the development of S&T.

Third, the research culture in Indonesia has recently improved considerably. At the macro level Bappenas, the National Development Planning Board, now allocates research spending through various competitive grants (known as RUT, RUK and RUSNAS) and other programs with an important research component. Habibie himself plays a significant role in ensuring government support for research activities. The Ministry of Education and Culture is also promoting the development of research capability by, for example, establishing interuniversity centres in several leading universities.

There is, fourth, the issue of BPPT's status as a government institution. BPPT is subject to the same administrative standards as any other government institute, including those relating to salaries and budget disbursements. Although take-home pay is much higher in BPPT than in similar government institutes, such as LIPI, BPPT's scientists are still poorly paid ($200–800/month) compared with those in the more advanced and richer East Asian countries. As a result, the work ethic in BPPT is relatively poor, and the management system less efficient than in similar institutes abroad.

Finally, organisational changes have often been discussed in BPPT. In the first STAID program, great attention was paid to institutional and human resource development. Institutional development included improving S&T capabilities, developing policies to strengthen S&T mechanisms, and providing analytical support for S&T related to industry development. The program also focused on strengthening S&T infrastructure, with the following objectives: to strengthen R&D planning and coordination by the DRN and Menristek; to raise the efficiency of the LPND (particularly BPPT and LIPI); and to accelerate the development of the science-based city at Serpong, as a key element in the country's industrial development efforts.

The integrated technology development project, which is concerned with institutional development, has the following objectives: to guide and support the process of organisational change as S&T institutes become private sector-oriented technology service organisations; to implement and make operational core functions for business development and the management of technology; and to implement and make operational core management information systems, specifically including finance and inventory management. If it is to meet the needs of industry, BPPT will need to be reorganised to become more private sector oriented. It must develop the technology and innovations that will be required by national industry in the future, as well as serving its current needs. It should provide support and leadership to national industry. To this end, BPPT must endeavour to promote linkages with industry in accordance with its available expertise.

CONCLUSION

In assessing the role of BPPT in Indonesia's technological development, it is important to look both at its development projects and at other national projects. Since these development projects cover a broad range of sectors, consistent with BPPT's assigned tasks, the agency is responsible for the assessment and application of technology in all areas, especially in the implementation of S&T policies. This involves the integration and utilisation of foreign technologies, followed by the implementation of R&D, and basic research. In carrying out its assigned role, BPPT is constrained by the fact that only a small portion of its budget is allocated to direct research. This makes it difficult for it to carry out the assessment and application of technology to international standards, and has created the perception that the role of BPPT in technological development is not significant. To counter this perception, the agency should be reorganised into a private sector-oriented technology service organisation, as planned under the technological service provisions initiative, as soon as possible.

9

THE HABIBIE APPROACH TO SCIENCE, TECHNOLOGY AND NATIONAL DEVELOPMENT

Robert C. Rice

INTRODUCTION

This chapter presents and discusses the approach to science, technology and national development of Professor Dr Ing. B.J. Habibie, whose thinking greatly influences Indonesian development policies. Professor Habibie has served as Indonesian Minister of State for Research and Technology since 1978. Most of the elements in his approach have become Indonesian government policy, as can be seen by comparing the discussion below of these elements with the country's management of technology as described in other chapters in this book (see also Luhulima 1996). In a previous paper (Rice 1990), I have compared Habibie's approach with that of other Indonesian intellectuals. In this chapter I present Habibie's view of development and discuss elements in his approach, compare the Habibie and Porter (1990) approaches to competitive advantage, and offer a critique of the Habibie approach.

HABIBIE'S VIEW OF DEVELOPMENT AND ELEMENTS IN HIS APPROACH

Like the Indonesian government, Habibie has a broad view of development that embraces increasing real per capita incomes and achieving a more even income distribution, together with broader objectives of nation building. In human resources he includes six items: belief in God, mastering science and technology (S&T), possessing a national culture, putting the national interest before personal/group interests, endeavouring to raise general levels of prosperity based on Pancasila and the 1945 Constitution, and endeavouring to maintain eternal unity and integrity (Rakasima and Linrung 1995, p. 62). One way to achieve the ideals of Pancasila is through development, which can spur an increase in national capability and living standards to levels comparable to those in advanced nations. Habibie further argues that Independence in its true meaning must be accompanied by the capability of a nation to stand on its own feet economically, and to be able to defend both its cultural and political integrity (Habibie 1995, p. 9).

Another characteristic of development is the capacity of a nation to use S&T to confront challenges and avail itself of opportunities. Habibie is concerned that developing countries will be unable to catch up with developed countries—that they are rapidly increasing their productivity without participating actively in the S&T revolution, which requires capable human resources. If developing countries are not part of this revolution, they will be excluded from the circle in which important decisions affecting global futures are made (Habibie 1995, p. 373).

Owing to the importance of an even income distribution, Habibie argues, development strategies that emphasise economic growth and comparative advantage will need to be reconsidered. GDP is just one of several indicators that can be used to evaluate a nation's potential for development. Habibie proposes another, national performance productivity (PPN), and its growth (Habibie 1995, p. 45). PPN is the total of the performance productivities in a nation's enterprises.[62] PPN and growth in PPN are more precise than GDP in evaluating the capability and future prospects of a nation or firm. A country with a high GDP growth rate but low PPN may eventually be surpassed by a country with a high PPN, because the latter is superior in developing and producing new products. A country with a high GDP growth rate but low PPN will experience

---

[62]Habibie apparently measures PPN by the ratio of value added from production to the value of the raw materials or intermediate inputs used in the production process.

difficulties in marketing its products and its growth rate will suffer (Habibie 1995, p. 45), because growth in demand for new products will be greater than that for old, well-established lines.

Habibie considers investment in human resources and advanced technology to be critical in expediting economic development, because human resources do not depreciate as quickly as physical capital, and because they are less mobile internationally than financial capital. He advocates utilising Indonesia's large domestic market and the state to facilitate human resource and technological development. He generally thinks of high technologies as being in manufacturing industries, but also includes advanced agricultural, medical and biological technologies (Habibie 1995, p. 113).

The Minister contends that Indonesia needs an alternative development strategy oriented towards the optimisation of human resource capability to utilise technology and available resources. He hopes that this strategy will guarantee continuing growth with high value added, which in turn can be used to promote community welfare. Habibie calls this strategy 'Development Oriented to-wards Value Added' (Pembangunan Berorientasi Nilai Tambah) (Habibie 1995, p. 46). The objective is not to adopt sophisticated technology for its own sake, but rather in order to contribute to national welfare through the value adding process (Habibie 1995, pp. 112–113).

I now discuss other elements in Habibie's development strategy, based on his views as presented and paraphrased in Habibie (1995). An important element in the Minister's strategy is for Indonesia to specialise in the production of products—mainly technology and human capital intensive—for which world demand is increasing rapidly. One attraction of high-technology products is that, due to a lack of competitors, producers determine their prices (Habibie 1995, p. 90). Habibie advocates making Indonesia's domestic market available to these new products until production can become internationally competitive. He insists that Indonesia will not be able to continue to depend on its comparative advantage in products based on natural resources, footloose industries and low-cost labour, and says that it must instead develop a competitive advantage in new, higher technology products before its comparative advantage in labour-intensive industries is lost. This is necessary because future world market prospects for primary commodities and simple labour-intensive commodities are not favourable, owing to the development of substitutes for these products in the richer countries and increasing competition among the poorer countries that supply them.[63]

Habibie has advocated a strategy of 'picking winners'. Thus the government has become directly involved in specific production sectors, by promoting the establishment of enterprises grouped together under the umbrella of the Agency for Strategic Industries (BPIS). The primary mission of BPIS is to anticipate a shift in international business from a resource to a knowledge base, which further involves a shift in emphasis from comparative to competitive advantage (Habibie 1995, p. 564). Achieving this goal requires an increase in human resources and S&T capability; the latter in turn requires practical application. Habibie also emphasises human over physical capital, because human capital appreciates with its use in production while physical capital depreciates.

The Minister foresees the rich nations having surplus financial capital once profitable investment opportunities in their own countries diminish. He believes that Indonesia may be able to attract this surplus capital through the development of complementary and nearly immobile factors of production, such as economic, human resource and S&T infrastructure, including human resources highly skilled in S&T (Habibie 1995, pp. 91–92, 99). This is another reason for his emphasis on the expansion of human resources and technological capabilities. Like Johnson (1964), he has a broad concept of capital that includes human resources, ideas and relations, as well as economic, S&T and other types of infrastructure (Habibie 1995, pp. 92, 103). Rather than attracting capital through foreign direct investment, Habibie stresses build, operate and transfer (BOT) agreements, together with cooperation between Indonesian firms and government agencies, and foreign entities (Habibie 1995, pp. 363, 376–377).

Habibie divides industries into three categories according to the density of their capital (broadly defined): low, medium or high

---

[63]Raul Prebisch and Hans Singer formulated the thesis in the early 1950s that the commodity terms of trade of primary commodity exporting countries would decrease over time because of slow rates of increase in world demand for such commodities. There is still considerable disagreement about the validity of a broader version of this thesis which, with the widespread adoption of export-oriented strategies by developing countries, includes labour-intensive manufactures. Some notable recent participants in the debate are Bleaney and Greenaway (1993), Hughes (1992), Helleiner (1989, 1990) and Cline (1982).

TABLE 9.1 Industry Characteristics by Density of Capital

Criterion	Low Density	Medium Density	High Density
Education	Elementary	High school + 3 to 12-month course	High school + 4-year course
Technology	Low density	Medium density	High density
Return on investment	Fast	Medium	Slow
Industry movement	Quick to move to a profitable country	Quick to move	Permanent
Risk	Depends on the management	Depends on the management	Depends on the management

Source: Habibie (1995), p. 61.

(Table 9.1). Garments, textiles and footwear are typical low-density industries; electronic consumer goods is a medium-density industry; and the aircraft, maritime and telecommunications industries are examples of high-density industries. It is easy for investors to move low-density industries from one country to another, but difficult for them to transfer high-density ones. Low-density industries are based on comparative advantage, medium-density industries on comparative and competitive advantage, and high-density industries on competitive advantage (Habibie 1995, pp. 60–62).

According to Habibie (1995, pp. 209–210) there are five principles to be heeded if the desired technological and scientific transfers, applications and developments are to be achieved.

(1) Education and training needs to be carried out in S&T fields that are relevant to national development needs.

(2) A concept of technological and scientific advancement that is clear, realistic and capable of implementation needs to be developed. The technologies required will not always be the simplest; often they will be among the most advanced in the

world. A technology is considered to be appropriate if it can solve the real problems confronting a nation.[64]

(3) The most important principle is that technologies can only be transferred, applied and developed further if they are truly applied to solving concrete problems. Technologies cannot be understood, let alone developed, in the abstract.

(4) A nation which wants to develop itself technologically must be determined to solve its own problems, and cannot remain a net importer of technology for ever.

(5) In the early stages of technological transformation, a nation must protect the development of its national technological capability until it is able to compete internationally. Each nation has to plan to achieve that capability in the shortest possible time.

According to Habibie, many strategies could be used to apply these five principles in a concrete manner. However, conceptually any such strategy would need to describe the phases of, and 'vehicles' for, its implementation.

Development Phases

In the process of becoming advanced, technologically and industrially, countries pass through four overlapping development stages (Habibie 1995, pp. 211–215, 1986, pp. 44–45).

(1) Management and production technology is used to add value to raw materials and semi-processed goods—usually by transferring technology from abroad and producing under licence. During this phase, the capability to understand designs from overseas, together with more advanced techniques and production methods, is augmented, and both production capability and managerial and organisational competence increase. This technology transfer process can be expected to be long and arduous.[65]

---

[64]The emphasis on S&T to solve concrete problems is also found in the Broad Outlines of State Policy (GBHN) for 1993 and Repelita VI.

[65]Pack and Westphal (1986) hold a similar view. They consider much technology to be tacit in nature, with the result that importing countries need to acquire a technological capability in order to choose wisely which technologies to import and how to apply them efficiently.

(2) Existing technologies are integrated into the design and production of completely new products. The designs and blueprints thus produced introduce a new element of creativity and encourage the further development of design skills, leading to the design of more new goods.

(3) Existing technologies continue to be improved and new technologies developed. Whereas in stage 2 existing technologies are still used, in this phase completely new technologies must be created, within the framework of planning future products.

(4) Large-scale basic research is undertaken, including the development of new theories.

To give one example, in the case of Indonesian aeroplane technology, production of the NC-212 aeroplane took place in stage 1 and manufacture of the CN-235 Tetuko in stage 2. In stage 3, several scientific disciplines were combined to create new technology, resulting in the production of the N-250 Gatotkoco. Now, in stage 4, the N-2130 project is spearheading research into how the aircraft industry will develop after the year 2000 (Habibie 1995, pp. 216–217; see also Mursjid, Chapter 11, this volume).

Habibie emphasises that the first three development phases are the most relevant for developing countries, although basic research directly connected with the national interest should be carried out in Indonesia by Indonesians. Such research needs to be oriented towards the resolution of basic scientific problems that are specific to Indonesia's cultural, social and national conditions (Habibie 1995, pp. 168–169).

According to Habibie, not all concrete programs for the design and production of goods are equally effective in bringing about the transformation of technology and industry in a developing country. To achieve this goal and make best use of available funds, priority must be given to the transfer and development of technology that contributes to the value adding process and that fulfils the following two necessary conditions.

The first necessary, but not sufficient, condition is that there be a *progressive production plan* to facilitate the penetration of related technology among national producers in a step by step process. Consequently, as production increases, so will the percentage of domestic value added in an industry. The objective of the progressive production plan is to increase the level of technological capability of related firms to the point where the percentage of the value of the product accounted for by domestic value added is the same as that of similar firms in advanced countries. In the case of aircraft companies, this would be 40–60% of value added for an aeroplane.

The second condition, which is both necessary and sufficient, is that the goods or group of goods produced must *fulfil market demand*. Developing countries with large domestic markets will need to produce goods incorporating sufficiently advanced technology that they can compete with imports in the domestic market, while developing countries with small domestic markets will need to be able to compete in export markets. It is appropriate that developing countries with large domestic markets—such as Indonesia—be oriented towards domestic market needs when determining and implementing programs for technology transformation. It would not be prudent for Indonesian firms to carry out production programs for goods for which there was not a domestic market, even though this would be quite feasible for countries like Singapore and Hong Kong. The availability of protected markets in Indonesia gives national producers the opportunity to progress to the point where they can become competitive in international markets, and eventually export their products. Habibie supports exporting to finance the import of goods that Indonesia cannot produce itself economically (Habibie 1995, pp. 208, 218–220). He recommends that Indonesian firms actively seek out an industry segment or product in the domestic market that is in great demand locally, but which is not being adequately supplied by foreign firms (Porter 1990, pp. 677–678). A medium-body passenger aeroplane would be one such product.

Vehicles for Transformation

According to Habibie, determining which products or groups of products fulfil the conditions to be vehicles for the transformation of technology and industry depends also on a consideration of the country's geographic conditions (including location), natural wealth, rate of economic development and domestic market size. In Indonesia the sectors that fulfil these conditions, while meeting the need to augment political and economic unity, are the transportation equipment industries (aircraft, shipbuilding and maritime, railway and automotive, electronics and telecommunications).

Some other industries are also attractive as vehicles for transformation. As incomes rise there is an increased demand for energy, which makes this industry (turbines, boilers, generators, heat exchangers) a candidate. Demand for machinery and equipment in such sectors as sugar and palm oil processing, petrochemicals and cement can be expected to increase, and thus the engineering industry is important in this schema. The development of extensive agriculture in the areas off-Java requires a high level of mechanisation, making agricultural equipment an attractive activity. Security and defence are significant industries in view of Indonesia's strategic location and abundant natural wealth. A final category comprises industries that will benefit from the expansion of demand for their products through backward and forward linkages with the vehicle industries enumerated thus far. Examples include road construction, housing, food production, agro-industry, pharmaceuticals, aircraft propellers and landing gear, railway cars, hydraulic systems and associated services (Habibie 1995, pp. 222–224; Rice 1990, p. 54).

Habibie argues that Indonesia must utilise its limited funds and forces optimally. They must be used to transfer and develop technology in value adding processes in agriculture and industry with significant multiplier effects through backward and forward linkages (Habibie 1995, pp. 162–163, 181). He identifies the following virtuous circle: provision of basic needs (such as health and education) and economic infrastructure brings about an increase in the capability to produce sophisticated high-value commodities; this results in a higher income, part of which can be invested to fulfil basic and infrastructural needs; this in turn leads to further improvement in basic needs and economic infrastructure (Habibie 1995, pp. 67–68).

THE HABIBIE AND PORTER APPROACHES TO COMPETITIVE ADVANTAGE

Habibie's approach to developing competitive advantage is similar to that of Michael E. Porter, author of the famous book, *The Competitive Advantage of Nations*. First, Habibie agrees with Porter (1990) on the four broad attributes that shape the national environment in which local firms compete, and that affect the creation of competitive advantage (the so-called Porter 'diamond'). These are factor conditions; demand conditions; related and supporting industries; and firm strategy, structure and rivalry (Porter 1990, p. 71). Porter considers vigorous domestic rivalry in the production of a product to be a very important factor in the creation of competitive advantage, and believes that state enterprises (especially state-owned monopolies) are usually ineffectual in developing competitive advantage (Porter 1990, pp. 117, 377, 474,

694–697). Habibie seems to have a different view on this latter point, as is evidenced by the extensive support provided to Indonesian state-owned vehicle industries, some with few or no domestic rivals. Indeed, in 1993 he stated that the only way to protect the aircraft industry was to have a monopoly in the domestic market (Rakasima and Linrung 1995, p. 44).

Second, both men advocate protection of the domestic market to assist infant industries, although Porter restricts the use of this policy to 'developing nations without a strong base of industries, in industries where foreign competitors are already well established' (Porter 1990, p. 665). Both see potential in industry segments where domestic demand conditions are favourable, especially those ignored by other countries, and consider it important that industries eventually become internationally competitive.

Third, Habibie's first three development phases for a society's industrial and technological advancement are similar to Porter's first three stages of national competitive development. Habibie places greater emphasis than Porter on investing substantial resources to induce the vehicle industries to move into the third development phase.

Fourth, unlike Habibie, Porter is very cautious about government assisting firms directly to develop competitive advantage, especially in the third phase:

> Many of the ways in which government tries to 'help' can actually hurt a nation's firms in the long run (for example, subsidies, domestic mergers, supporting high levels of cooperation, providing guaranteed government demand, and artificial devaluation of the currency) (Porter 1990, p. 681).

Finally, both Habibie and Porter advocate direct and indirect targeting of industries by governments. However, Porter is more restrictive in his support of direct targeting, stating that it is only likely to succeed when a nation has investment-driven national advantage (his second stage, which is similar to Habibie's second phase) (Porter 1990, pp. 673–675).

CRITIQUE OF THE HABIBIE APPROACH

In this section I assume a government effectively dedicated to development—a condition only partially fulfilled in Indonesia owing to blatant government–business collusion, corruption and rent seeking behaviour (Thee 1996). The presence of this 'government failure' weakens the case for selective interventions in Indonesia.[66]

I agree with Habibie's views on the high mobility of international capital—and the resultant need to focus on the development of factors of production complementary to it—and on the advantage that human resources have over physical capital. Also important is the attention Habibie gives to Indonesia's changing comparative advantage, and his cognisance that some technological development processes are cumulative.

However, the Habibie approach also has important weaknesses. Indonesia still has abundant unskilled labour and natural resources, and these will continue to affect its comparative advantage significantly for at least the coming two decades. Given its shortage of human capital, would it not be more efficient for Indonesia to move into more medium-technology industries, as South Korea and Taiwan did in the 1980s? Moreover, history has shown that the prices of high-tech goods—not just those of low-tech goods—can fall rapidly.

Habibie pays little attention to the importance of the various trade-offs between alternative types of investment in applied S&T development. He appears to have placed a high priority on the transformation of technology and industry through the development of the eight vehicle industries without having examined alternative applied S&T development needs, their priorities and the resources required. Just as there are internationally tradeable and non-tradeable goods and services, so too are there tradeable and non-tradeable knowledge and technology. Much technology can be imported because it is embodied in goods (Rice 1990, pp. 57–58). A large part of the research in the Technology and Research National Matrix must be carried out within the country, and therefore needs to be given a high priority. Examples include basic human needs; energy and natural resources; philosophical, cultural, economic and social problems; and, perhaps to a lesser extent, the industrial, defence and security priority areas in the Matrix. Most vehicle products could be imported if necessary, even if the imported product did not meet Indonesia's distinctive needs as well as a locally manufactured one. It could be that because of the particular circumstances (physical, social and economic) of a country such as Indonesia, no imported product will be well suited to local conditions (Evenson and

---

[66]Some other critiques of the Habibie approach are discussed in Rice (1990, pp. 60–62).

Westphal 1995, p. 2,250). However, the development programs in the S&T chapter of Repelita VI, prepared under the direction of Minister Habibie, do include considerable research and technology development resulting in knowledge and other outcomes which cannot be imported (*Repelita VI, 1994/95–1998/99* (1994), Book 2, Chapter 14).

There is also the question of the efficient allocation of resources to support national technological development. The resources of the Agency for the Assessment and Application of Technology (BPPT) and the National Centre for Science and Technology Development (Puspiptek) appear to have been directed mainly toward promoting the 10 strategic industries under the aegis of BPIS, rather than privately owned and other state-owned industries. This has very likely resulted in reduced funding for the nine sectoral R&D institutes, five industrial R&D and testing centres, and 10 regional testing laboratories under the Agency for Industrial Research and Development (BPPI), Department of Industry and Trade, which have the responsibility to support manufacturing industries in general. Because of poor funding and staffing, these agencies have been unable to serve their clientele effectively (Thee 1996, pp. 22–23).

Habibie emphasises the importance of moving swiftly to the third phase, of integrating and improving on existing technologies and developing new technologies, with an apparent stress on creating new products. This emphasis means that fewer resources are available for accessing and adapting imported technologies that could eventually enable Indonesian products to compete with those of firms from industrialised economies. A World Bank evaluation report of Indonesia's 1985 S&T Training Project stated that, at Indonesia's stage of industrial development, industrial technology development should consist of assimilation of imported technologies rather than more innovative design and development work (Najmabadi and Lall 1995, p. 81). Such an approach could result in greater productivity than if scarce resources were invested in the development of new technologies and products. Other studies have found that the results of government promotion of some vehicle industries have been disappointing—especially in terms of these industries' commercial viability, low level of international competitiveness and weak forward and backward linkages—although in the aircraft industry important progress has been made towards achieving technological competence (McKendrick 1992; Chapman 1992; Braadbaart 1996).

Habibie recognises the problem of unemployment, but does not explicitly consider the possibility that large investments in human capital and technology-intensive activities (including related industries) may exacerbate it. The drain on resources caused by his policies means that there may be insufficient investment resources left to employ directly some of the unemployed, or to make the investments needed to attract sufficient international capital inflows to create employment.

Moreover, the opportunity costs of making large commitments to the development of high-technology industries at the expense of other sectors will be greater the more limited are the available technological resources. That these resources are very limited in Indonesia has been demonstrated by Ray (1996), Ramelan (1997) and Hill (1995). Hill claims that Indonesia's technological capabilities and human resources are low relative to its neighbours, not to mention developed countries. He concludes that there is scope for greater government involvement in the promotion of broad-based technological development, but that 'it is doubtful whether ambitious "high-tech" investment projects contribute significantly to efficient, broad-based technology development, particularly when the underlying research and education infrastructure is still rather weak' (Hill 1995, pp. 117–118). Our greatest concern about the high-technology industrial development programs is that Indonesia's human resource and organisational capabilities will not be sufficient to design and produce the new products. A second serious concern is that the costs of production will be well above the world prices of close substitutes, resulting in very low or negative economic rates of return.

Finally, if we relax the assumption at the beginning of this section, the widespread existence of collusion, corruption and rent seeking behaviour renders the successful implementation of the Habibie approach—including, for example, the government's attempt to pick winners—even more problematic. In addition, in view of insufficient private sector interest in developing vehicle industries, the government has made large investments in state enterprises whose performance has often been poor (Hill 1996b, pp. 103–106).

In summary, there are some attractive features in the Habibie approach. These include his emphasis on human resources, the close relationship between technological advancement and production, the focus on changing comparative advantage, and Habibie's dynamic approach to technological and industrial development.

However, there is a danger that the commitment of large resources to the development of high technology will result in large inefficiencies in resource use, higher unemployment and slower economic growth. This is especially so if industries are pushed to advance quickly to the third stage of creating new technologies and products, before having fully realised the first and second stages of technological advancement in labour-intensive sectors. Furthermore, Habibie presents only a partial theory of technological and scientific development, in contrast to the more comprehensive theory developed by Dosi, Pavitt and Soete (1990), for example.

Our analysis has been mainly in the context of achieving higher per capita incomes. In fairness, it should be pointed out that Habibie's broader concept of development partly explains his commitment to some policies and programs which may not be economically efficient.

10

TECHNOLOGY AND HUMAN RESOURCES: ARE SUPPLY-SIDE CONSTRAINTS HOLDING INDONESIA BACK?

Chris Manning

> The development of a larger, better and more specialised base for industrial skills is universally recognised as a top priority for Indonesian industrial success ... The lags are particularly great at higher levels of technical education ... The vocational system needs to be greatly improved. In-house training by industrial enterprises needs to be raised and improved, and SMEs [small–medium enterprises] need to be assisted by specialised training institutes (Lall and Rao 1995, p. 124).

> As we have seen, most firms engage in little formal training, because of appropriability problems and short time horizons. To stimulate more training the government could impose some kind of training levy ... It could allow such training to be tax deductible ... It could also encourage such training at industry level to avoid 'free rider' problems ... [T]he current scheme of partial tax deductions for training costs is probably the most appropriate strategy (Hill 1995, p. 111).

INTRODUCTION

The need for greater investment in human resources to support development is accepted as an article of faith in many countries. Indonesia is no exception. Few policy makers, academics or private sector investors disagree on the need for more investment in human capital to underpin structural change and promote technological upgrading. There is more debate, however, on the form that investment should take, and on the role of the government in supporting improvements in the education and skills of the workforce. For example, what levels of basic education should receive most attention? What is the desired mix of public and private sector investment in vocational education and training? And to what degree should the government invest directly rather than play a more indirect role in encouraging private initiatives? Most would agree that overall levels of investment need to be increased. However, different policies will inevitably mean a reallocation of scarce public sector funds and administrative support within the education sector.

The fields of vocational education and training have received substantial attention from the Indonesian government during the sixth Five-Year Development Plan (Repelita VI), which began in 1994. This marked a shift of emphasis away from efforts mainly directed to improving basic education. The change was based largely on the belief that private firms have underinvested in the skills of their workers, and training institutions have not been sufficiently in touch with the needs of industry. It is also frequently argued that high unemployment among the educated occurs because school leavers lack the skills necessary for productive employment.

The two quotations above—from economists who have written widely on problems of industrial development in developing countries, including Indonesia—state the case for greater government involvement in investment decisions related to vocational education and training. In a recent report on industrial development in Indonesia, Lall and Rao (1995) assert forcefully that skill shortages are one explanation of Indonesia's failure to move out of labour-intensive exports and compete with neighbouring countries in more skill-intensive products. Hill (1995) draws attention to under-investment by the private sector in training. He views this as partly related to a failure to capture fully the 'external' benefits from firm-sponsored outlays. While Hill makes it clear that he is not a supporter of greater public intervention in industrial development in general, 'market failure' in the case of training efforts is seen to justify special government policies.

However, other indications cast some doubt on these conclusions. Indirect evidence that skill shortages may be exaggerated comes both from Indonesia's rapid non-oil manufacturing growth over the past 25 years—over 10% growth in output per annum—and from diversification of exports in the 1990s (James 1996; Manning and

Jayasuriya 1996). The capital-intensive structure of several fast growing key industries, such as cement, fertiliser, pharmaceuticals and other chemicals, and pulp and paper, all of which require significant numbers of professional manpower, would also seem to suggest that skill shortages may not be a major problem. The number of approved foreign workers—just under 60,000 for the economy as a whole in 1996—is not large by the standards of a rapidly developing and now quite large modern industrial sector.

This chapter examines the evidence for 'market failure' in training, the alleged scant supply of skilled manpower, and the social benefits of government support programs in this area. The next two sections discuss some general considerations for analysis of the supply and demand for skilled manpower, in the context of economic and labour market change in Indonesia in the past decade. The fourth section presents the case for skill shortages and government policy initiatives to overcome the problem. In the fifth, I raise some questions about the general proposition of skill shortages and point to some of the potential costs of extensive and rigid government intervention. The final section discusses priorities for investment in training in the context of human resource development.

The chapter concentrates on the supply and demand for vocationally trained manpower at secondary and post-secondary school level. To focus the discussion, emphasis is given to formal sector manufacturing, although there is no presumption that this is the only significant issue related to vocational education and training in Indonesia.[67] The chapter is not based on detailed analysis of empirical data, seeking rather to raise general issues which might be taken up in further research.

THE PROCESS OF SKILL ACQUISITION:
SOME GENERAL CONSIDERATIONS[68]

Evaluation of the private and social benefits derived from vocational education and training is a complex and multifaceted

---

[67]Of course, processes of skill acquisition in small-scale industry and in other sectors (agriculture, construction, trade and services) are no less important.

[68]See especially Behrman (1990) and Middelton, Ziderman and van Adams (1993) for a general discussion of these issues, and Manning (1979), World Bank (1991) and Godfrey (1987) on processes of skill acquisition in manufacturing and manpower planning in Indonesia.

task. For our discussion of Indonesia, it is useful to take account of some basic issues related to the processes of skill acquisition. As in other countries, many technical skills are acquired in the workplace through on-the-job informal processes. Some of the costs of acquiring skills are specific to individual establishments, or industrial processes related to a given technology, and hence are not readily transferred if a worker moves from one firm or one industry to another. In practice, most skills are unlikely to be completely specific, but rather to represent a continuum from general to specific; employers pay more of the training costs for more specific, scarce skills and less (if anything) for more general skills.

Workers with general skills, including machine operators, workers with basic trade jobs (welders, plumbers, fitters and turners), and those in more modern occupations such as computer operators, generally acquire their expertise through a variety of channels. These include short training courses run by the firm, on-the-job guidance from other workers, various kinds of apprenticeships, and formal training in vocational schools. As opposed to training in specific skills, the costs of which tend to be borne jointly by firms and individuals, general skills are normally paid for by workers themselves, either through lower wages during periods of apprenticeship or through fees paid to training instructors or institutions. Since such skills can be applied in a range of work environments, workers tend to be more mobile and employers are loathe to invest in their training. However, inability on the part of workers to finance general education, in particular because of imperfections in capital markets which preclude borrowing against future income, can represent a major obstacle to skill attainment and justify government subsidies for education.

What sort of training is likely to be undertaken to acquire the various skills associated with changing technology? It is useful to distinguish several different patterns at the lower and upper ends of the skill range. At lower skill levels, three contrasting processes of skill acquisition can be distinguished for machine operators and machinists, which are among the most common semi-skilled occupations.

First, in traditional industries such as handloom weaving, batik making and *tahu* [tofu] or tempe manufacture, many operators learn their skills entirely on the job, by observing other workers and filling in for absent employees. Sometimes they receive guidance from friends or family engaged in the same activity.

Second, in modern industries such as electronics, featuring 'fordist' type, continuous assembly lines, short training courses introduce workers to company practices and basic repetitive skills. Such skills are often quite general to the industry, and are largely paid for by the employee in the form of lower starting wages. Employees tend to be mobile, moving from firm to firm in response to personal preference and differences in pay and conditions.[69] While employers sometimes complain of labour turnover, in reality there may be a large pool of readily employable people who can be recruited through existing employees, especially under conditions of relative labour surplus such as Indonesia experienced during much of the 1970s and 1980s (Manning, forthcoming).

Labour conditions can alter, however, as became apparent in parts of Indonesia towards the mid 1990s when shortages of unskilled and semi-skilled labour emerged. In the short term, firms need to adapt wages and conditions of work to keep labour. In the longer term, they will need to consider other options for dealing with higher labour costs, such as adopting new technology, shifting investment to less labour-intensive processes and modifying employment systems.

Finally, firms which utilise equipment in more specialised processes tend to carry out more intensive training (often involving equipment suppliers). In such cases, the firm may bear some or all of the cost of training. Skills may not be directly transferable to other firms or industries, and hence employers will have a strong incentive to maintain employment stability. They are likely to pay higher post-training wages to achieve this end.

The process of skill acquisition is fundamentally different for higher level professional occupations. Formal educational requirements are minimal (aside from basic literacy and numeracy) in both traditional and 'fordist' factory environments, although they are likely to be higher (at least lower secondary school) in firms which have introduced specialised equipment. Formal educational qualifications are a prerequisite, however, for employees entering white collar and professional occupations, such as book-keepers, secretaries, accountants, engineers and managers. The general and applied skills acquired through formal education underpin any

---

[69]This has been true of export-oriented firms in Indonesia in major industrial centres in recent years. See, for example, Lok (1993).

subsequent post-employment training, which tends to be broad in scope and transferable, in fields such as general management and financial management. Under conditions of buoyant demand for professionals, such as experienced in Indonesia in recent years, firms have little choice but to fund at least part of such training, even though the enterprise will be vulnerable to poaching from competitors.

In summary, the strategies which governments and private firms adopt to improve the skill base of workers will depend on both the kinds of skills required (partly related to the nature of the production process and technology) and general labour market conditions. Especially under conditions of increasing international competitiveness, all these are likely to change quickly. Government policies towards vocational education and training need to be sufficiently flexible to enable job seekers and private sector firms to adapt to the changing needs for skilled manpower.

RECENT ECONOMIC AND LABOUR MARKET CHANGE

In the context of recent rapid economic change, three major economic developments are relevant to the discussion of training.[70] First, economic growth and labour demand have remained high, especially in manufacturing and non-tradeable sectors, despite the recent currency crisis. Technology and the associated skill mix of employment began to adjust rapidly in Indonesia, particularly following deregulation in the mid 1980s. Second, there are signs of considerable diversification in manufacturing away from traditional labour-intensive industries, a pattern also reflected in the composition of exports, as noted above.[71] Finally, the reduction of tariff barriers and removal of many non-tariff barriers have forced much of the private sector to become more competitive in international markets (Fane and Condon 1996). All three factors, accompanied by Indonesia's commitments to AFTA and the WTO,

---

[70]See Manning and Jayasuriya (1996), Tubagus (1997) and Sri Mulyani's analysis in Chapter 2 of this book.

[71]Nevertheless, although labour-intensive exports faltered in recent years, the realignment of currencies as a result of the Thailand–Malaysia–Indonesia exchange rate crises will probably enable Indonesia to maintain competitiveness in many of these industries in the short term.

mean that private sector adaptation to changing market conditions is likely to be much greater than in the past.

These developments have important implications for the labour market. First, as noted, the labour market for unskilled workers tightened and real wages rose in the 1990s. This was associated with a historic turning point in agricultural employment, which began to decline quite steeply in absolute numbers for the first time (Manning and Jayasuriya 1996). Both slowing labour supply growth and continued rapid growth in labour demand encouraged a shift away from labour-intensive industries and exports in the middle of the decade.

Two other labour market developments are worthy of mention. Government direct intervention in the labour market increased in this period, primarily through minimum wages set by the central government for each province. The increases were substantial, a more than doubling in nominal terms and an approximately 50% increase in real terms from 1993 to 1996. Owing to labour market tightening in unregulated labour markets, the backwash effect on employment from minimum wage increases was probably quite small. But the new policy has injected a greater degree of inflexibility in the more regulated, modern sector labour markets. Such policies are likely to encourage employers to conserve on unskilled labour to an even greater extent than in the past.

Second, senior high school and tertiary graduate unemployment rates are particularly high among young school leavers in Indonesia in the 1990s, a pattern which has been a peculiar feature of the labour market for some time (Manning and Junankar, forthcoming). Urban unemployment rates were close to 30% among young senior high school leavers aged 20–24 years—both for academic and vocational school graduates. They were lower (but still around 20%) among tertiary graduates aged 25–29 in 1994 (Figure 10.1).[72] Some commentators have attributed high unemployment to the lack of skills and entrepreneurial capacities of young graduates, and have used these figures to advocate greater efforts in vocational education and training. We return to this subject below.

---

[72]This pattern was still apparent in 1996, according to the National Labour Force Survey (Manning and Junankar, forthcoming). While senior high unemployment rates have remained fairly stable among young people since the 1970s, they have risen significantly among young tertiary graduates in the past decade.

**FIGURE 10.1 Unemployment among Urban Males and Females
by Age and Education, 1994
(%)**

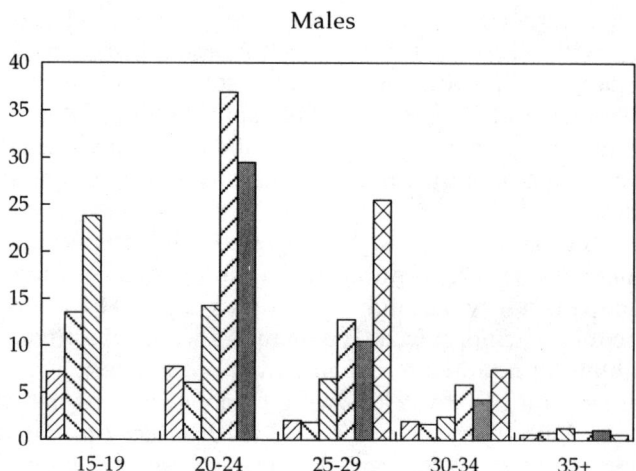

Males

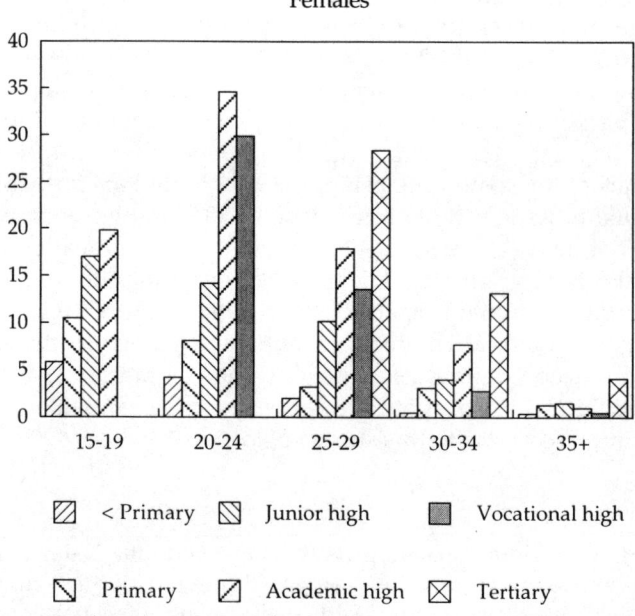

Females

Source: BPS, Sakernas (1994).

High unemployment among the educated has been associated with a rapid expansion of schooling, especially at secondary and tertiary levels, from the 1980s. Low levels of schooling when rapid economic growth began in the late 1960s, together with substantial government and private sector investment in education, both help to explain large annual increments in the number of senior high and tertiary graduates. Growth was in excess of 10% per annum throughout the 1980s and first half of the 1990s, and has raised major concerns about the availability of jobs for graduates (Keyfitz 1989). Much of this expansion occurred in the private sector, and could only have been achieved at the cost of some deterioration in standards (World Bank 1997a).

These rates of growth in the numbers of educated workers were as fast as anywhere in the developing world. They were superimposed on an economy in which over two-thirds of all educated manpower was absorbed by the public sector. By the 1990s this situation had changed quite dramatically, as public sector employment growth slowed and the private sector began to absorb the majority of new senior high and tertiary graduates.

THE CASE FOR EXPANSION OF VOCATIONAL AND TECHNICAL TRAINING

We now turn to vocational education and technical training. I examine the arguments for expansion of public sector participation first, and then look at the policies adopted in Repelita VI (1994–99). Later I raise some questions about the nature and extent of government involvement.

Much support for an expanded government role in the provision of vocational education and training comes from rather scant data collected through a survey of manufacturing establishments undertaken in 1989 by the World Bank (1991).[73] This is backed up by the findings of some tracer studies of graduates, the above-mentioned high rates of youth unemployment, and international experience in both developed countries (especially Germany and Australia) and developing countries (mainly Korea).

---

[73]The survey was conducted by a team of consultants for the World Bank and involved 132 manufacturing establishments. Another establishment survey of training activities and needs was conducted in 1992 (Dhanani 1994a).

Several pieces of evidence from the World Bank survey appear to support the argument that there has been underinvestment in training: relatively few establishments undertook formal training, and expenditure on training was a relatively small proportion of the firms' total payroll outlays. The survey (see especially pp. 30–33) found that formal training was undertaken only by a minority of firms. Most training was informal and on the job, or involved more highly educated professional and managerial workers rather than technicians and skilled workers. This was especially true of export-oriented firms and smaller establishments, although formal training efforts were limited even in higher technology firms. Expenditure on training amounted to around Rp 30,000 per employee per year only, and few firms had full-time training instructors or facilities to undertake training.

Despite this situation, however, over one-third of all firms reported that shortages of skilled manpower were a problem, and vacancy rates were reported to be quite high (around 10%) for skilled tradespeople and supervisors or technicians (World Bank 1991, pp. 27–28). Although evidence is scanty, poaching of skilled workers was also mentioned as a concern, and affected over one-third of enterprises (World Bank 1991, Annex Table 3.10). These data confirm the anecdotal evidence that '... rising levels of labour turnover discourage firms from undertaking extensive in-house training' (Hill 1995, p. 107).

Other data have been interpreted to suggest that not only do private establishments underinvest in training but school leavers suffer from a lack of technical skills. Unemployment surveys and tracer studies detail long durations—commonly 6–12 months—of job searching among unemployed high school graduates (Dhanani 1994b; Manning and Junankar, forthcoming). These high rates and long durations of graduate unemployment have been one of the driving forces behind the government decision to expand vocational education and training.

Finally, reference to international experience has also encouraged the government to consider expanding vocational education and training. The German-trained Minister of Education has been particularly attracted to Germany's 'dual' system, where there has long been close cooperation between the private sector and vocational institutions in the training of apprentices. On the assumption that lack of skills certification has been one factor reducing the returns to investment in vocational education and training, technical assistance has been sought from Australia, where skills testing and certification is a central part of the National Training Reform Agenda. In addition to this, reference has been made to the expansion of vocational education in Korea, in particular as evidence of how governments need to tailor educational systems to the needs of industry.

Government Policies to Expand Vocational Education and Training during Repelita VI

Based on some of the problems identified above, the Indonesian government has moved decisively to develop vocational education and training on a national scale. There is a strong belief, especially in the Ministry of Education during Repelita VI, that education must be made more vocationally oriented and relevant to the needs of industry, with certification of skills at all levels.[74]

There are two major prongs to the push for greater improvements in and expansion of vocational education. First, a considerable rise in vocational school enrolments (close to 50%) is envisaged during Repelita VI. A new policy of 'link and match', fashioned on the German dual system, was to involve 250 vocational schools (senior high level) in cooperation with 6,000 companies in the initial stage from 1994. Students will spend part of their time in formal study and part as apprentices in the private companies.

Second, apprenticeships in private companies are to be linked to training provided through Vocational Training Centres (Balai Latihan Kerja), to cover just on half of all existing 153 centres by the end of Repelita VI. This scheme is administered by the Ministry of Manpower, whereas responsibility for the vocational schools lies with the Ministry of Education and Culture. One production supervisor will be allocated to trainees, who will be taken on for a period of 1–3 years and receive either company or nationally approved certificates. An advisory team from the German agency GTZ established a training advisory unit within the Ministry of Manpower in 1994 and planned to contribute DM 9.5 million for experts, equipment and counterpart funding.

The exact cost of the two schemes is not clear. But there will be substantial government involvement in underwriting basic equipment and physical facilities in the schools, in paying instructor salaries

---

[74]See especially speeches by the Minister of Education made at various conferences on the subject in the mid 1990s (Djojonegoro 1994, 1995).

and in training instructors.[75] The estimated cost of developing examinations for private training courses during Repelita VI alone amounts to Rp 133 billion. Given that costs per student in the vocational schools are approximately 40% higher than in the academic (general) schools, the budgetary allocations are likely to be substantial.

Two features of the revamped policy on vocational education and training deserve special mention. First, a national scheme for skills certification is an important, albeit secondary, aspect of the scheme. The setting of standards with support from industry is to provide the basis for developing teaching materials, assessment and certification of skills (Ministry of Education and Culture 1995). Envisaged initially to cover selected industries, it is anticipated that the scheme will extend across all industries, including small-scale establishments.

Second, the initiative has largely been taken by the central government rather than private industry, and like other national programs will be managed, coordinated and planned from Jakarta. While the rhetoric stresses a shift from a 'supply' to a 'demand'-driven system, many elements of the former remain evident. It is envisaged that a substantial new and complex bureaucracy would be established, including such organisations as the proposed National Vocational Education and Training Council, the Vocational Standards Implementing Agency, and national and provincial/regional industry training bodies. Private sector participation is envisaged, and is indeed central to the reforms, but policy leadership and substantial budgetary control will continue to remain with the central government.

EXPANDED VOCATIONAL EDUCATION AND TRAINING: SOME QUESTIONS

The argument for substantial commitment of public sector funds to vocational education and training ultimately rests on two simple questions. First, is the lack of skilled manpower a clear case of market failure, which therefore justifies substantial government

---

[75]The Ministry of Manpower plans to train 19,500 instructors during Repelita VI, and the Australian government has offered to assist with the national certification plan.

intervention? And second, will the new programs make a significant contribution to lowering unemployment and raising incomes of trainees and school leavers? I will address these two questions first before looking at organisational and other aspects of the proposed schemes.

Underinvestment in Training in the Private Sector?

There are a number of reasons to question the *general* proposition that there is a shortage of skilled and technical manpower, which arises partly because firms underinvest in skills. To take the general proposition of shortages first, in the 1989 World Bank survey, a shortage of skilled manpower was reported as an issue, but ranked low among the problems firms mentioned as constraining output and profits. Much more important were failure to meet deadlines, low-quality products, high levels of absenteeism, and the need for worker supervision (World Bank 1991, Annex Table 3.6). The 1992 survey (Dhanani 1994a) came to similar conclusions: relatively narrow wage differentials for semi-skilled and skilled operators did not suggest a significant shortage of the latter. Indirect support for this conclusion comes from the narrowing differentials between male vocational high school graduates—those most likely to be employed as technicians and skilled workers—and other groups of educated manpower from the early 1980s (Table 10.1).[76]

With respect to training efforts undertaken by private firms, the evidence for underinvestment is remarkably thin. According to several surveys, training costs are around 1–2% of total payroll costs in Indonesia, a figure not out of line with common practice internationally (World Bank 1997b). Neither the prevalence of on-the-job training nor the concentration of formal training efforts among professional and more educated employees is evidence of underinvestment or misdirected training priorities. On-the-job training is widely accepted as the most prevalent means of skill acquisition in small and medium establishments in Indonesia, and is particularly suited to industries where technological change is

---

[76]The decline was less marked among females, and occurred in the 1990s rather than during the earlier decade. It is possible that increasing demand for teachers, many educated through the now disbanded vocational schools, may have contributed to high relative wages paid to vocationally trained females in the 1980s.

TABLE 10.1 Indices of Wage Income Differentials in Manufacturing
by Level of Schooling and Gender, 1982–94
(primary schooling = 100)

	Males				Females			
	1982	1987	1990	1994	1982	1987	1990	1994
< Primary	82	78	88	94	68	83	87	86
Primary	100	100	100	100	100	100	100	100
Lower secondary	129	121	122	120	132[a]	143	136	135
Upper secondary								
General	205	174	167	161	210[a]	248[a]	197	188
Vocational	251	164	171	160	192[a]	226[a]	196	163
Tertiary	472	340	447	400	[b]	520[a]	439[a]	471

[a]Only 10–99 cases in category.
[b]Less than 10 cases in category.
Sources: BPS, Sakernas (1987, 1990, 1994); Susenas (1982).

gradual, adaptive and linked to past technologies.[77] That is not to argue that more formalised systems may not need to be introduced as firms expand and the rate of technological change speeds up. However, many firms surveyed in the early 1990s were not in this position. Similarly, emphasis on more formal training among professional and managerial workers is not unusual; as noted above, it is precisely this group which is likely to gain most from building on knowledge gained in formal education through more structured courses.

There is considerable evidence that turnover in particular of operatives and several categories of professional worker (accountants, managers and secretaries) increased in the 1990s. One should be careful in interpreting this as a constraint on investment in

---

[77]See Manning (1979) for a discussion of on-the-job training in small and medium-scale enterprises in contrast to more formal training courses undertaken by larger, and especially foreign, firms.

training. Tightening labour markets at the lower end of the skill spectrum and excess demand for high-quality, general professional skills at the upper end are likely to have been much more important.[78] Failure to retain workers is likely to be related to a lag in adjustment of wages and salaries to changed circumstances—many firms continue to function as if they are still operating under labour surplus conditions. Thus adjustment of remuneration levels and packages would help firms to retain workers. It is instructive that businesses that invest heavily in human capital, such as PT Astra in the automotive industry, are also renowned as high-wage firms. Firms which invest in the skills of their workers, especially to the extent that the new skills are general, will have to adjust their wages to minimise the risk of turnover.

In sum, there is little evidence that underinvestment in training is a general problem facing Indonesian industry or a major constraint on skill upgrading. This is not to say that it is not an issue in some firms and industries where technological change has been particularly rapid. As a corollary, the general case for government taxes (levies) or subsidies imposed according to investments in training is not strong, especially if it is to be used to build a large centralised bureaucracy to administer training programs. I return to this issue below.

Vocational Education and Training to Improve Job Access?

Even though firms may invest sufficiently in training, public subsidies and promotion of vocational education and training may still be justified if it significantly improves access of job seekers to employment, especially where there is significant youth unemployment, as experienced in Indonesia in recent years. However, it does not appear that a lack of vocational training has been a major factor associated with high unemployment. Unemployment rates were similar, although slightly lower, among both male and female vocational graduates compared with academic stream secondary graduates in the mid 1990s (Figure 10.1). Indeed, one survey found that unemployment rates among all graduates from public economic high schools (SMEA) and technical high schools (STM) were considerably higher than those from academic public high schools

---

[78]It is instructive, however, that labour turnover was not mentioned as one of the major production problems faced by firms in the 1989 survey.

two years after graduation in 1990 (Strudwick and Cresswell 1994, p. 15).

Second, as Table 10.1 suggests, the earnings of vocational graduates differed only slightly from those of graduates of academic schools, despite the much higher cost of the former. Rate of return studies have not found that investment in vocational training is more profitable from either a private or social point of view (Clark 1983; Boediono and McMahon 1991; World Bank 1997b). Although mean earnings were comparable, social rates of return were lower among vocational graduates, partly because of higher costs of tuition and partly because of higher rates of repetition and attrition.

Nevertheless, there were some differences in the employment experiences of different groups of vocational school graduates. Strudwick and Cresswell (1994, pp. 18, 20) contrasted the job experience of the two largest groups of vocational graduates: STM and SMEA graduates. Job seekers from both public and private STMs appeared to have more job choice, and a much higher proportion of their graduates were employed in manufacturing. They recorded earnings higher than those of graduates of SMEAs or academic high schools.

Overall, however, there is little evidence that high levels of unemployment are *generally* the result of a lack of employable skills on the part of school leavers. There is no doubt that some graduates from academic streams lack the skills demanded by employers, and undergo training at personal expense in order to move forward in unemployment queues. But most are prepared to wait and search for better opportunities in a segmented labour market, rather than accept low-wage options soon after graduating. Substantial wage differentials between sectors and industries, partly associated with the major role of mining and relatively protected capital-intensive manufacturing industries, are a key explanation for graduates opting to spend long periods in unemployment queues (Manning and Junankar, forthcoming).

Will a closer linking of skills acquired in formal institutions to demand in the workplace make a significant difference to the employment prospects of young graduates? There is no doubt that there are readily identifiable cases where vocational training can be more closely linked to industry needs. This is especially so where there is a heavy concentration of particular industries in certain regions. For example, it makes good sense to involve private enterprise in planning courses and training students in vocational training centres oriented to demand. Examples include meeting the needs of tourism in Bali, textile design schools and courses in selected locations in Java, and maritime and fisheries courses in parts of Maluku and Sulawesi. A much less compelling case can be made for public funding of industrially oriented training centres which have only tenuous links with industry in the immediate vicinity. This is the case in many urban centres where manufacturing and service industries are quite diversified.

Skills Certification: Is a National Program Needed?

Is the lack of nationally certified skills a constraint to industrial development and the deployment of skilled manpower? There are three main arguments for a certification scheme. First, to improve product quality and competitiveness, industry needs some benchmark for assessing the availability and level of skills. Second, labour markets work more smoothly where employers can select from a pool of skilled workers who have achieved a similar, measurable standard of competence. And third, workers can confidently apply for jobs and plan careers once their skill competence is graded and publicly recognised.

These arguments apply particularly to the upper end of the skills range. It is essential for safety that the minimum level of skills required for an aircraft engineer or skilled technician in the aircraft industry be identified with precision. Craft and industry organisations also sometimes provide certification through apprenticeship standards or industry testing mechanisms. Certification systems have thus grown up in traditional industries and as modern industry has sought to recognise and standardise levels of competence in high-tech industries.

In the Indonesian case, there is a strong argument for government support of industry initiatives to set up systems of training, testing and certification. The case for extensive government involvement in certification across all industries and major occupations is much less obvious in such a rapidly industrialising country. The wide range of technologies, even within industries, means that certification may apply only to small subsectors and certain occupations. There is a persistent tendency to undervalue the efforts of skill providers in the private sector on the basis of low 'quality', despite demonstrated high demand for basic skills in such areas as computers, language, crafts, tourism, business studies and home economics (World Bank

1997b, p. 85).[79] The government has a role to play in providing information on the level of skills provided by the private sector, and in countering fraud and misinformation. But in the Indonesian context this has more frequently led to a desire to control all activities, on the grounds that standards do not meet often unrealistic minimum requirements suitable only to employment in a small segment of the modern sector.

The broader the coverage of a national scheme, the greater the danger that the program will be hijacked by bureaucrats and involve planning and implementation increasingly divorced from the needs of individual employers and workers. Typically, it is this latter kind of national program which is being planned in Jakarta. Despite seemingly appropriate reference to a 'demand-driven system' and industry involvement, the initiative for the system has been taken by the central government, and not by industry leaders.

AN ALTERNATIVE ROLE FOR GOVERNMENT IN VOCATIONAL EDUCATION AND TRAINING

What is the appropriate level and type of government assistance for vocational education and training, given our argument that the *general case* for direct government involvement in the funding and administration of training is weak? In our view, the most important contribution can be made through two avenues: provision of information to industry, and coordination of industry-specific initiatives. Direct involvement in training will, of course, be required in public sector or public utility activities where social costs and benefits are not always reflected in standards set by private trainers—for example, the training of air controllers, nurses, teachers and fire fighters. But by far the greatest amount of training will continue to be undertaken by private firms in industry and firm-specific skills. Government, working closely with particular industries, can play a major role in identifying what skills might be required in the future and in setting up a framework for industry coordination to help meet those needs. Ideally, in line with

---

[79]The World Bank (1997a, pp. 75–88) draws attention to the dramatic growth (over 7% per annum) in private training centres, despite questionable quality from a modern sector perspective. It also draws attention to the high social and private returns from such growth and recommends less regulation given the high costs associated with extensive government controls.

government attempts to achieve greater decentralisation in the planning and execution of public programs, these needs and strategies could be identified and planned at the regional level (and on occasion at provincial or *kabupaten* [district] level).

Given the sometimes costly investments that have already been made, there may be a case for upgrading and transforming some of the existing publicly funded vocational schools and training centres to meet industry-specific demand, or for providing more general trade skills. It is not clear, however, that all—or even most— existing schools should be supported as publicly funded vocational institutions. 'Privatisation' or conversion to other uses (academic schools, or vocational schools providing skills that are in greater demand) would seem a high priority. Even in those schools which provide skills which are in demand, a targeted program for increased private sector management and accountability would seem desirable if the goal of 'demand-driven' supply of skills is to be achieved.

To what extent should the government encourage training through taxes (levies) or special tax rebate schemes? We agree with Hill (1995) that a training levy such as that established in Singa- pore is unlikely to be administered efficiently to benefit those firms most in need of assistance. Should firms receive tax rebates based on their investments in training? Such a policy can actually result in *overinvestment* in training, if firms are induced to undertake activities that they might not have considered, merely to obtain the rebate. More useful would be efforts directed at investigating why some firms invest more and others much less in training, in the context of the internal and external market for skills in which they operate. It may be that the principal difference is in the form training takes rather than enterprise commitment to training per se; that is, large firms tend to undertake more formal, identifiable activities, while small firms concentrate more on on-the-job training. A rebate is thus likely to be biased in favour of larger firms.[80]

Two final points are worthy of mention. First, if one assumes that the total size of the education budget is broadly predetermined, significant funds allocated to vocational education and training

---

[80]Of course, regardless of some of the above considerations, a rebate might be seen as the best way of inducing some training which would not have been undertaken otherwise, if it is demonstrated that all firms consistently underinvest in skill upgrading.

through the education budget does have implications for funds available for other educational purposes, especially raising the quality and coverage of basic education. Given the pressing need for raising the quality of basic schooling, a strong argument can be made that the opportunity cost of funding vocational education and training through the public budget is likely to be extremely high. This is especially the case given that government investment in basic education is a pressing need on both efficiency and equity grounds (World Bank 1997a).

Second, in many countries there is a strong—and not entirely disinterested—lobby group in favour of greater vocational education and training. This appears to be the case in Indonesia as well. For some, a questioning of public allocation of funds to this area is regarded as near heresy. One should note in this context that much of the assistance to vocational education and training in Indonesia has come through official international assistance programs. It is in the interests of the recipients, and consultants and 'experts' from the aid donor country, for programs to continue and to expand, sometimes with little attention given to their cost in terms of diversion of national resources (not only funds but also scarce administrative skills). This is in no way to argue *a priori* against the upgrading of technical skills, which is obviously central to modernisation and development. But I am suggesting that the social benefit–cost ratio of government allocations to vocational education and training might be more carefully evaluated in light of the pressing need for public investments in other areas to promote a more prosperous and just society.

PART IV

TECHNOLOGY:
CASE STUDIES

11

THE FUNDING OF PT DSTP,
A HIGH-TECHNOLOGY PROJECT:
LOFTY GOALS, DEFINED TASK

Saadillah Mursjid

The application of technology in the industrial progress of a nation is a critical although sometimes controversial aspect of economic development. This chapter is written in my capacity as President Director of PT Dua Satu Tiga Puluh, or DSTP as it is commonly called. This company is the financier of Indonesia's passenger jet aircraft, the N-2130. It is my hope, as President Director of the company, that I might convey a sense of the vision, indeed share some of the excitement, that motivates the leadership and management of DSTP. It is also my job as President Director to thoughtfully acknowledge and discuss the many challenges we face as a firm. These are the two aims of this contribution.

The debate about the proper role of the state in the process of industrialisation is a longstanding one and far from settled. All too often this debate takes place on an abstract level, governed by economic theory and coloured by ideology.

What is the economist's view of industrial policy? As a non-economist, my understanding is that economists start by assuming that markets for products and inputs exist, that they work efficiently, that information is widely shared and known with certainty, and that the market consists of a large number of firms. Given these conditions one reaches the unassailable conclusion that no intervention by government can improve on the market-determined outcome. And yet the picture of a perfectly competitive, efficient market does not fit the real world. In the real world, market failures exist and efficiency is not the only goal. Theory may suggest that the state should play only a minimal role. But can one truly believe that, without some state steering, the economy will evolve in an optimal manner and with sufficient speed? As the noted economist Jagdish Bhagwati recently wrote, 'only a romantic belief in the virtues of laissez-faire combined with a cynical disbelief in the capacity of useful state action would lead one to answer these questions in the affirmative' (*New Republic*, 19 May 1997, p. 41).

Like most developing countries, Indonesia began its economic history with a strong predilection for state intervention. Over time we learned, as did the economics profession as a whole, that the state was not always a good promoter of economic growth, and that all too often state interventions resulted in substantial welfare losses. For this reason we are now relying increasingly on the private sector as our main economic engine. There are, of course, economic activities that clearly call for government intervention. Activities that are characterised by market failures or substantial techno-logical spillovers are widely accepted as providing a legitimate rationale for government involvement.

The government of Indonesia saw the development of aircraft manufacturing capacity as such an activity. It has made substantial infrastructure and training investments through the establishment of PT IPTN (Indonesian Aircraft Industry), a state-owned enterprise founded in 1976. IPTN is presently entering the third stage in acquiring aerospace technology with the development of the N-250 fly-by-wire aircraft, which is now undergoing international certification and for which IPTN is already taking orders. However, having invested public sector funds in the start-up of the aircraft industry, the government made the decision to use private funding for further development of the sector. It is for that purpose that PT DSTP was established.

LOFTY GOALS BUT A NARROW TASK DEFINITION

Our enterprise was formally established in February 1996. Our initial and continuing purpose is to fund the development of the N-2130, a regional twin-engine jet in the 100–130 passenger range. It was determined from the outset that the N-2130 was not to be funded from the public purse. Rather, it was to be a commercial venture in high technology, fully financed by the private sector. Few if any such aviation projects have been financed anywhere in the world without some form of public subsidy, and so this approach must be recognised as being both bold and innovative. In the US, the development of commercial aircraft has enjoyed substantial cross-subsidies from the development and sale of military aircraft to the US government. Elsewhere in the world, aircraft firms have been the direct recipients of full or partial government funding. In the case of the N-2130 we are aiming for a new approach: broad-based private sector financial support.

At the firm's first major shareholders' meeting on 29 January 1997, we adopted the name PT Dua Satu Tiga Puluh, and set for ourselves ambitious goals and a sometimes intimidating agenda for action. Our acts of incorporation outline our mission as:

> ... the establishment of a mechanism for promoting innovative activity in the development and application of technology, as well as for the purposes of making short-term, medium-term and long-term investments for the mastery and application of technology in the fields of aerospace affairs, maritime transportation and communications ...

These are lofty goals. It is important to recognise, however, that we are first and foremost a financing firm—not an aircraft manufacturer, or shipbuilder, or communications provider. Indeed, our immediate task is even more narrowly defined. Our job is to raise the funds needed for the development of the N-2130 passenger jet, and then to disburse those funds to IPTN. In this sense, IPTN is our premier client, with whom we share considerable, but not unlimited, risk.

Our financial obligations in this project terminate upon full certification of the N-2130 by Indonesian and international aviation authorities, which is anticipated in the year 2008. One important aspect of the contract between DSTP and IPTN that might be worth emphasising is that all the prototypes produced, and all the intellectual property generated by this undertaking, will be owned by DSTP. IPTN's responsibility is to produce and market the aircraft, and DSTP will receive a royalty for every aircraft sold. In other words, DSTP shareholders will be entitled to the flow of royalties generated by sales of the N-2130 aircraft by IPTN.

The N-2130 will incorporate superior technology and will be offered in the market at a competitive price. By careful analysis, and in consultation with IPTN, we have estimated the cost of development to be $2 billion. This estimate includes initial project development costs, construction of the prototypes, and operational costs incurred in the certification of the aircraft by international aviation authorities. It is very similar to the $2.2 billion anticipated cost for the development of an 80–130 seat passenger jet by a joint venture of Aviation Industries of China, Singapore Technologies and non-German partners in Airbus. The European AI(R) consortium is also engaged in the development of a regional jet, and estimates development costs at just $1.1 billion; it expects to achieve substantial savings through the incorporation of already well-developed technologies from various partners (*Air Transport World*, March 1996, p. 25).

Our immediate task, therefore, is to raise the necessary $2 billion. Our secondary task is to disburse those funds to IPTN on the basis of a thoughtfully determined timetable and cautious disbursement scheme.

RAISING THE FUNDS

Where do we stand in the effort to raise this $2 billion? In an initial private placement, 100,000 shares were issued, valued at approximately $1,000 each, thus raising approximately $100 million (Rp 2.3 billion).[81] These priority shares (deemed Class A shares) were purchased by 15 individuals, two firms and eight foundations (Table 11.1). We consider these shareholders to be our founding members. We were pleased with these results, but still faced the formidable task of raising another $1.9 billion. We were unable to obtain funds through a formal public offering followed by listing on the Jakarta or Surabaya Stock Exchanges, as we are not yet an established firm with a documented performance record, and are therefore unable to fulfil the various requirements for a public listing. At this stage, we want the project to maintain a national character, and have thus far not explored the possibility of raising additional funds in international markets. We may consider such a step in the future. We therefore needed to come up with an innovative plan to fulfil our goals.

---

[81]Throughout this paper a conversion rate of Rp 2,300/$1 is used, as this was the rate prevailing prior to the recent currency turmoil; it is used in our current published annual statement and most recently audited financial statements.

**TABLE 11.1 Composition of Share Ownership of PT DSTP
as at 29 January 1997**

No.	Name of Shareholder	Share Holding (No.)	Total Shares		
			Par Value (Rp '000)	Total Par Value (Rp million)	Owner-ship (%)
	Class A: Registered Stock				
1	Soeharto	50	2,300	115	0.05
2	Umar Wirahadikusumah	10	2,300	23	0.01
3	Sudharmono	10	2,300	23	0.01
4	Saadillah Mursjid	10	2,300	23	0.01
5	Rahardi Ramelan	10	2,300	23	0.01
6	Giri S. Hadihardjono	10	2,300	23	0.01
7	Anthony Salim	5,000	2,300	11,500	5.00
8	Eka Tjipta Widjaja	5,000	2,300	11,500	5.00
9	Henry Pribadi	5,000	2,300	11,500	5.00
10	Prajogo Pangestu	5,000	2,300	11,500	5.00
11	Rachman Halim	5,000	2,300	11,500	5.00
12	Sudono Salim	5,000	2,300	11,500	5.00
13	Sudwikatmono	5,000	2,300	11,500	5.00
14	Usman Admadjaja	5,000	2,300	11,500	5.00
15	Putera Sampoerna	5,000	2,300	11,500	5.00
16	PT Tugu Pratama Ind	5,000	2,300	11,500	5.00
17	PT Nusantara Ampera Bakti	5,000	2,300	11,500	5.00
18	Bank Dagang Negara Employee Welfare Fund	6,140	2,300	14,122	6.14
19	Bank Eksim Pension and Welfare Fund	6,140	2,300	14,122	6.14
20	Bank Rakyat Indonesia Employee Welfare Fund	6,140	2,300	14,122	6.14
21	Bank Rakyat Indonesia Danar Dana Foundation	6,140	2,300	14,122	6.14
22	Bank Tabungan Negara Employee Welfare Fund	6,140	2,300	14,122	6.14
23	Bank Bumi Daya Pension Fund	6,140	2,300	14,122	6.14
24	Bank Indonesia Employee Welfare Fund	7,160	2,300	16,468	7.16
25	Purna Bakti Pertiwi Foundation	900	2,300	2,070	0.90
	Total Series A Registered Stock	100,000	2,300	230,000	100.00
	Class B: Registered Common Stock				
26	Holders of Series B Common Stock, consisting of 175 individuals/legal entities	291,466	2,300	670,372	100.0
	Total Series A & B Common Stock	391,466		900,372	100.00

The founders adopted what we believe to be a unique and exciting vision. We determined that additional funds were to be raised from citizens throughout Indonesia, and so a certain number of shares were to be denominated in units of just Rp 5,000 per share, making the purchase of shares accessible to millions of Indonesian citizens. In that manner, when the N-2130 is airborne, numerous Indonesian shareholders will be able to take pride in their participation in this important national project.

Our previous capital plan called for the sale of 100,000 Class A shares; 1,065,864 Class B shares (at full face value of $1,000 per share); and an additional 400,000 Class B shares (to be split into 460 units each for sale at Rp 5,000 per split share). In summary, with the full approval of the Ministry of Justice and by letter of approval from Bapepam (the Stock Exchange Supervisory Agency), our authorised capital currently stands at over $1.5 billion (Rp 3.6 trillion). We have applied to increase the authorised capital to $2 billion (Rp 4.6 trillion), which would allow us to fund fully the development of the N-2130 in accordance with current budget estimates. Specifically, this authorisation would allow us to float an additional 434,136 Class B shares when paid-up and issued capital reaches $650 million (Rp 1.5 trillion). In line with the standard approval process required of any company wishing to float shares, we will seek Ministry of Justice approval for our projected funding effort.

As already discussed, the 100,000 Class A shares were fully subscribed upon founding of the enterprise. By the shareholders meeting of 29 January 1997, we had placed 291,466 Class B shares, valued at approximately $290 million, with 175 individuals, foundations, cooperatives and other legal bodies. Our registration with Bapepam became effective on 4 March 1997, and our 'Evergreen Public Offering' began on 11 March 1997. This offering will remain open until all two million shares are sold. In the public offering structure, the 400,000 Class B shares are each divided into 460 split shares valued at Rp 5,000 per share. Since each share of $1,000 carries with it the right of a single shareholder vote, individuals purchasing split shares may, in time, trade 460 split shares for a single Class B share and thus obtain full voting privileges.

Starting in September 1997, the distribution and sale of these Class B shares is being administered by Bank Pembangunan Daerah (the Provincial Development Bank) in all 27 provinces. In the first week of September, when our 'roadshow' through the 27 provincial capitals ended, total shares issued stood at 400,500, valued at just over $400 million, and with shareholders spread over 17 provinces. This is sufficient to meet our near-term budgetary commitments, but continued development of the N-2130 will require raising substantial additional funds.

DSTP AS VENTURE CAPITALIST AND HIGH-TECHNOLOGY PARTNER: A DUAL CHALLENGE

It is perhaps best to view DSTP as a venture capitalist partner in a high-technology project. As a finance company, we face a dual challenge. Our first challenge as venture capitalist is that of a company managing a financial relationship with its principal (indeed sole) client, and that of a worthy steward of shareholder interests. Our second challenge as partner in a high-technology endeavour is to face the competitive and technological challenges of a firm engaged (albeit indirectly) in a sophisticated and globalised industry. Let me deal with the various issues that must be dealt with as a venture capitalist firm first. Consider as a starting point the relationship with our client, IPTN.

We are confident that we have met the first challenge as venture capitalist in a spirit of partnership with our client and through the institution of various financial controls. Our relationship with IPTN is governed by a cooperative agreement that incorporates various checks and balances. Funds are to be disbursed to IPTN only on the basis of regular quarterly reporting by IPTN. These quarterly reports must thoroughly document expenditures for development of the N-2130. Only following our review and acceptance of these accounts will additional funds be disbursed. In return, DSTP will receive royalties in two phases. The first phase of royalty payments will start with the sale of the first commercial aircraft by IPTN, and continue until the break-even point. The second phase involves 'pure royalties'—which are the payments by IPTN for the utilisation of DSTP's intellectual property rights over the N-2130. If IPTN fails to fulfil its commitments contained in the cooperative agreement, it is required to return all remaining funds, plus any materials and equipment already purchased with DSTP funds. Likewise, if we at DSTP fail to meet our commitments, IPTN is permitted to seek funds from a third party, and DSTP is required to relinquish its intellectual property rights and all future royalties vested therein.

Next, we need to consider the relationship with our shareholders. We recognise that we are asking shareholders to undertake
somewhat unusual risks by participating in this scheme. Indeed, we
are calling upon shareholders to act not only out of immediate
economic interest, but also to look towards a proud, technological
future in a spirit of mutual assistance and national commitment. As
the stock does not trade on the Jakarta Stock Exchange, many issues
of pricing, liquidity and governance remain untested. We have
consulted regularly with Bapepam in preparing this 'unlisted public
offering', but we acknowledge that few regulations exist as yet to
fully govern and protect shareholder interests in such cases.
However, Bapepam will be drafting more such regulations, and in
the meantime we are fully governed by existing applicable laws
such as the Limited Liability Companies Act and the Capital
Market Act.

Shareholders can expect no return on their investment in the
near future. No dividends are to be paid until at least 2003, when
IPTN is scheduled to produce the first N-2130 and thus begin royalty
payments to DSTP. Until market success appears imminent and more
certain, share prices are likely to remain stable at or near the initial
offering price. Once royalty payments begin, we project high
operating profits, as DSTP itself is a very simple and lean
organisation employing less than a dozen individuals. We concede
that DSTP shares, judged on the basis of a traditional net present
value calculation, do not yet compete favourably with most publicly
listed firms, or even time deposits at the bank, but we are convinced
that committed shareholders will reap clear benefits in time.

The stock is legally tradeable, but we recognise that there is no
readily accessible public forum for trading, and therefore the stock is
likely to remain illiquid—at least until such time as returns appear
more certain and a trading mechanism develops. Transfers of shares
will be governed by applicable laws that require the reporting of
transfers to the Board of Directors for subsequent registration. These
changes, in turn, will be fully reported by DSTP to Bapepam as
required by law.

The offering price of shares has been set by DSTP and to date, in
the absence of active trading, the price of shares remains at its
initial nominal value of $1,000 per share. The initial offer was to
remain open for a period of 10 years, although the offering price is to
be adjusted every six months. Price adjustments will have to be
managed carefully in order to provide an incentive for individuals
and firms to invest now rather than later.

At this stage, it is very difficult for an investor to estimate a future value or stream of income deriving from an investment in DSTP, even allowing for a very high level of risk. Continued sale of shares will therefore require a concerted and innovative public relations and sales effort.

DSTP as High-Technology Partner

Consider the issue now from the perspective of DSTP as a high-technology partner. We face the second challenge as any firm does in a highly competitive business: we analyse the strengths and weaknesses of our design, production and marketing systems, as well as those of our product, and review the competitive opportunities and threats in the marketplace. Let me focus first on IPTN as a designer, producer and seller of aircraft.

The product, the N-2130, will be designed, developed, produced and marketed by IPTN. As the high-tech partner, our future success is heavily dependent on the demonstrated competence of IPTN. IPTN's launch into the passenger jet market is founded on an improving track record. It has been engaged in the manufacture of aviation equipment for somewhat more than 20 years. IPTN's product line is now quite extensive, and includes a range of helicopters developed under licence from Messerschmidt Bolkow Blohm (in Germany), Aerospatiale (France and the UK) and Bell Helicopter Textron (the US), propeller aircraft made under licence from and in cooperation with CASA (Spain), the manufacture of aircraft components of various types, plus the provision of a variety of aircraft services.

The two most recent additions to the IPTN product line are the CN-235 Maritime Patrol Aircraft (MPA) and the N-250. Both are on the road to commercial success. The CN-235 MPA is a turboprop aeroplane produced by IPTN in long-term cooperation with Spain's CASA aircraft manufacturer. International sales of the CN-235 MPA continue to grow. The N-250, a 70 seat twin-engine turboprop, is the first fully Indonesian-designed aircraft, and successfully completed its maiden flight in August 1996. It is the first electronically controlled, propeller-driven commercial aircraft incorporating IPTN's unique fly-by-wire electronic system.

The N-250 is expected to receive clearance from the US Federal Aviation Administration (FAA) in the very near future, paving the way for an intensive production and marketing effort. IPTN promoted the N-250 heavily at the most recent Paris Air Show, and eight countries have expressed strong interest in purchasing it. There are now firm commitments for more than 30 aircraft, from such countries as Pakistan, Sweden and Colombia. IPTN has learned much from these prior ventures, and we see this as a firm foundation on which to develop the N-2130. Like the regional jet aircraft market, the 30–70 passenger turboprop market (serving routes up to around 500 miles) is a dynamic market with expectations of high growth. Experience in the production and sale of the N-250, and the establishment of an international service organisation for the aircraft, provides an enormously important basis for success in the next generation aircraft, the N-2130.

What can one say about the particular product—the N-2130? The N-2130 is planned as a 'regional jet' designed to serve distances in the 1,600 nautical mile range and to fly at a very fast 461 knots. IPTN plans to produce both a 100 and a 130 passenger seat version. The projected success of the N-2130 is expected to rest on three foundations: superior technical specifications, compelling operational features, and a relatively low acquisition price and operating cost. The N-2130 will be distinguished by the use of a full fly-by-wire control system, digital flight deck and cockpit, and digital avionics. It will perform at high speed and at high cruising altitudes over medium-range distances. The aircraft will feature a wide cabin offering superior passenger comfort. Finally, Indonesia's lower labour and managerial costs will allow the aircraft to be produced relatively inexpensively, and marketed at a highly competitive price. IPTN has elected aggressive pricing for the aircraft in order to penetrate the regional jet market (Table 11.2).

A clear schedule for product development has been specified by IPTN. The plan calls for conceptual development to be completed by December 1997. Then, over the next five years, IPTN will develop and design the N-2130, with prototype construction to be completed by early 2002; its maiden flight is also anticipated in 2002. Preliminary certification and testing is to be accomplished by 2005, with full certification to be achieved by 2008. Commercial production is to begin in 2004, with expectations of reaching the break-even point of 326 aircraft by 2013. We at DSTP see this as an ambitious, but not overly optimistic, schedule.

There remain two final issues: the growth of the market, and the threat of competition. The competitive opportunity lies in a rapidly growing market for small jets in the 50–130 passenger range. The growth potential of this market is widely recognised in industry

TABLE 11.2 Aircraft Selling Prices
($ million)

Aircraft	Passenger Capacity	Base Selling Price		
		Low	Mid	High
F-70	79	20.5	22.1	23.2
RJ-70	80	18.1	20.5	21.5
RJ-85	98	20.2	22.7	24.1
N-2130-100	100		20.5	
F-100 IGW	107	20.6	23.1	24.3
RJ-100	112	20.9	24.1	25.4
B-737-500-IGW	122	23	25.7	27.1
N-2130-130	130		26.65	
A-319 Std.	134	32	33.3	34.6
B-737-300 IGW	140	25	28.5	31.1
B-737-400 IGW	159	30	35	40.0

Source: *Indonesian Capital Market Journal*, March 1997.

journals.[82] The opportunity is reinforced by IPTN's confidence that it will be able to offer a technologically superior product at a highly competitive price. Boeing's most recent *Current Market Outlook* forecasts that approximately 1,400 units in the 91–120 seat range will be delivered between 2004 and 2014, with growth slowing in the second half of that period. IPTN, using very similar data, provides a longer term projection showing demand for passenger aircraft in the N-2130 class to be 3,237 units in the 20 years between 2005 and 2025. The N-2130 program calls for reaching the break-even sales point of 326 aircraft around 2013, requiring IPTN to capture a substantial share of the market between full certification in 2008 and break-even point in 2013. No decline in demand for aircraft in this range is expected until around 2025. In this somewhat longer time frame IPTN's projections of a 25% market share appear more realistic.

---

[82]See, for example, Robert W. Moorman's article, 'Order up!' in *Air Transport World*, May 1996.

There is no doubt that reaching break-even by 2013 is an ambitious goal, and IPTN may have to re-estimate its break-even time frame projections. Still, it is clearly a robust market, and there is some suggestion in press reports that published projections underestimate the demand from smaller airlines and emerging economies. In addition, Indonesia's market position is strengthened by expectations of vigorous growth of domestic demand in our expanding archipelagic economy. We anticipate that at least half of N-2130 sales will derive from the domestic market.

Let me turn now to the issue of competition. Growth expectations are indeed high, but competition in this market is expected to be fierce. We face a competitive challenge from firms with august reputations and long histories in this business: McDonnell-Douglas, Boeing and Airbus. These leaders in this field offer competitive aircraft, although, as we are all aware, Boeing and McDonnell-Douglas are now in the process of merging. We also face challenges from Avro, Aerospatiale, British Aerospace and Alenia, which have formed a consortium, AI(R), to service and manufacture an extensive family of jet aircraft, including development of a regional aircraft in the N-2130 class. Avro already markets a competitive regional jet and will bring that experience to the consortium. In addition, the Chinese have recently initiated a joint venture between Aviation Industries of China, Singapore Technologies and the non-German partners in Airbus to produce jets in the 80–130 seat class. (An earlier joint venture between South Korea's Samsung and Aviation Industries of China to produce a 100 seat jet appears to have stalled due to partnership disagreements. But South Korea continues to have strong aspirations to aircraft manufacture, and may well introduce new plans for a regional jet.) Also, the dramatically improved efficiency and competitiveness of Bombardier (of Canada) and a revitalised Embraer (of Brazil) pose future challenges, as they, too, market comparable regional aircraft. These are challenges we recognise.

IPTN is a relative newcomer to the global aircraft industry, albeit with a growing reputation. Successful development of jet aircraft by emerging firms has generally occurred in partnership with more experienced manufacturers. DSTP will encourage the establishment of joint ventures or other strategic partnerships by IPTN if we decide that the N-2130 project would be best served by such relationships.

CONCLUSION

I have outlined some of the risks, acknowledged the competitive threats, and discussed the many uncertainties inherent in this approach to high-technology development. I ask that DSTP be judged not by the loftiness of our ambitions nor by the novelty of the financial strategy. I have asked myself, and will ask myself again, many of the same questions that are raised in public discussions of the initiative. There are many untried aspects of this plan—issues of governance, shareholder interests, stock liquidity, the production and marketing experience of our client and so on. We managers and leaders of DSTP will no doubt continue to spend many late hours fine-tuning our plans, perhaps scrapping some of what we initially thought to be our best ideas. We are convinced, however, that this endeavour is worth a try. We do not believe that we are imposing undue levels of financial risk on investors—indeed, we believe we are providing a viable opportunity to participate in national development and to gain financial returns. We therefore ask to be judged, not today, but a number of years from now, when the N-2130 is flying commercial routes throughout the Indonesian archipelago, and indeed throughout the world. Then we can, will, and indeed must ask, whether the investors in PT DSTP, in the N-2130, have received an adequate return for their visionary participation.

12

SOME COMMENTS ON 'THE FUNDING OF PT DSTP: A HIGH-TECHNOLOGY PROJECT'

Ross H. McLeod

Many economists are concerned about the way the Indonesian government has pursued its objectives relating to the development and application of high technology. There is certainly room for debate about whether the Indonesian people are best served by policies that try to accelerate the pace of technological advancement—given the heavy demands this makes on scarce skills in the areas of engineering, management, computer science and so on—at the expense of other sectors of the economy. But whatever policy direction is chosen, there can be no argument that everything possible should be done to ensure that available funds are not used wastefully. This is where the establishment of PT Dua Satu Tiga Puluh (DSTP) appears to mark a highly significant advance on past policies relating to high technology.

Specifically, by making DSTP a private company, subject to the Companies Law,[83] PT IPTN becomes fully accountable under its agreement with DSTP for all of its actions in relation to the development of the N-2130 jet aeroplane. All expenses must be properly documented and justified before payments are made; the schedule of work must be laid out in advance and adhered to; and the

---

[83]Law No. 1 of 1995 on Limited Liability Companies.

developer must work within a reasonably hard budget constraint, not only in relation to the total cost of the entire project, but also in relation to the expenditure of funds quarter by quarter. This kind of approach is essential if costs are to be kept under control and the project is to be completed on time.

The new arrangement contrasts strongly with those for the development of the CN-235 and N-250 (Mursjid, Chapter 11, this volume), which seem to the outsider to lack any semblance of financial control. Almost nobody knows what the development budgets were—or, indeed, if there *were* any budgets—much less whether budgets have been exceeded and, if so, by what margin. It was precisely this circumstance which led to the unplanned diversion of reafforestation funds to these projects,[84] creating a negative and lasting impression (both domestically and internationally) of an apparent lack of financial control by the government over these activities.

Notwithstanding these comments, there was considerable doubt among Update Conference participants as to whether the rules of the game have truly changed. Many seemed to doubt that the creation of DSTP would in fact permit the costs of developing the N-2130 to be kept under control and that any excess would be borne solely by the shareholders of DSTP (quite apart from concerns as to whether the development of a jet aircraft represents a wise use of productive resources in Indonesia). Many also doubted that DSTP's efforts to raise the full $2 billion needed for development of the aircraft by way of voluntary share purchases would be successful.

It is widely believed that many of the private sector founding shareholders did not participate voluntarily in DSTP. The fact that the employee welfare and pension funds of the central and state banks are the largest shareholders is also striking. In principle, these funds should be managed in the best interests of past and present employees of those banks, but many of the funds' beneficiaries will have died long before DSTP begins to earn a profit (if it ever does), and thus contributes to their pensions. As well, there have been cases in which overzealous officials, failing to understand the true nature of DSTP, have taken it upon themselves to raise funds for the purchase of DSTP shares, such as in the recent

---

[84]Presidential Decree No. 42/1994 on Loan Assistance to the State Limited Liability Company PT IPTN, dated 2 June 1994.

case of teachers in Central Java having deductions made from their salaries for this purpose (*Jakarta Post*, 9 September 1997).

Another concern of participants was that future sales of the N-2130 might not be genuine, arms-length market transactions, but might contain hidden costs for the Indonesian people. One possibility is exploitation of the captive market in Indonesia, in the form of airlines owned by the government or dependent on it for their operating licences; they would probably not be in a position to decline to purchase the N-2130 if requested to do so by the developer. It is widely believed that many of the sales contracts already entered into in relation to the CN-235 and N-250 fit into this category (*Asian Wall Street Journal*, 17 June 1994). Such contracts artificially boost the apparent success of these aircraft development projects, and hide their true costs by spreading them across airline users and the government (by way of higher airline costs and consequently lower profits and tax payments).

A second possibility participants may have had in mind is sales to other countries in the form of barter transactions. The concern with such sales is that their value to Indonesia might be less than the nominal sale price, as many feared in relation to the 'planes for sticky rice' transaction with Thailand not long ago (Suryodiningrat 1995). This seems unlikely with the N-2130, however, since DSTP's shareholders will only be interested in cash royalty payments from IPTN, not shiploads of rice and the like.

One of the interesting features of the contract between DSTP and IPTN is that, for the time being at least, foreigners cannot become shareholders in the former. But with the recent relaxation of the rule that previously prevented foreigners from buying more than 50% of the shares of companies listed on Indonesian stock exchanges, DSTP may also be giving consideration to the possibility of allowing foreign shareholdings. If we acknowledge that it is likely to be quite difficult to issue $2 billion worth of shares domestically, this may seem an attractive option.

On the other hand, it is hard to imagine that the foreign investment community would be particularly attracted to DSTP shares, given the problems mentioned in Mr Mursjid's contribution to this volume—such as IPTN's lack of experience with jet aircraft design and manufacture, and the intensity of competition from long-established players in the world market. It would be an embarrassment to Indonesia if foreigners showed no interest in the offer of DSTP shares. On balance, therefore, it would probably be more sensible to maintain the present exclusion on foreign ownership of shares in DSTP. It seems wise also for DSTP to continue to rely entirely on equity finance (not debt), since any debt issue presumably would need to be guaranteed by the government; this would conflict with the principle of pure private sector risk bearing for development of the N-2130.

As Mr Mursjid put it, the establishment of PT DSTP 'must be recognised as being both bold and innovative'. While it does nothing to allay economists' concerns about using scarce resources for the development of a jet aircraft, it is without doubt a very significant change in the manner of financing such projects, imposing strong financial discipline by moving the ownership of the project to the private sector.

13

INNOVATION AND DIFFUSION IN RURAL DEVELOPMENT: CONTRASTING CASE STUDIES

Colin Barlow

INTRODUCTION

New agricultural innovations have much enhanced productivity and economic returns in parts of the Indonesian rural sector. They have been generated through a country-wide network of government research institutes, with workers effectively devising techniques themselves and modifying others from abroad. They accordingly comprise a menu of methods suitable for most kinds of farming. More background on Indonesian agricultural research is given by van der Eng (1996).

These innovations have much enhanced outcomes from large farm agriculture, comprising 'estates' run commercially by hired managers and workforces. With their access to capital and managerial expertise, such estates have been well placed to adopt profitable technologies. They have sustained substantial economies of scale, including the ability to review and select appropriate new methods. They have therefore been forward in adoption, with new methods being employed in all their spheres of activity.

This chapter focuses on new technologies for small farms, however, since these dominate Indonesian agriculture and seem likely to continue to do so. In 1995, 33 million persons worked on such farms,[85] which still comprise the largest sector of the Indonesian economy after manufacturing and trade. Small farms likewise pose some of the greatest problems in national economic advance, and therefore deserve special attention.

Innovations have sometimes been widely adopted in small farm agriculture, as seen from Table 13.1. This shows large yield increases and steady area expansions in 1950–95 for the key food crops, paddy, maize and soybeans. Innovations have also been important in cattle production[86] and in food, livestock and tree crop processing. They have substantially affected engineering aspects of the rural economy, especially irrigation and transportation. In the latter area they have chiefly had an impact on road construction and vehicle and shipping designs, reducing previously high costs.

But small farm adoption of innovations has been poor in other spheres. Table 13.1 shows that tree crops have gained little in *yield*; the large expansions in area under cultivation have continued to be based on innovations coming to Indonesia in the early 1900s or earlier. Farmers have learned to handle old technologies more efficiently, but that is nearly all on the production side. Large technological advances are needed to raise the viability of enterprises and help them match their counterparts in comparable countries.

In some regions all small farms have failed to adopt new methods. Those in most of Eastern Indonesia—notably the two Nusa Tenggaras, East Timor, all the Sulawesis, Maluku and Irian Jaya—have low yields of both food and tree crops (BPS 1995, 1996), while poor technological levels obtain in most aspects of life. Such deficiencies are key constraints on per capita incomes in those provinces.

This chapter first sets out paradigms covering the generation, diffusion and adoption of innovations in rural improvement. It then

---

[85]This is the 'number of persons aged 10 years or more that worked during the previous week' in 1995 in agriculture, forestry, hunting and fishing (BPS 1996), adjusted downwards by BPS's estimated 1,240,000 labourers involved in estate agriculture and 1,000,000 individuals engaged in fishing.

[86]These crop and livestock innovations have done far more than raise productivity. They have, among other things, secured better environmental adaptation, improved resistance to disease and adverse weather, higher palatability and superior processability. Details are given in the scientific literature.

TABLE 13.1 Area, Production and Yield of Small Farm[a]
Agricultural Items, 1950–95

Item	1950	1970	1990	1995
Wetland paddy[b]				
Area ('000 ha)[c]	4,963	6,837	9,378	10,081
Production ('000 mt)	11,495	21,280	42,825	46,806
Yield (kg/ha)[d]	2,316	3,112	4,567	4,643
Dryland paddy				
Area ('000 ha)	979	1,470	1,125	1,358
Production ('000 mt)	1,332	2,121	2,534	2,939
Yield (kg/ha)	1,361	1,443	2,092	2,165
Maize				
Area ('000 ha)	2,272	2,939	3,158	3,652
Production ('000 mt)	2,115	2,826	6,734	8,246
Yield (kg/ha)	931	962	2,132	2,258
Cassava				
Area ('000 ha)	884	1,398	1,312	1,324
Production ('000 mt)	8,282	10,478	15,830	15,442
Yield (kg/ha)	9,368	7,495	12,100	11,700
Soybeans				
Area ('000 ha)	360	695	1,334	1,477
Production ('000 mt)	281	498	1,487	1,680
Yield (kg/ha)	780	717	1,115	1,137
Other food crops ('000 ha)[e]	506	962	844	968

ha: hectares; mt: metric tons; kg: kilograms; na: not available.

[a]Not including estates (farms with areas over 40 ha). In 1995, the latter constituted just under a quarter of the agricultural and forestry land of Indonesia.

[b]Including rainfed and irrigated paddy. Van der Eng (1996) classifies the whole paddy area as 'irrigated'.

[c]Harvested area for food crops, and planted area for tree crops. With wetland paddy there may be more than one crop and the same area may be counted twice.

[d]Yields are usually expressed in terms of the dry weight of harvested items, including dry paddy (paddy) and dry copra (coconut). But with cassava the weight of fresh roots is used.

TABLE 13.1 (cont.)

Item	1950	1970	1990	1995
Coconut				
Area ('000 ha)	1,500[f]	na	3,308	3,574
Production ('000 t)	na	na	2,298	2,542
Yield (kg/ha)	na	na	695	711
Rubber				
Area ('000 ha)	1,302	1,813	2,639	2,920
Production ('000 t)	na	na	913	1,159
Yield (kg/ha)	na	na	346	397
Coffee				
Area ('000 ha)	182[f]	na	1,014	1,099
Production ('000 t)	na	na	385	427
Yield (kg/ha)	na	na	380	389
Other tree crops ('000 ha)[g]	na	na	2,334	2,945
All crops ('000 ha)	12,442[h]	na	26,446	29,398
Livestock ('000)				
Cattle[i]	4,261		11,210[j]	11,550
Goats	5,032		12,062[j]	13,309
Buffalo	2,734		3,342[j]	3,112
Pigs	1,201		8,135[j]	7,825
Sheep	na		6,235[j]	7,169

[e]Sweet potatoes and peanuts.

[f]In 1940.

[g]Including palm oil, cocoa, tea, cloves, kapok, cashew, candlenut and arecanut.

[h]Excluding areas of 'other tree crops', which are unknown but probably less than 1,000,000 ha.

[i]Not including milk cows.

[j]In 1992.

Sources: Creutzberg (1975–80); Barlow, Jayasuriya and Tan (1994); BPS (1995, 1996); van der Eng (1996).

analyses the cases in Indonesia of rice, rubber, goats and water closets, each involving groups of new technologies. It compares these experiences, reaching conclusions on ways in which diffusion of improved methods to family farms and communities might be better achieved.

SUITABLE PARADIGMS

The generation of technologies is usefully considered in the induced technological innovation framework posited by Hicks (1932) and later demonstrated for agriculture by Hayami and Ruttan (1985). This supposes that changes in the relative prices of land, labour, capital and other factors of production induce researchers to produce innovations that economise on scarce and expensive factors and use those that are abundant and cheap. Changes in output prices likewise encourage the development of technologies in spheres that promise added returns.

The adoption of innovations by firms, households or individual farmers is helpfully viewed as primarily determined by net benefits, calculated according to relevant resource endowments, production opportunities and skills (see, for example, Anderson and Hardaker 1979). How much will it cost to marshal the required land, labour and capital? Are climates, topographies and expected output prices favourable to the production opportunities being appraised? Can complex new technologies be handled, given current levels of skills? The benefits of innovations estimated in this way are always perceived from the subjective viewpoints of producers and others concerned.[87]

The fact that resource and output prices, geography and skill levels differ greatly throughout the Indonesian archipelago, and even sometimes between neighbouring localities, means that many new innovations are needed to suit varying requirements. But a good

---

[87]Perceived risks informed by low assets and variable climates are also often quoted as affecting estimated benefits. But agents largely recognise them by discounting estimated yields and output prices pertaining to production possibilities. Preferences related to cultural backgrounds are likewise thought important, but agents usually include these in their estimated opportunity costs of resource endowments. Individual decision making under such circumstances is addressed by Becker (1976).

menu is available, as indicated, and beyond this Indonesian research workers have been flexible in developing techniques as needed. It might thus be thought that little else needs to be done, with 'the market' ensuring suitable technology adoption takes place. That is broadly true in limited regions adjoining centres of active economic development, notably in parts of Java. But it does not hold for tree crop producers, rural dwellers in Eastern Indonesia and numerous non-adopting pockets elsewhere.

In these places, 'incomplete markets' put up critical barriers to the adoption of innovations. Weak infrastructure and services, resulting in high transaction costs for inputs and outputs (and sometimes even preventing trade), are a feature of areas with incomplete markets. Information is lacking, and also asymmetric in the sense that it differs between participants. Even persons in apparently similar situations consequently react divergently, securing lower economic returns than would be obtained with greater knowledge. Such incomplete markets characterise the 'traditional sector' in the dual market structure postulated by Myint (1985). The well-integrated and smoothly functioning 'modern sector' includes Indonesian estates, which have strong market linkages including access to full and relevant information.[88]

It is reasonable to postulate that sparse adoption of new innovations by Indonesian small farmers is mainly due to such incompleteness. People do not adopt because information is insufficient or misleading, while input prices are too high and output prices too low. They have limited access to capital, especially for longer term investments. But since many innovations would prove profitable given better linkages, the question arises as to how adoption might be achieved.

The paradigm of 'interveners' specifically addressing and overcoming such problems and helping to promote innovations is crucial in this context. Government and non-government organisations from Myint's modern sector may be viewed as economically justified in seeking to rectify incomplete markets by providing public goods; these comprise information, infrastructure and key services, all of which are profitable for society as a whole, but either not supplied

---

[88]Myint's dualism, unlike that of Boeke (1952), postulates links between the two sectors. There is hence the possibility of steadily better attachments to modern markets being secured by traditional producers.

or only produced suboptimally by private agents.[89] Once such goods are in place, however, the modern intervener of private business is much more attracted to helping to enhance input and output markets. Given that new innovations are economically viable, these interventions set in train autonomous adoption and growth beneficial to all.

These paradigms are now used to explore the four cases. The scene is basically one of potential adopters seeking net benefits from innovations, but being able to adopt them only if the reigning incomplete markets are rectified through suitable interventions by government and other parties.

CASE STUDIES

Rice

Rice is Indonesia's pre-eminent crop, in terms of both area and value added (BPS 1995, 1996). Over three-quarters is produced on farms of below one hectare in size in Java and Sumatra; other centres of cultivation include South Sulawesi. While rice improvement efforts go back hundreds of years, modern scientific seed selection in special gardens commenced in Indonesia only in 1914. Robust *Indica* rices that cope well with drought and water stress were brought from other countries and selected for improvement; they spread widely in Sumatra and Kalimantan. Local *Javanica* rices, which mature faster, yield more under good growing conditions and are more palatable, remained dominant elsewhere. *Javanica* rices also cross-pollinate easily, allowing farmers to generate varieties that suit their specific growing conditions.

A big breakthrough—sometimes called the 'green revolution', even though that had been under way for half a century—occurred with the arrival of new rice varieties in the late 1960s. Most of the new rices came from the International Rice Research Institute in the Philippines. They were short-stalked, which meant that they

---

[89]Intervention may be further justified on the 'infant industry' argument, where farmers previously refrained from otherwise profitable activities owing to deficient knowledge and consequent uncertainty. Interveners assist introduction of new technologies on the grounds that doing this will eliminate previous barriers, also encouraging other farmers to adopt technologies autonomously without assistance. The benefits of the latter repay costs of initial intervention.

responded well to fertiliser and gave higher yields, and had relatively brief fixed maturity periods, making it easier to fit two crops into a limited growing time.[90] The palatability of the Philippine rices was improved by crossing them with local types to secure *Pelita* and other varieties, which tasted better while retaining the superior traits of imported cultivars.

Although massive attacks from grassy stunt fungus and brown planthopper severely cut *Pelita* yields in the 1970s, the breeding in of resistance enabled new varieties to continue spreading. The Indonesian wetlands moved from no new rices in 1967–68 to almost 100% in 1980–81 (Barker and Herdt 1985); in the 1990s yields were virtually double those of the 1950s (Table 13.1). Researchers producing these rices provide a good illustration of induced innovation: they generated plants able to yield more on increasingly expensive land; in developing varieties needing purchased inputs, they made more use of capital, which was becoming relatively cheaper than other factors; and they took measures to overcome losses incurred through disease and pests.

But research was not the only factor in securing the huge increase in rice production that was of such major significance for the Indonesian political economy; heavy government intervention in at least two other spheres was also vital. One of these was irrigation, which had in fact been emphasised by Indonesian governments since the mid 19th century (Booth 1988; van der Eng 1996). This initiative was now vastly enhanced by the New Order, which added nearly three million hectares to the stock of land from the late 1960s to the mid 1990s. While these new areas varied enormously in quality, with one-half being tidal and swamp developments, they usually had enough water for one crop and sometimes two. Importantly, these irrigation improvements were supplemented by the better roads and the improved health and education services now provided by government throughout Indonesia. The roads, together with the new construction and transportation technologies mentioned

---

[90]The long-stalked 'traditional' varieties common in the 1950s and 1960s chiefly responded to fertiliser by adding to their leaves, consequently falling over and undermining grain quality as well as being more difficult to harvest. The short-strawed varieties when fertilised mainly increased their grains, and are rarely lodged. The shorter fixed maturities entailed about 100 days, compared to 135–140 days for older photo-period sensitive rices. Growing time for rice is customarily dependent on climate and the availability of water, but with high-class irrigation is at most about 180 days.

previously, substantially reduced forwarding costs, while enhanced communications and education levels meant that rural people became far more aware of new possibilities.

The other important official initiative was extension and credit, vigorously pursued through all irrigated rice regions to encourage farmers to adopt new varieties. This began in West Java in the early 1960s with the famous Bimas rice intensification program, which from the mid 1980s targeted *unit desa* [village units] of up to 1,000 hectares and 2,000 farm households. Under the scheme, government field officers, together with bank representatives and certain villagers, provided 'contact farmers' with a standard package of high-yielding rice varieties, fertiliser and pesticides, backed up by information and low-interest credit.[91]

This extension program has been crucial in assisting the spread of high-yielding rice varieties. Given such help, farmers have perceived net benefits from participating and acted accordingly. Without the program, prevailing incomplete markets could have delayed the advance achieved by the 1990s for several decades. Yet it is also significant that, even with intervention, farmers took over 10 years to become proficient in handling the complex new technology, only reaching that stage by the mid 1980s. The program was then increasingly superseded by private market facilities and enhanced business participation, reflecting improved linkages and signalling the start of autonomous rice development. Many producers began to select the input combinations they judged most economically appropriate, purchasing materials and obtaining credit from private traders who were now happy to supply these (Birowo and Gondowarsito 1990).

The original failure of traders to provide such facilities illustrates a not uncommon deficiency on their part; in the rural areas of many developing countries, such agents—despite furnishing efficient output markets—hardly ever introduce new technologies or supply the requisite inputs for them.[92] It is likewise pertinent that rice farmers with improved linkages in their new autonomous

---

[91]The fertiliser has also been provided at what until recently has been a large subsidy for all users.

[92]The failure of private traders to promote improved technologies in the small farm sector is a worldwide phenomenon, catalogued for global tree crops by Barlow (forthcoming). The returns to be gained from such activity are too uncertain within the dominant incomplete markets, and in the case of tree crops, too long term.

situation increasingly applied their experiences to adjacent food crops, helping to explain the productivity rises for maize and soybeans shown in Table 13.1.

A third area of government intervention was the regulation of rice prices through the Bulog logistics agency. Bulog used purchasing and stocking operations to reduce fluctuations in the price of rice, and sometimes to increase Indonesian prices above world levels (van der Eng 1996).[93] While this measure may have helped farmers, in circumstances in which a well-developed private output market could almost certainly have handled the crop efficiently, its economics have been hotly debated. Bulog in fact now seems likely to be dismantled, leaving traders to operate largely on their own. This initiative is the least important of the government's actions on rice, at least from the producer's viewpoint. It should be noted that all rice interventions have been justified not only by economic perceptions, but also by broad-based political support for what is indisputably Indonesia's key crop.

Rubber

This is still the most significant product marketed by the small farm subsector (BPS 1995, 1996). Rubber production is concentrated mainly in Sumatra, with smaller output in Kalimantan. Rubber trees need high rainfall and year-round humidity for successful cultivation. Smallholding rubber ventures cover larger areas than paddy enterprises, usually involving 1–3 hectares of land in scattered parcels.

Even today, most small rubber farms grow unselected seedlings of provenances brought from Brazil in the late 19th century, and most replanting is still done with these materials. Such seedlings come into tapping after 10 years and remain in production for some 20 years thereafter. Yet Indonesian research led the way in rubber improvement, with the first rubber clone being produced in the Bogor Botanic Gardens in 1916. Given capital to invest and reasonable management, this spectacular advance allowed yields to be doubled over the low levels of the time, while subsequent research enabled them to be tripled by the end of the 1930s (Dijkman 1951). The new technologies were actively adopted by rubber estates, which by 1941

---

[93]Bulog has maintained regional food depots and at times handled a large proportion of the rice crop. It has also been responsible for importing rice when this has been judged necessary.

had replanted almost one-third of their plantations with high-yielding varieties. They resumed this thrust when political conditions improved in the late 1960s, and by the early 1980s had virtually all plantings under advanced provenances (Barlow, Jayasuriya and Tan 1994). The rubber innovations, although suitable for estates, were more land and labour saving and more capital intensive than was initially appropriate for small farms.[94] From the 1970s they became increasingly suitable for smallholdings, and even better varieties were developed in subsequent years. Levels of adoption remained minimal, however.

The typical conditions of small rubber farms made improvement almost impossible without outside assistance. Most were situated in remote places, which up to the 1980s were connected to the outside world by river or rough tracks. Their low cash incomes meant that they could not afford the substantial costs of replanting with high-yielding trees, and they did not in any case know much about them. Even given sufficient knowledge, they could not have secured private long-term credit for an investment that would take 10 years to show a net benefit.[95] Markets in the areas where rubber was grown were even more incomplete than those facing rice, with no apparent chance of autonomous improvement. Private traders again failed to enhance production technologies, although they supplied efficient output markets and were forced by government to introduce improved processing methods. Small wonder that nearly all smallholders continued replanting with seedlings selected from the best low-yielding trees, following century-old practice.

Government intervention to assist small farm rubber production was much less than was the case with rice. Other things apart, this was a crop grown in remote locations by relatively few people, and had little political significance. However, the New Order regime

---

[94]This was not surprising, in that research generating these technologies was largely financed by estates.

[95]Unselected rubber seedlings are conventionally planted under the swidden (shifting cultivation) system, with food crops being grown among them for 1–2 years. Under the *karet hutan* [jungle rubber] system, cultivation is abandoned for a further eight years until the low-yielding trees can be tapped. Any moneylenders wanting to invest in rubber loans—and there were none—would certainly anticipate a gap of 10 years before positive but low net revenues could be secured (Barlow 1997). Actually, new rubber varieties could, if managed properly, be expected to come into tapping after seven years.

did provide well-organised research through two major rubber institutes in North and South Sumatra, and rubber growing areas increasingly benefited from the general improvements that were being made in infrastructure and services.

Although government supplied little or no extension to the great majority of rubber farmers,[96] it made some direct efforts to promote intensive improvement in selected locations—supplying, for example, high-yielding rubber packages of a similar nature to those provided for rice to limited groups of farmers attached to nucleus estates. Some of these ventures were very successful, and mainly explain the 15% rise in smallholdings under better trees by the mid 1990s (Barlow, Jayasuriya and Tan 1994).

Yet this small achievement must be compared with the successes of Thailand, Malaysia, India and Sri Lanka, where comprehensive official extension of packages improved large proportions of total smallholder plantings and greatly enhanced national rubber productivity. These interventions raised producers' incomes substantially, and in Thailand and Malaysia were estimated to have earned economic rates of return of 25–30% from official help (Barlow, forthcoming). In Indonesia, where most small farms did not receive direct official help, rubber production expanded chiefly through steady increases in planted area with minimal yield changes (Table 13.1).[97]

---

[96]Thus the typical situation with most crops and livestock, excluding rice and special projects, is one in which a lowly paid extension officer, sometimes without even a motorcycle, is responsible for perhaps 10,000 farmers in a big district. Although many officers are highly motivated, it is manifestly impossible for them to make much progress. Similarly it is hard for back-up specialists to have much impact under such circumstances.

[97]But it is interesting to record that in the 1990s, 80 years after the Bogor development of high-yielding rubber, the private sector in Sumatra is finally responding to the availability of better material. This development has been pioneered largely by Javanese ex-rubber estate employees, operating small planting material nurseries and working through traders who have usually gained their experience in public sector improvement schemes. The establishment of such enterprises has been greatly facilitated by lower transport costs and much improved information in the 1980s. Large numbers of rubber smallholders are buying materials from these nurseries, with the main difficulty being quality control over sales of inferior stock which may not achieve the promised yields. The development has been analysed by Barlow (1997).

Goats

Goats are the most numerous of the larger animals in Indonesia (Table 13.1), and are chiefly kept in groups of 10–20 by individual small farmers throughout the archipelago. They are potentially prolific: after five months' gestation, females can bear offspring at 12 months; they can have up to two gestations a year over the next 6–8 years, bearing 3–4 young annually and producing excellent quality milk. Goats are usually corralled continuously in Java where land is scarce, but elsewhere roam freely by day to be brought in and fed at night. Contrary to widespread perception, they are kind to the environment, leaving a basic cover of vegetation that would be stripped by cattle.

Goats can be marketed for meat or for breeding when they are only six months old. Farmers can also profit from the sale of goats' milk. The benefits flowing from the initial purchase of a young female can accordingly be secured quite early, like rice but unlike rubber. However, as goats are susceptible to disease and birthing problems, the outputs just quoted are only achieved with excellent management. Theft of free range animals is also common. The benefits of keeping goats are hence subject to high uncertainty, discouraging investment in them.

Much Indonesian research has been undertaken to improve goats. This has focused on breeding, sometimes for milk and sometimes for meat production. Superior selected types from abroad have been introduced to secure crossbreeds well adapted to Indonesian conditions. Other work has concentrated on better management, including feeding regimes to achieve higher growth rates, and improved methods of disease control (Sarwono 1990). The adoption and effective application of these approaches could greatly enhance the net benefits from goat husbandry.

However, as with rubber, the diffusion of goat technologies to small farmers has been disappointing. For the most part unimproved animals are still being managed in the traditional manner, with indifferent results. Little effort has been made to improve goat extension, with most farmers seldom having contact with this service or with back-up veterinary services. Even where improved animals have been released in village communities, their qualities are soon diluted in what Indonesian livestock specialists recognise as the all too familiar pattern of the best meat animals being sold quickly for immediate gain. While understandable, this soon destroys the impact of introduced stock. Farmers have also not understood how to handle the special inputs needed to get the most out of improved animals, even if they could find the money to purchase them. Although non-government organisations (NGOs) have sometimes intervened, their characteristic lack of technical expertise (Riddell and Robinson 1995) has often undermined their attempts to facilitate long-term improvements. Traders, while keen to purchase goats, once again have not promoted new technologies or supplied the inputs needed to support them.

Water Closets

While there remain large remote areas where the inhabitants still follow traditional approaches to waste disposal,[98] water closets are now common in rural Indonesia. Although several technologies have been tried, the most popular system comprises a bowl connected by pipe to a septic tank, with minimal water employed for flushing.[99] Such closets have beneficial health effects, and studies in Eastern Indonesia, where diarrhoea is common, have indicated additions of 15 days per year to household working time following installation (Dinas Kesehatan 1997). Closets are also more conveni-ent than traditional approaches, especially for females, and this too has had a positive impact on community health.[100]

Unlike some other innovations, closets have not involved much Indonesian research and adaptation; well-developed international systems have proved suitable for direct national application (Pacey 1980). While they entail the purchase of cement, piping and other inputs, their positive health effects give them perceived net benefits soon after installation. They are simple in design and can be used with minimal training. This has meant that NGOs have been

---

[98]In parts of Eastern Indonesia where pigs are common, for example, waste enters a 'natural cycle', being consumed by these animals in their roving activities.

[99]Such flushing systems can cause adverse accumulations of soil nitrates under some conditions (Fox et al. 1993), albeit rarely recognised by those involved. It is hard to judge how dangerous these accumulations are, and the fact that the dangers may not eventuate for many years reduces their perceived significance. Other systems involving composting methods and not having such effects are available, but have not proved popular.

[100]The expansion of individual household toilets apparently helps prevent community-wide infections, thus raising general levels of health (Pacey 1980).

able to initiate projects involving water closets without being limited by their own lack of technical expertise.

Agencies and other groups introducing closets have provided materials and limited training to selected households; this has usually set in train rapid diffusion, helped by rotating credit[101] but also characterised by widening autonomous adoption by all but the poorest households. This outcome, which contrasts with difficulties over rubber, goats and even rice, can be attributed to the simplicity of the innovation, accompanied by relatively low initial cash expenditures and early high perceived benefits.[102] It has also been encouraged by 'bandwagon effects' from high surrounding adoption, as well as by pressure from *kepala desa* [village heads] following prompting by health authorities.

CONCLUSIONS

The four cases discussed above involve innovations that would prove profitable at prices in the modern sector, but which are rendered less attractive by incomplete markets characterised by sparse information and limitations on input supplies. Each has available to it a good menu of suitable technologies springing from research induced by modern sector prices, with the difficulties mainly residing in diffusion. While rice, goat and closet innovations can, if adopted, secure early returns, the introduction of higher yielding rubbers produces benefits only after many years, and then with much uncertainty.

The new rice, goat and rubber technologies are complex, with successful application often requiring long periods of adjustment. This may involve tackling unanticipated difficulties, as for example when researchers had to impart resistances to rice varieties to

---

[101]Households receiving credit are expected to pay back the value of materials purchased on their behalf over a period of, say, two years. The *kelompok* [group] generally organises such projects and collects loans, enabling others to get closets in turn.

[102]These elements assure a high perceived present value of net benefit, with any uncertainty being further reduced as more closets are established in the surrounding area. It is interesting, however, that attempts to introduce communal toilets shared by several households have never proved popular, with persistent difficulties over maintenance leading to the facilities falling into disuse.

counter pests and diseases, or to modify goat control regimes to overcome emerging health problems. Farmers in turn have had to learn how to handle the new technologies available to them. While the cases I have discussed demonstrate the broadly effective responses of research mechanisms to challenges, they also show how difficult it has been for people to adopt innovations. Hence even with strong official support for the introduction of improved rice strains, it took over 10 years for farmers to reach the stage of autonomous cultivation in conjunction with private traders. Such long periods of adjustment naturally impose substantial costs.

All cases demonstrate the usefulness of intervention in promoting the diffusion and adoption of new innovations in Indonesian rural development, thus helping people move from traditional to more technological approaches. Here government seems to have two key roles, one of which is to supply the public goods of infrastructure and services, including research, irrigation, roads, health and general education. The Indonesian authorities have in fact progressively improved facilities in rural areas, successfully financing research to generate technologies and using other measures to enhance market integration.

Government's other key role is to undertake direct targeted extension programs for particular innovations, providing information, training, limited credit and other support to promote adoption. So far it has only done this comprehensively for rice, with its political imperative, but there is a need for intensive programs in other spheres.[103] Little has been achieved where levels of assistance are low, as demonstrated by the persistence of traditional methods in the rubber and goat sectors. However, quick autonomous adoption was secured with less extension in the case of water closets, a simpler and more straightforward technology. Although NGOs can assist with targeted actions, to be successful they need better technical skills and improved linkages with government support services.[104]

---

[103]The need for lengthy and intensive extension involvement for smallholders was recognised in Thailand and India, where government advisory services followed targeted programs for even longer than was the case in Indonesia for rice. Indeed, in India such programs continued for over five decades. But the high level of expertise now evident in the largely autonomous situation of Kerala rubber farmers is very impressive.

[104]This is not easy, owing to both the small size and highly independent nature of NGOs. The difficulties are explored by Riddell and Robinson (1995).

A third government role, that of intervention in output markets, has more doubtful justification given the usually efficient and competitive nature of private output trade. Questions have accordingly arisen over Bulog's regulation of rice prices.[105]

Private business is a vital intervener, despite its seemingly characteristic failure to transfer new production technologies to rural peoples. It has an especially important role to play in supplying the inputs needed for new technologies, including credit, seed, fertiliser, and pest and disease control items. The case of Indonesian rice shows that private agents will supply these inputs once improved infrastructure is in place and farmers have acquired the necessary skills; the same proved true in Thailand and India for rubber.[106] But in Indonesia a lack of inputs from private sources has seriously constrained advances in rubber and goats, chiefly owing to the failure of government to provide the foundations for autonomous development. Private enterprise can also provide improved output processing facilities, although the techniques necessary to underpin their success may once more have to be supplied by government.

Some Indonesian rural development will certainly occur even without government and other interventions, and some will be fuelled by growth of the national economy. But the slow pace of development, together with the high rates of return that could be achieved on some initiatives, suggest that more active policies are needed in all cases except rice. This argument is reinforced by the manifest impact of economic improvements in raising the miserable living standards of many regions. Technologies can be critical in economic progress, but require facilitation where markets are incomplete. As incomplete markets still predominate in most rural areas, more positive interventions should undoubtedly be considered.

---

[105]Even the blanket Indonesian subsidy for fertilisers, mentioned earlier, has doubtful economic justification, appearing to result in gross overapplication of nutrients in many instances. The adverse consequences of output price interventions in other countries are highlighted by Barlow (forthcoming).

[106]Once the infant industry is launched and demand for inputs well established, it becomes profitable for private agents to supply them.

14

BALI'S GARMENT EXPORT INDUSTRY

William Cole

INTRODUCTION

One of the factors that will make or break Indonesia as an industrial power in the early 21st century will be its success in assimilating new flows of technical information in the manufacturing process.[107] In addition to capital, regulatory and physical technology constraints, small and medium-sized manufacturing enterprises face a range of information-related problems. Among these are poor management skills, poor workmanship, inefficient organisation of production and transport systems, and limited access to innovative product designs and to other market-related information. From the perspective of the individual firm, addressing these problems is a matter of overcoming the high cost of accessing and incorporating new information. For those hoping to raise Indonesian industrial productivity quickly, this leads to two central questions: where will this knowledge come from, and who will pay for it?

---

[107]The chapter is based on data from several sources: research interviews and personal experience (the author lived and worked in Bali for several years between 1978 and 1991); policy research conducted on the Bali garment industry by Dr David Wheeler and the author in 1988; preliminary findings of an update on the industry in 1997 by Dr Thee Kian Wie; and a series of industry interviews conducted by the author in early 1997.

There are three potential answers: the government, the independent sector (non-government organisations) and the private sector. Government clearly has a role in supporting industrial development. Yet the scale of the problem is so large, the time before AFTA and other agreements come into effect so short, state resources so limited, and the instrument of delivery (the bureaucracy) so blunt that direct state assistance on a firm-by-firm basis is unlikely to achieve much. The programmatic solutions currently being considered by the Indonesian government will not have an impact on the mass scale required over the next 5–10 years. Non-government organisations (NGOs) have had better success in direct firm-by-firm assistance, but they work on a very small scale with virtually no possibility of scaling up for mass impact. The chronic constraint for NGOs is not just funding, but the short supply of managers and workers willing to make the personal sacrifices necessary to ensure that this solution really works. That leaves the private sector. Direct assistance at the enterprise level by private sector consulting firms is a partial solution, although it runs into many of the resource and quality limitations inherent in government-supplied assistance.

A more effective private sector solution would be 'strategic alliances',[108] or the transfer of knowledge as a natural part of cooperative long-term business relationships. In the context of such relationships, buyers of products and vendors of technology and capital often provide information-related assistance to less developed firms as a normal part of doing business. Such transfers are driven by long-term profit motivation and have nothing to do with welfare. To work, knowledge transfer through strategic alliances has to be entirely voluntary and must provide enough returns for the knowledge provider to cover the costs and the risks involved.

Can this solution work for Indonesia? In fact, provision of information through strategic alliances is almost certainly where the vast majority of technical knowledge transfer and the diffusion

---

[108]The closest Indonesian term is *kemitraan*. However, in Indonesian policy circles this has come to refer narrowly to assistance/subcontracting linkages between very large conglomerates and small traditional producers. Such linkages tend to be justified in 'welfare' terms, and their establishment requires heavy pressure and direction from the state. *Kemitraan* in this definition is an aberrant subset of the common phenomenon of voluntary, for-profit strategic alliances among businesses at all levels.

of innovations is now occurring.[109] This mechanism receives little attention because such transfers are private, incremental, highly dispersed, and hard to detect and measure.

Among the industries in which this pattern is important is export manufacturing, where relatively small-scale producers are linked to final markets through a network of assembling, finishing and trading firms. Well-known examples are the Jepara wood industry, the Ceper machine parts industry, some components of the fashion batik industry around Yogyakarta, and the Astra auto and motorcycle industries. In many of these cases, the flows of information and interfirm assistance originate with foreign buyers, who are seeking higher value products for export and who are willing to give assistance to producer firms to get those products.

BALI'S EXPORT INDUSTRIES

The impressive growth record of export industries on the island of Bali provides several cases of industrial development based on information inflows through networks of strategic business alliances. Over the past 20 years Bali has become the site of innovative production in several specialised products: garments, silver jewellery, wood carving, quilting, leather products, bamboo furniture, ceramics, stone carving and textiles. Each of these industries is now producing competitive products for international markets, based on highly flexible small-batch production, quick turnaround times and a capacity for rapid adjustment to new designs. Some of these industries had their origins in traditional Balinese crafts, and some did not. But in every case the production system quickly developed the capacity to adapt to new designs and methods that have no roots in traditional Balinese or Indonesian culture. The process of absorption and replication of these designs has been so thorough that the casual observer in Bali often has the impression that he or she is looking at indigenous Balinese goods.[110]

---

[109]Foreign direct investment is, of course, important at the high end of technology transfer, but the sheer number of strategic alliances in which most private sector firms are engaged argues for this being the greatest path of information and innovation flow.

[110]Examples of new products introduced over the past 20 years are embroidery (from Tasikmalaya in West Java); sequins, applique and silk screening in garments; virtually all silver jewellery design (earlier mostly of Indian origin, later 'generic ethnic'); all quilts (from the US); the vast majority

The positive social characteristics of the Bali industries are of particular interest from a policy perspective and are precisely what the government of Indonesia has been attempting to produce:

- Most production is based on thousands of rural village work groups and small/micro enterprises networked through small and medium-scale assembly/finishing firms.

- There has been a rapid growth of indigenous Indonesian (*pribumi*) entrepreneurship, particularly in the numbers of *pribumi*-owned assembly/finishing firms.

- Capital for expansion is largely self-financed through retained earnings, while until the 1990s working capital was primarily covered by downpayments from foreign buyers.

- No specific government subsidies or protection were involved— the growth rates and growth characteristics noted above were neither anticipated nor planned by government.

The last point bears special emphasis. With the provincial government's attention focused primarily on the development of tourism and agriculture, the initial establishment and early growth of most of these industries went all but unnoticed by the state.

Among the industries on Bali, the rise of the garment export industry is typical. Because the Bali model of garment production has no roots in any indigenous craft or production pattern, this industry shows how far it is possible to go with minimal initial endowments when conditions are favourable. The lack of a direct role for government in the start-up and expansion of the industry may provide insights into what role the government could and should play in support of similar industries elsewhere.

GROWTH CHARACTERISTICS OF THE GARMENT INDUSTRY

One of the most striking characteristics of the Bali garment industry is the rapid expansion of total output, with exports being the primary market from the outset. While actual production figures are

---

of painted softwood carving (inspired by modern American and European designs); most ceramic designs; all modern bamboo furniture designs; and all leather clothing and most leather accessories.

FIGURE 14.1 Bali's Garment Exports, 1975–95 ($ million)

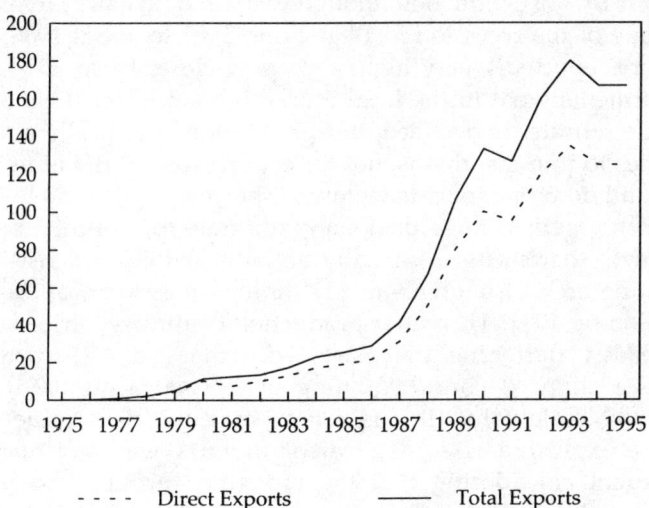

- - - - Direct Exports ———— Total Exports

Notes and Sources: Direct export data are based on estimates provided by the provincial authorities. Total exports are the sum of direct exports and tourist purchases. The latter are based on various fieldwork estimates. All data are approximate, especially for the period 1988–95.

not available, estimates can be made based on export figures.[111] From a minimal export base in 1975, exports grew to over $10 million in 1980.[112] After a brief drop due to the recession in 1980, exports quickly resumed growth and rose rapidly to $31 million by 1987. Thereafter garment exports rose very fast, but in a steadily flattening arch, to around $125–135 million in the mid 1990s, with a serious but temporary drop during the recession of 1990–91 (Figure 14.1).

Garment production intended specifically for Bali's tourist retail market was relatively limited until the late 1980s. However, export rejects have throughout the industry's history been dumped on

---

[111]Up through 1988 the figures presented are in 1988 US dollars adjusted for inflation. More recent figures are not inflation-adjusted.

[112]All export and production figures are free on board (fob) values reported at the point of export. Until the mid 1980s all exports were by air through Ngurah Rai Airport.

the tourist market, and this provides a rough estimate for tourist retail consumption of Bali garments. In the 1970s rejects were rarely more than 10% of production, though with an unusually large dump at the time of the recession in 1980. From 1981 to about 1986, reject rates were reportedly very high, averaging closer to 20–30%. All of these garments went to the local tourist market. From 1986 on, the reject rate reportedly declined, but production specifically aimed at retail sales to tourists (that is, not for export) rose as the numbers of foreign and domestic tourists increased sharply.

Taken together, these data suggest a trajectory of total production growth that starts at virtually nothing in 1974 and rises in an accelerating arch through about $11 million in 1980 on up to about $45 million by 1987. Thereafter production continues to rise fast but in a steadily flattening trajectory, with the end of growth and possibly a slight decline beginning in 1994 at around $160–185 million wholesale value (that is, fob export equivalent value).

This explosive rise in garment exports was a remarkable achievement considering that the industry, through most of its history, was built on simple foot-treadle sewing machines and part-time, seasonal, rural village labour. Moreover, the industry had to compete for export quotas with larger Indonesian producers using more advanced mechanical technologies and based in the industrial centres of Bandung, Jakarta and Surabaya. Quotas in Indonesia were largely allocated on the basis of *past performance*. To expand exports at the high rates achieved, each year the Bali industry had to outbid other Indonesian producers for additional new and unused quotas.

One of the most striking characteristics of this growth history is the explosive rise in unit values (US dollar value fob per kilogram of garments) that began in 1981 and ended around 1986 (Figure 14.2). The mean figure moves from about $1 to $3 per kilogram over this period. The 25th percentile for all shipments actually moves from about 30 cents to $2.50 per kilogram. This rise is all the more impressive when one considers that dozens of new assembly/finishing firms and hundreds of new village work groups were entering the industry during this period. These new entrants, each starting low on the learning curve, would naturally draw the overall unit value averages down. Almost as striking as the onset of this rise is the rapid flattening of unit values after about 1985.

Taken together, these total value and unit value data suggest three phases of growth up to the early 1990s. The industry began

FIGURE 14.2 Unit Values of Bali's Garment Exports, 1979–87
($/kg)

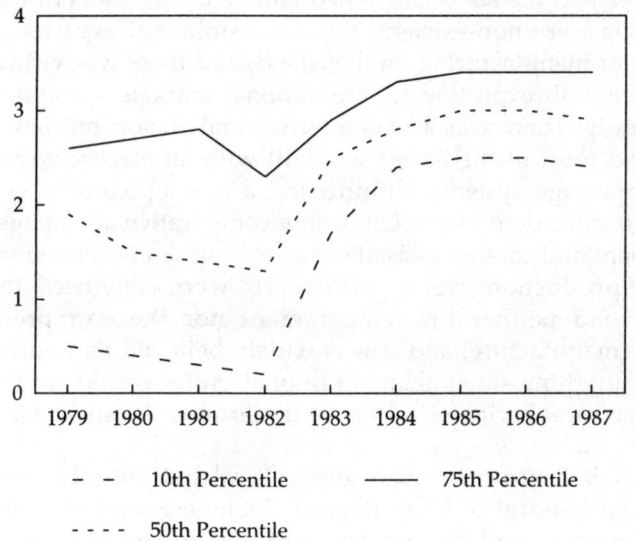

- - - 10th Percentile   ——— 75th Percentile

- - - - 50th Percentile

Notes and Sources: As for Figure 14.1.

with the establishment of a system capable of rapid *extensification* of production at fairly low unit values. Around 1981, rapid quality and design improvements produced increasingly higher unit values in what might be called an *intensification* of production. Beginning in 1986, though progress up the technical innovation curve faltered, a *new phase of extensification* began, with faster growth in volumes making up for flattened growth in unit values. Throughout this third phase the rate of increase was declining, and by 1994–95 total production appears to have plateaued.

Initial Conditions and the First Phase of Growth

What drove this explosive rise in export garment production? The endowments of southern Bali in the mid 1970s were not promising. Land and labour were cheap, but not much space is required for garment manufacture and most Balinese were, and still are, averse to factory work. There were no garment making skills aside from the very low-quality traditional tailor shops. Labour laws made it extremely difficult to employ more than about 40–50 workers on one site before owners struck regulatory difficulties with the Ministry of Manpower and the state-sanctioned labour union. Modern management skills were non-existent, there was minimal experience with any kind of manufacturing on the island, and there was virtually no experience with exporting to international markets. Capital was in short supply. There was a new international airport but few paved roads, and most of the island was still without electricity, running water or sewage systems. Culture and a rich agricultural environment were considered to be Bali's only comparative advantages, and government and foreign assistance were focused entirely on tourism and rice production. Nearly all experts were convinced that the Balinese had neither the temperament nor the entrepreneurial skills for manufacture, and it was widely believed that encouragement of anything other than traditional crafts would destroy the island's cultural heritage and hence undermine the sustainability of tourism.

What Bali did have were two critical but entirely unnoticed elements: a handful of local 'Balinised' Chinese with experience in the tourist retail trade, and a group of low-budget tourists, mostly Australian surfers, living for months at a time over several years in Bali. These, combined with a nearby airport and an untapped market for cheap beach wear in Australia, were the only ingredients necessary for explosive growth—all that was needed was the spark.

In the early 1970s the local beach market retailers discovered that the few tourists visiting Bali in those days would pay handsomely for old Javanese sarongs and embroidered *kebaya* [women's traditional blouses]. Some retailers began collecting these items, first in Bali and then around Surabaya, for resale in this market. The young Australians soon picked up on the opportunity to carry quantities of these goods home in suitcases for resale on Australian beaches. Mark-ups were many times over cost. For most, the objective was simply to make enough money to finance a low-cost lifestyle in Bali. However, by 1975 about half a dozen of these foreigners were pursuing the trade seriously, and soon the demand for used traditional garments outstripped supply. At that point some of the beach market retailers entered manufacturing, using small-scale, more or less traditional tailor shops to produce simple beach wear. Initially, the tailoring units consisted of no more than 5–10 workers making simple, low-quality goods from domestically produced batik, polyester and cotton textiles. As demand in Australia surged and the foreign entrepreneurs learned how to export larger quantities legally, they began pressing local producers to increase supply. There was a natural resistance by local counterparts to investing in expanded production given the unknown risks inherent in a new line of business and a regulatory environment adverse to manufacturing. Foreign buyers overcame this resistance by offering high down-payments for products, no-interest loans (or outright gifts) for machines and building modifications, and no-cost advice on design and manufacturing logistics.

Garment designs through this period were simple. A skilled Italian designer living in Bali was experimenting with new designs based on domestically produced jersey and simple dyeing methods. Together with a local counterpart, he began producing in very small quantities for the New York fashion market around 1975. By about 1977, two long-term American residents began to create simple but innovative designs and new wave colour schemes (for example, the well-known 'Jak-Pak'), but they produced in very small quantities for sale locally to tourists. Scaled-down, low-quality versions of these designs, the new methods and the use of jersey were all copied for export by the Australian entrepreneurs and their local producer counterparts.

The beautiful old embroidered Javanese *kebaya* had always been a favourite of tourists. At some point around 1977, the largest of the Australian entrepreneurs established a link to traditional embroidery producers in Tasikmalaya, West Java. Initially, they imported raw embroidered cloth for garment making in Bali. Even-tually, however, they hired embroidery experts from Tasikmalaya to train Balinese women in this work. Though the quality was poor, embroidery skills added a new element to the garment options available for export out of Bali.

Quality control during this period was minimal, but preshipment reject rates were reported to be about 10% through this first phase of growth. Though the workmanship was generally of poor quality, demand in the Australian beach market was for simple designs at very low cost. One Australian entrepreneur joked that his quality control operation consisted of pitching the garment in the rubbish dump in Australia if local retailers thought it was of too poor quality to sell.

Probably the single most important innovation during this early phase was the establishment of the 'put-out system', which later evolved into full-scale subcontracting networks. With the Australian market absorbing virtually everything that could be produced, foreign entrepreneurs were putting tremendous pressure on local producers to increase volumes. With the rapid expansion of exports, skilled tailors were in short supply. Many of the new tailors were young village women who would return home after a short period of work in the new workshops. The producers began to entice these women into continuing to work in their rural villages, providing them with machines. Clusters of new village workers then grew up around these trained women. A new position of 'go-between' emerged in the producer firms to distribute raw material to the village work groups and collect completed output for final assembly and finishing at the core shop. Over time these core shops became small factories with 30–50 workers, networked to dozens of village work groups. As problems of logistics, worker training and basic quality control were worked out in close collaboration with the foreign entrepreneurs, the industry developed a capacity to expand easily and at an accelerating rate.

During this early phase the government played no role as a provider of information, subsidies or protection. The local representative of the Ministry of Trade did play an important role in streamlining processing of small-batch exports: whereas in around 1981 a small-batch shipment out of the Surabaya office in East Java was reported to take 5–7 days, the same could be achieved in Bali in 24 hours. Other than this, the government appears to have had little direct effect on the industry. In fact, the main contribution was a critical decision apparently made very early on by the Bali governor, on the recommendation of the Ministry of Trade representative, not to intervene to shut down the Australian entrepreneurs after receiving complaints that they were engaging illegally in business. From that point on, the difficulty of distinguishing between foreign entrepreneurs and tourists meant that the entrepreneurs could operate with a degree of freedom unheard of at that time in other locations in Indonesia.

By the end of the decade a garment production system was in place that was highly flexible, easily expandable, and able to handle increasing quantities of production with fairly quick turnaround times (at least fast enough to handle at least one round of reorders per season). Though the foreigners were in the beginning heavily involved in active management of production, over time they were able to move out of day-to-day involvement. Once the basic pattern for production and production expansion was jointly established, volume could expand quickly with minimal further technical inputs *as long as no further quality increase was necessary.*

As is often the case with 'first movers', the returns for the dozen or so main foreign entrepreneurs were enormous, but they also bore the greatest risk. The 1980 recession caught them entirely by surprise—most had no knowledge of business cycles (which underscores how little sophisticated business knowledge mattered in the early stages of the industry's start-up). In that year, the largest entrepreneurs had made huge expansion investments in raw material to get an early start on production to meet expected mid-summer reorders from Australia. The collapse of the market left them with large stockpiles of cloth and garments in Bali which subsequently had to be dumped on the tourist market at far below cost. To their credit, most of the Australians made good on their commitments to their Indonesian producers. Though the recession was devastating to the foreign entrepreneurs—none regained a major share of exports in the years that followed—their willingness to shield their local counterparts, in many cases from ruinous loss, probably helped the industry to survive this early disaster.

Second Phase of Growth: Intensification of Production

The garment producers were shocked by the impact of the recession on their foreign counterparts. Though they had moved a long way up the manufacturing learning curve, they were all still heavily dependent on individual alliances with the foreign entrepreneurs for production assistance and access to foreign markets. The collapse of many of the first phase foreign entrepreneurs left them ready for new foreign buyers. The gap was quickly filled by a new type of foreign entrepreneur with a more sophisticated set of business, production and design skills.

Throughout the 1970s tourist flows had been steadily increasing. Among these tourists were young Americans from the West and East Coasts with a heightened sense of design and looking for entrepreneurial possibilities. Picking up on the existence of the basic small-batch production system in place on the island, many returned with new design ideas or design skills, eager to penetrate small-scale boutique markets in the US. Success stories fuelled a rising flow of foreign buyers from the US and, to a lesser extent, Europe.

Creative, higher value designs required better workmanship and experimentation with new methods to achieve acceptable quality and quicker turnaround times. A great deal of precisely targeted consultation specifically tailored to a given firm's needs was therefore required. The new group of 'buyer–consultants' was willing to work closely and for months at a time with the producer firms, providing whatever information and assistance inputs were necessary to get their new designs transformed into a saleable product of acceptable quality and cost. High returns and the very pleasant social, cultural and physical environment in Bali made it worth their while to spend several months of the year working as 'unpaid consultants' to get a few hundred or a few thousand creative, high-value boutique garments produced. The high returns were, however, accompanied by high risk. Before any sales in the foreign markets could begin, the buyer–consultant had to expend the full up-front costs for product purchase—typically involving a down-payment, but always with full payment before export from Bali—and consulting (the living costs in Bali plus the opportunity cost of the time spent working with the producer). Success was never assured—few foreigners had more than two or three good years in a row before facing a very bad year, or even total loss. Producers, on the other hand, could steadily improve their scale and production efficiencies with minimal risk.

One of the most striking characteristics of the buyer–producer relationship during this phase was the intense application of tough post-manufacturing quality control standards. Despite the heavy involvement of the buyer–consultants in the production process, rejection rates at the last stage were commonly as high as 20–30%. The producers could tolerate this level because they could always dump export rejects on the rapidly expanding Bali tourist market. In fact, by the mid 1980s many of the larger producers had opened their own large retail shops in Kuta and Sanur for just this purpose. The producer firms would also, however, seek to absorb fully the improvements or innovations recommended by the buyer–consultant to bring reject rates down. As a firm progressively improved its capacity for workmanship and service, it would eventually 'out-grow' the individual buyer–consultant to which it was temporarily allied. When a new buyer–consultant entered with a higher value design, offering higher unit prices to produce it and willing to supply the more sophisticated skills needed to get the product out, the producer could, and often did, switch buyer–consultants. The overall effect, both at the firm and at the industry level, was a very rapid rise in the capacity of firms and a sharp upward rise in export unit values.

As the design and quality mix moved upscale, a process of firm specialisation occurred. Paralleling their assistance to garment producer firms, the foreign buyers became heavily involved in encouraging and advising the spin-off firms handling textile dyeing, and print specialists, including silk screening operations. A sub-industry of specialised and innovative button making also emerged in this phase.

One of the most important specialties that emerged was the packer/shipper cum exporter, who took on the task of expanding and allocating export quotas, especially for the US market. By the late 1970s a small number of local firms were specialising in packing and shipping, initially to serve tourists. Some had worked with the Australian entrepreneurs before the recession. They were expert at cutting through the complex Indonesian business export documentation and at dealing with the bureaucracy.[113] It was these firms that came to handle the equally thorny problem of US and European import restrictions. A quota allotment was required to ship garments legally to these markets. Each year the quota was allocated on the basis of a firm's export volume in the previous year (that is, *past performance*). Because the Bali industry was built around small-batch orders under highly flexible production scaling, producer firms had great difficulty maintaining and expanding their own quota levels. Instead, exporters served as the firm of record for documentary purposes. By stringing together many small orders for many foreign buyers under its own name, an exporter could assure itself access to at least that much garment quota allocation in the following year, thereby ensuring a market share for itself in that year. As an exporter's existing quota allotment for a given type of garment ran out each year, it would bid for additional quota in the quota markets in Surabaya and Jakarta. With unit values higher in Bali compared to what was being exported from the factories in Java, the Bali exporters were nearly always able to outbid non-Bali exporters. Given the easy access to export licences and the intensely competitive environment, exporting firms could not restrict trade and extract monopoly rents.

A remarkable array of innovative designs and techniques was introduced and widely copied during this creative second phase of growth, including the use of sequins and applique, combinations of textiles and leather in apparel and accessories, and experimentation with new colour schemes. It was during this period that many products now seen as 'traditional' Bali garments were created.

---

[113]The Indonesian state was not geared to facilitate manufactured exports during these years, and certainly not exports in numerous small batches.

As Bali's reputation as an easy and service-oriented production site grew, new buyer–consultants with increasingly higher value designs poured in. As competition among them increased, 'knocking off' the design innovations of competitors became a serious problem. The most critical point for foreign buyers was the one or two month lag between development of a prototype, production in an initial small batch for market testing, and full-scale order and reorders. As turnaround times got shorter, it became possible to copy an innovative design and actually beat the designer to the market with a full-scale order, if one moved quickly enough. When this happened, the producer firm gained a reputation as a bad risk.

Cooperation between buyer–consultants and local producers to address the problem took several innovative forms. A typical response was to break production down into several steps, with final assembly occurring in the core factory under tight supervision. More important from a social perspective, however, was the geographic spread of production. To hide design innovations, producers engaged rural village work groups more distant from the centre of the industry in southern Bali. By the mid 1980s production had dispersed all over the island. There were several unintended results. Employment opportunities were distributed more equitably. A new independent specialty in the form of 'collectors' (*pungumpul*) evolved out of the in-firm go-between role; their job was to handle links to distant village work groups. New logistical innovations and skills essential for managing dispersed production over great distances, while maintaining quality and delivery schedules, were developed. This made it possible by the mid 1980s to expand production into East Java with its virtually unlimited labour supply.

From an Indonesian policy perspective, one of the most important aspects of the rapid growth experienced in the Bali garment industry during this period was a rapid increase in *pribumi* ownership among producer firms. As the put-out and subcontracting networks expanded, the Chinese Indonesian owners of the original firms increasingly came to depend on employees who filled the middle manager function of linking village work groups with the core factory. This position was almost always filled by indigenous Indonesians. In the early 1980s, as a producer firm's reputation grew, new foreign buyers would come looking for a production source. The firm's owner would almost always be fully occupied with its one or two main buyer–consultants. Rather than lose the business to a competitor, however, the new buyers would be passed on to middle management. After working closely for a period, the buyer sometimes offered to assist the *pribumi* middle manager to set up his or her own manufacturing operation. The middle manager already knew most aspects of the business by then, and the new buyer–consultant would promise to supply whatever skills were missing. The buyer–consultant would then have access, at least temporarily, to a source of product responding specifically to his or her needs. The original producer firm's owner might be livid, but could do little. This pattern recurred frequently enough that the net effect was a rapid increase in new producer firms, nearly all of which were *pribumi*-owned. From 1977 to 1987, the share of total value added in the Bali garment industry generated by *pribumi*-owned producer firms rose from almost nothing to nearly 50%. Not surprisingly, the fastest rise occurred during the period of intense buyer consulting, from 1981 to 1986.

During this phase the Bali garment industry (and in fact other similarly organised industries on the island) matured into a somewhat unique *Bali model of production*. This model was based on small-batch production, a highly flexible labour pool using simple mechanical technology, a strong service orientation toward foreign buyers, and almost 100% local ownership.

Third Phase of Growth: A Return to Extensive Production

Maintaining a rise in unit values in the garment industry would have required at least three things: (1) continued inflow of innovative design inputs (that is, based on a better understanding of rapidly evolving consumer preference); (2) higher quality workmanship and better management; and (3) better quality material inputs (especially textiles). Since the Bali garment industry has primarily produced in response to orders from buyers, higher value design inputs were not a problem as long as the system could produce garments that were of acceptable quality and quantity and low enough cost. Steady improvements were being made in the quality and design of domestically produced textiles throughout the 1980s. The key factor in the industry's continuing ability to attract ever higher value design inputs for foreign markets would, therefore, be its ability to maintain steady improvements in the quality of products and service to buyers. Producer firms obviously could and did learn on their own, but that process was slow. Rapid improvement required types of information inputs that the firms could neither identify nor easily access on their own. The buyer–consultants were the only source of accurate knowledge about what quality improvements mattered most to foreign consumers, and how exactly to achieve those improvements given the producer firm's existing capabilities.

At the start of the second phase of growth around 1981, the returns to the buyer–consultant for small batches of unique 'boutique market' garments were reportedly very high. They were certainly enough to attract a growing number of increasingly sophisticated new entrants willing to invest months of consulting time with producers in Bali. However, as the system matured through mid decade, the ability to squeeze out higher quality through incremental technical, managerial and logistical improvements in the basic Bali system was getting harder and more expensive. The rise in costs for a buyer could only partially be compensated by an increase in the volume of orders, since the buyer, with limited time, could only provide a finite amount of quality control oversight.[114] The result was a slow squeezing of foreign buyer profit margins from below.

At the same time, producer firms were getting better at serving new buyer entrants. As barriers to entry fell and the word spread among garment producers, the flow of new buyers increased dramatically. Rising competition among foreign buyer–consultants for attention from the best producer firms weakened their bargaining power, and their profit margins began to be squeezed from above.

It is possible that constraints on the quality of raw materials placed a ceiling on unit values by mid decade, though this was probably not the case. Responsiveness of the large textile producers in Bandung and Jakarta to the needs of the Bali export industry was generally poor. However, these manufacturers were producing an increasing variety of better quality textiles in response to the demands of large garment producers on Java, and these new textiles spilled over into Bali. By the end of the decade it was also possible to access foreign textiles, though at high costs due to import

---

[114]In some cases, in the garment as well as other manufacturing sectors, foreign buyer–consultants established fixed base operations in Bali and trained local workers to do quality control under foreign management. In the 1980s, however, this option was risky since it meant establishing a workplace and hiring and training local workers for an operation that could at any time be interpreted by officials as a 'foreign-owned business'. The costs of continuing at that point would rise substantially. The great reduction in the cost of setting up a legal PMA (foreign investment) business since 1994 may become a factor in any renewal of the industry in the late 1990s.

tariffs.[115] Lack of access to high-quality textiles on the international market was therefore probably not a key factor constraining unit value ceilings in Bali (though it may explain in part why, in the 1990s, Bali has had difficulty in breaking into a higher value cycle of design, workmanship and material).

In any case, as margins thinned the ability of buyer–consultants to absorb the costs of remaining in Bali for months at a time to help firms produce their designs declined. Without the flow of consultant inputs, quality and workmanship improvements halted. Between 1985 and 1987 the steep rise in unit value came to an abrupt halt at an average of around $3 per kilogram (1988 dollars). Adjusting for inflation, this average export unit value has not risen substantially since. From that point on, there have been few real innovations in the garment designs coming out of Bali.[116]

---

[115]In theory, the duty drawback system in force in the late 1980s became available for use with textile imports. However, the time costs due to bureaucratic complexities, and the fact that refunds were much delayed and reportedly always well below the amounts owed, reduced its value tremendously. Moreover, the general lack of public knowledge about the duty drawback scheme and how it worked meant that a few larger volume firms could quietly use the system to reduce their costs and improve their competitive position, while the many medium and small-scale firms could not.

[116]The proximal cause for the exit of many American buyer–consultants in 1985, and therefore part of the explanation for the sharp flattening in unit values after that year, was a problem with the Immigration Office. Most buyer–consultants were entering Indonesia on business or social visas. To work officially as a paid employee one needed a permanent stay visa (KIM/S) and a work permit. In 1985, new immigration officials arrived in Bali and immediately discovered that by interpreting the foreign buyer's status as 'consultant', an opportunity for extracting large rents was created. Under that interpretation, the absence of wage compensation notwithstanding, the foreigner did not have the proper visa and work permit. Faced with the choice of very large and immediate payments or immediate deportation, most had no choice but to comply. This practice was ended less than a year later and the officials responsible were transferred. But the damage had been done and many of the American buyer–consultants did not return the next year. The problem was widely discussed among expatriates in Bali at the time, and the news may well have discouraged other would-be buyer–consultants. While this development was very negative, it is important to keep in mind that buyer–consultant profit margins were thinning anyway. Thus the large and unexpected payments to immigration officials simply hastened the end of information flows that would probably have ended anyway within another year or two.

As noted in the section on export growth, this flattening of unit values did not slow growth in total export values, at least initially. This was because the rush of new foreign buyers to Bali and the expansion of production into East Java ensured that the *volume* of exports continued to rise. With minimal ability to increase quality, this volume increase involved garments with relatively minor variations on designs and techniques introduced before 1987. Since these were no longer unique and exciting new designs hitting higher end markets for the first time, retail prices on the importing country end inevitably began to fall. Foreign buyers naturally adjusted to thinning margins by increasing order volumes. It was not long before discount retailers from the US (such as the Price Club) were placing very large orders for what had become 'traditional' Bali garments.

As producer firms pursued larger volume orders from distant buyers, quality control eroded and a non-virtuous cycle began. By the early 1990s Bali was acquiring a reputation for poor quality and unreliability that further eroded the willingness of high-value designers to chance working with the local industry. Compounding the problem, larger buyers began to insist on the release of letter of credit (LC) payments on arrival, and inspection in the country of import rather than at the point of shipment (fob) as had always been the practice in the past. This was a critical shift in the burden of risk, and some producers began to experience rejection of product on the import side, with buyers citing poor quality as the cause. Some producers claimed manipulative practices and breach of promise on the part of foreigners. There had, of course, always been some friction in individual relationships, but the decline in extended, face-to-face cooperation between Indonesian producers and foreign buyers seems to have opened the door for more unethical practices. At the very least, the ease with which orders could be placed by foreign buyers with no knowledge of local capacities or limitations must have increased the frequency of misunderstandings. Suspicions peaked with the onset of the US recession in 1990, when the number of rejected shipments was reported to have risen dramatically and several Bali garment producers were severely hurt.

A few Bali-based producers tried to address the problem of thinning margins and increased volume, while avoiding problems of quality decline, by moving to large factory production using high-speed industrial stitching machines. It is not clear why this approach did not start a new cycle of rising unit values, but the most likely answer lies in the substantial increase in unit costs and the inflexibility of production, which failed to attract higher end foreign design inputs. First, it required heavy fixed capital investments in construction and imported machinery, and imposed high variable costs for electricity and fuel for production, neither of which was necessary under the put-out system where each worker used his or her own low-cost, low-technology, foot-treadle machine. Second, it required the hiring and training of a permanent, inflexible workforce which, under Indonesian labour law, could not be laid off during the off-season. Since Bali industry production has always been for a seasonal market (spring and summer clothing), factory production incurs much higher total labour costs per unit than the dispersed village work groups working under short-term contracts.

One development that seemed to promise the start of a renewed cycle of new designs, better quality and higher unit values was backward integration into manufacturing by Indonesian garment designer cum tourist retailers. By the mid 1980s, several Balinese and an increasing number of Javanese and Jakartans with creative design skills were opening retail shops for the tourist market in Kuta.[117] Garment production typically took place in a back room of the shop. Direct feedback from tourist consumers led to more marketable designs. As these shops were 'discovered', they might eventually get orders from buyers for larger quantities for export. While this source of Indonesian-generated new design was important and provided high returns for some entrants, it ultimately failed to start a new cycle of rising unit values across the industry. In part, this was because the Bali industry by the mid 1980s required design inputs of a level of sophistication and international market savvy that was beyond what local designers could produce. More important, the indigenous designer–entrepreneurs did not have the skills needed to move the quality of production significantly higher. Ultimately, this new source of innovative design failed to produce anything fundamentally innovative enough to excite the international boutique markets, as the new designs had done in the early 1980s.

Starting in the mid 1980s, several government agencies (particularly the Ministry of Industry) began to provide sporadic assistance

---

[117]Many creative young people, particularly from the Yogyakarta/Solo area of Central Java, were discovering Bali in this period. The free-thinking, international and open-market environment, especially around Kuta, not only allowed a greater freedom of creative expression, but rewarded it well. The result was an unexpected explosion of entrepreneurship that equalled or exceeded the changes that were occurring among young Balinese.

to the industry, almost exclusively in the form of worker training. These programs probably involved a few hundred to a few thousand workers in very short-term instruction. By then garment production had spread all over the island, and so the impact on industry development was negligible—a point confirmed repeatedly by producer firms. In any case, no government-provided assistance addressed the central problem of how to generate new flows of high-quality information and assistance to producer firms to keep unit values rising.

In a subtle way, the Bali production model had shifted from the small-batch, high information input and high-service industry of the early to mid 1980s to a larger batch, low-information, limited service industry. In this it was becoming increasingly similar to garment manufacture in other low labour cost countries around the world. It is not surprising that, by the early 1990s, Bali producers were reporting an increasing loss of business to China, Vietnam and the Philippines, and even to large factory producers within Indonesia itself.[118] The head of steam built up through production innovations and design ideas by the mid 1980s drove the expansion forward for several years more, but this was largely fuelled by running down the value of Bali's 'design capital'. Continued extensification of the system at stagnant unit values could not be sustained, and growth gradually decelerated in the 1990s.

Maintaining profitability and managing risk under higher unit value production requires better control over inventory levels and input delivery schedules, improved production scheduling and overall coordination, efficient methods for training workers, improved protection of new designs from competitors, and improvement in quality control throughout the production process to reduce reject rates. To be successful, the Bali garment industry needed to make more or less simultaneous improvements on all these fronts. Making that leap in production sophistication would have required substantial new inputs in management, technical and logistical information, and innovation. The increased costs of this more advanced production system could only be covered by higher unit values, and these in turn would come only with a parallel leap to better design inputs and raw material for designer label markets. The Bali

---

[118]Very low-cost, low-quality copies of Bali-style embroidered apparel were being produced in factory settings in East Java by the late 1980s. This could only have the effect of hastening erosion of the appeal of these products in foreign consumer markets.

garment industry has never made it over that 'innovation hump', and consequently is foundering in the mid 1990s.[119]

SOME TENTATIVE POLICY LESSONS

One of the most striking characteristics of the Bali garment export story is its simple, unsophisticated origins. This case clearly shows how quickly a new, labour-absorbing industry producing for a highly competitive international market can become established and grow with very minimal initial endowments. The only necessary condition for start-up was that foreign entrepreneurs be in sustained and unsupervised contact with local entrepreneurs. When the foreign entrepreneurs discovered an opportunity for high returns in their home country market, they were able to draw on their relationships and knowledge of conditions in Bali to obtain product for that market. The ongoing interaction of the two sides in a relatively open market naturally started a cycle of technological improvements and learning that was self-replicating and largely self-financing, and which resulted in a spectacular rate of sustained growth over the next 20 years.

Information transfer and assistance provided by foreign buyers achieved a level of efficiency and accuracy unimaginable through any other mechanism. The specific assistance in the production process that was offered at each stage of a producer firm's development was *precisely* and *only* what was appropriate for improving production quality and quantity at that level. Translated into the language of business support programs: the assistance was provided on a for-profit basis; it was tied specifically to tangible product output results; the provider of the assistance received no compensation unless the assistance was successful; and the firms targeted for assistance were those with the best potential and a

---

[119]It is also possible that there are inherent limits on the dispersed, village-based production system that makes up the Bali garment industry. If so, no amount of incremental innovation would overcome these. However, a similar system forms the backbone of production for several advanced industries in northern Italy (including the high-fashion industry). These industries have achieved a high level of technical sophistication and their success suggests that there may be no *inherent* limits to movement up the technological improvement curve in the Bali production model. The problem in the Bali garment industry is that it has simply been unable to get 'over the hump' into the sustained cycle of improvement that Italy has achieved.

demonstrated willingness to absorb assistance inputs. There is surely no other source of assistance that would be more accurate and timely, and certainly no mechanism for delivering it that would involve more performance-based incentives for the provider.

Especially during the 1970s and 1980s, Indonesian observers often complained that the mark-ups on garments retailed in the importing country (up to eight times the purchase cost in Indonesia) were outrageous and unfair. Yet it was precisely these high returns that underwrote the technical consultation that buyers provided to producers and the assumption of nearly all risk by the foreign buyers. It is virtually certain that substantial intervention by the authorities to protect what might have been viewed as local interests, and to secure a higher share to total returns for the producers, would have killed the golden goose. Rapid growth was maintained precisely because government rarely intervened to 'correct' perceived problems. An open and highly competitive market was left to sort out opportunities, prices and incentives. A hands-off approach to conflict resolution allowed firms and individuals to work out long-term trust relationships, with information transfer and cooperative innovation being a natural consequence.

The history of the Bali garment industry suggests that there may be a point (or points) along the development path at which continued upward movement through a cycle of production experience, buyer-assisted learning and innovation, and higher unit value production may falter. With the decline in intensive, long-term, buyer-provided inputs after 1986, the industry had to fall back on self-generated learning and whatever assistance government programs had to offer. Neither source of new knowledge has been sufficient to push the industry forward to a new technological level and to higher unit values. A costly dose of new ideas, technology and innovation is required to help firms break through to higher value markets. The key questions are: who can best provide it, and how is it be financed?

There is no simple answer to these questions. Nonetheless, the experience of the Bali garment case tells us that: (1) the inputs needed to drive cycles of technological innovation are best provided by the private sector through networks of strategic alliances and on the basis of mutual profitability; (2) healthy cycles of growth under conditions of rapidly improving technology almost certainly require sustained inflows of information, technical knowledge and assistance with innovations derived ultimately from advanced economies; and (3) sustained information inflows channelled through private sector alliances take place only where markets are left as open and competitive as possible, and when interfirm negotiations and the building of long-term trust relationships are left unhindered and unsupervised.

What has been the role of government in all of this, and what is the appropriate role of government in similar light export industries elsewhere in Indonesia? Throughout the development of the Bali garment industry, the role of government has been critical, but limited and indirect. Steps taken by the local representative of the Ministry of Industry and Trade to streamline the export process were critical early on. Tourism, driven in part by government policy, obviously made initial contact with foreign entrepreneurs possible, and tourism markets were essential as a site for dumping export rejects. There is also no question that Ngurah Rai International Airport, built to facilitate tourism, made exploitation of the rapid turnaround, high-value order niche possible. The provision of a relatively stable macroeconomy and a stable sociopolitical environment was clearly the responsibility of government. There were a few points (visa problems in 1985 and problems with the duty drawback system) at which poor implementation of regulations actually created constraints to growth, but these instances seem to be less critical when viewed in the context of the industry's overall success.

Beyond these points, the role the government played seems more positive in its absence than in its actions. The forbearance of the provincial government (consciously intended or by default) was a critical factor in the success of the industry. Local government took an essentially tolerant, hands-off attitude to the development of the garment industry, despite what may have appeared to be highly undesirable characteristics during the early stages (the high-profile role of foreign entrepreneurs, the highest returns clearly going to the foreigners, minimal initial ownership by *pribumi* Indonesians). At any point government might easily have stepped in more forcefully to 'direct' growth of the industry or to correct perceived shortcomings or inequities, as has often happened elsewhere in the country (the most recent case being the deportation of 100–120 foreigners involved in the booming Jepara furniture industry).

Probably the strongest evidence for the importance of government forbearance is the fact that Bali is the only location in the country that has given rise to so many new export industries of this type, and did so beginning in the 1970s and 1980s. In the Bali garment case, heavy intervention would almost certainly have put an early end to the industry's spectacular growth. Perhaps the greatest challenge for government in its efforts to carve out a positive role in light industry development is to find regulatory and service solutions that stimulate cycles of self-generated technology creation within firms and industries, and that do not in any way hinder the inflow of foreign information and technology transfer that is essential for sustaining rapid growth.

15

TECHNOLOGY AND HUMAN RESOURCES IN THE INDONESIAN TEXTILE INDUSTRY

Rick van der Kamp, Adam Szirmai
and Marcel Timmer*

INTRODUCTION

The textile industry in Indonesia has experienced rapid growth over the past two decades. Between 1975 and 1993, gross value added increased over 16 times and the number of employees in the industry more than tripled. Since the mid 1980s, when the Indonesian government sought to lessen Indonesia's dependence on primary exports by promoting non-oil manufacturing exports, growth in output and value added in textiles has been exponential. The textile industry now ranks as the largest manufacturing employer. In 1994 it produced 16% of Indonesian manufacturing value added.

GDP per person employed in the industry increased by 17% per year between 1975 and 1993. In comparison with the US, the leading manufacturing economy, relative labour productivity in the textile sector increased from 7.4% of the US level in 1975 to 17.4% in 1993,

*This chapter is based on Rick van der Kamp's Masters thesis, 'Technology and Human Resources in the Indonesian Textile Industry', submitted to the Eindhoven University of Technology. Research for the thesis was carried out between September 1996 and April 1997 at the Centre for Research on Technology at the Institut Teknologi Bandung in Indonesia.

indicating a rapid process of catch-up and technological dynamism (Szirmai 1994).

This chapter investigates the role of technological progress in the Indonesian textile industry, using a 'narrow' definition of technology to refer to technological change embodied in machinery. Because of data constraints, the analysis has been restricted to spinning and weaving. Within these sectors, our focus is on the core technologies in use in the production process: spinning frames in spinning, and weaving looms in the weaving sector. Information on the machinery park is used to create a quality-adjusted index of capital inputs, which will also be used in an analysis of the role of technological progress in economic performance.

The Indonesian textile industry is increasingly integrated in the world market. It cannot indefinitely continue to rely on low-cost, labour-intensive production, as it faces competition from other producers with even lower wage costs. Continued export success will therefore require an increased emphasis on quality and the modernisation of production techniques. A corollary is an increased need for human resource development (HRD). Using secondary statistics on education, interviews and data collected on field visits to textile firms, we discuss trends in human capital and HRD. As in the case of technology and capital, these trends are quantified in indices and used in a growth accounting exercise.

PRODUCTION TECHNOLOGY

In the literature on production technology, there is a tendency to include virtually all aspects related to the production process in the definition of technology. For instance, UN (1989) distinguishes four dimensions of technology: hardware; 'humanware' (human resources); 'orgaware' (organisational processes); and 'infoware' (the control of information flows). However, there is no consensus on the broad definition of technology, and the measurement of changes over time in multidimensionally defined technology is very difficult. In this chapter we employ a 'narrow' definition of technology, referring essentially to the hardware part (machines), or to capital-embodied technology.

If technology is thus defined, it becomes possible to operationalise it with a relatively simple set of variables. Focusing on the machines used in the production process, technology can be measured by the maximum speed, or maximum physical output, per unit of time per unit of machinery. However, it is necessary to limit our scope to certain machines within the capital stock, because the machine park of an industrial sector comprises transport vehicles, construction machinery and other machines not directly involved in the production process. We have therefore chosen to focus on core machines in the production process. These core machines, or core technologies, are usually characteristic of a certain production process, and in many cases take up the bulk of the sector's investment in machinery. Pack (1987) showed that in spinning, the spinning frames account for 56% of investments in machinery. In weaving, this percentage is probably even higher, given the limited number of stages in the weaving process. We assume that non-core machinery and fixed structures have to move in step with core machinery.

Within an industrial sector, we are now able to measure 'technological progress' by examining the performance of the core machines (measured by maximum output, or productivity, per unit of machine) over time. Given that there are a number of alternative core technologies in a sector, the above-mentioned procedure introduces two distinctly different aspects of technological progress. The first of these is the changing composition of the machine park of an industrial sector. If different types of machines differ in their level of productivity, the productivity of the machine park will change over time as the share of each machine type in the park changes. This is referred to as the *shift effect*. The second aspect is the changing productivity of one type of machine over time due to technological improvements in its design, speed or other features. Younger vintages of machinery are more productive than older vintages. This will be referred to as the *advance effect*.

Using this framework to analyse and measure technology will result in an index of technological progress. By taking a sectoral approach, we have been able to look at the capital-embodied technology from a more technical point of view. In the next section, the technologies used in the Indonesian textile sector will be described, from both a technical and economic viewpoint. We will also attempt to construct a technology index that can serve as an indicator for the level of technology in the Indonesian textile sector, and which can be used in an analysis of the role of technology in economic performance. As stated earlier, we will confine our discussion to the two largest sectors in the Indonesian textile industry, spinning and weaving. These sectors are the ones most commonly used in research into the industry. In 1994, they accounted for 76% of value added and 69% of employment in the textile sector. The technologies they employ are described in more detail in van der Kamp (1997).

Spinning

The process of spinning refers to the transformation of raw fibres, natural or synthetic, into yarn. Historically, cotton has been the dominant natural fibre, with wool playing a smaller role. Over the last 20 years synthetic fibres have gained in importance, to make up almost half of the fibre consumed in the spinning sector in 1994.[120]

The transformation of fibres into yarn can be subdivided into a number of stages. During *opening*, the bales of raw fibres are opened up and cleaned, blended and separated. The *carding* stage attenuates the laps of fibre into slivers, removing short or immature fibres. During the *drawing* process, the slivers are drawn out to obtain longer, thinner slivers. *Roving* continues this thinning process, preparing the slivers for the spinning stage. During *spinning*, slivers are transformed into yarn by drawing out and twisting, with the thickness of the yarn called the yarn count. To obtain the various end products of different yarn count and structure, several yarns can be twisted together to increase strength during *twisting and winding*. Increases in the productivity and speed of many of the side technologies mentioned here have been impressive; UNIDO (1979) reports improvements of 400% and over for carding and drawing machines.

Yarn producers can choose among several spinning technologies: ring spinning, open-end rotor spinning, air-jet spinning and open-end friction spinning. At present only ring spinning and open-end (rotor) spinning are used on a large scale in Indonesia. (One mill has installed an air-jet system but is not yet producing yarn with it.) One reason for the failure to install the latest spinning systems—air-jet and open-end friction spinning—is that they are still being developed by manufacturers, and as such are not always 'ready to use' without further R&D.

Ring Spinning

Ring spinning is an old technology, first patented in 1828. Nevertheless, it is still the most popular spinning technology, in

---

[120]In this paper we exclude the production of synthetic fibres, which is part of the chemical products sector.

Indonesia and in many other parts of the world. The spinning is carried out on spinning frames, each of which may contain up to 500 separate spindles. Ring spinning machines further attenuate the slivers coming in from the roving process by roller drafting until the required draft is obtained. Simultaneously, the slivers are twisted to gain strength. This process requires two types of labour: spinners, who set up the roving supply (output of the roving process); and doffers, who remove the filled bobbins from the spinning frame and replace them with empty ones.

Although the basic technology has remained unchanged, improvements in productivity (mainly in speed) have been achieved over the years. UNIDO (1979), for example, reports an increase in speed of about 30%, although no time interval is given. It estimates a maximum spindle speed of 15,000 revolutions per minute (rpm), while Ishida (1991) gives an estimate of 25,000 rpm. These improvements in productivity are modest, however, compared with the speed increases achieved for the side technologies (opening, carding, drawing, roving) mentioned earlier.

Open-End Rotor Spinning

Open-end rotor spinning is one of the radically new methods developed in the 1960s and 1970s to obtain yarn. Rather than the whole bobbin being rotated, in open-end rotor spinning an open end of the yarn is rotated around the axis of the yarn, with the twist insertion being performed by a rotor. The main advantage of this system is that greater rotor speed can be obtained—in 1991 maximum speed was estimated at 120,000 rpm (Table 15.1). Another benefit of the technology is that the roving process can be entirely skipped.

An important disadvantage is the limitation on yarn count (yarn thickness). Open-end systems perform adequately for thicker yarn, but become increasingly vulnerable for yarn of larger count (thinner yarn). This has been an important factor in the limited use of open-end systems in Indonesia. In fact, most such systems are used in the production of 'jeans yarn', that is, yarn used to weave denim cloth. Use of open-end rotors in recent years has thus varied with the popularity of denim among garment producers.

Although open-end spinning is clearly much faster than ring spinning, this difference in speed cannot be used directly to compare the productivity of the two systems. Whereas the ring spinning speed refers to winding and twisting, the rotor speed in open-end

TABLE 15.1 Comparison of Spinning Technologies

	Ring spinning	Open-end spinning
Maximum speed (rpm)	Up to 25,000	Up to 120,000
Maximum output (index)[a]	100	442
Labour requirements (index)[a]	100	74
Productivity (output per labourer, index)	100	598
Investment costs (per spinning unit, 1980 $)	157	2,400
Lifespan (years)	30	30
Product specialisation	All yarn	Mainly lower count yarn

[a]Per spinning unit.
Sources:
Speed: Ishida (1991).
Output and investment costs: Acero (1984).
Lifespan and output specialisation: expert interviews.
Labour requirements: Acero (1984) and expert interviews.

spinning reflects twisting speed only. Since winding speed refers to the speed at which yarn is wound onto bobbins, this is obviously a better indicator of the productivity of both systems. Progress in speed and productivity has been made: 90,000 rpm was mentioned as a top speed in 1987, although in a different context to the 1991 estimate of rotor speed mentioned earlier.

Comparing Spinning Technologies

Table 15.1 compares the two technologies in terms of their speed, output, labour requirements, productivity, investment costs, lifespan and product specialisation. The difficulties in making this comparison become apparent when looking at product specialisation, which differs slightly. Maximum output data have been compared at a yarn count of 10s, since open-end performance deteriorates at higher (thinner) yarn counts while ring spindle performance remains relatively stable over a range of yarn counts. Also of interest is the large difference in ratio between investment costs on the one hand and productivity (maximum output) data on the other for the two technologies. Part of this difference can be explained by the fact that the data are from around 1980, when open-end rotors were, at least in developing countries, still rather new and therefore expensive. Failure to take into account the omission of the roving process with open-end rotors may also help explain their higher costs.

Spinning Systems in Indonesia, Past and Present

The spinning industry in Indonesia is not as long-established as the weaving sector. It was not until the 1930s that a small industrial spinning sector started to develop; and it was only at the end of the 1960s, when the New Order government facilitated the importation of capital goods, that the sector began to grow rapidly. The policy liberalisation packages of the mid 1980s provided a further impetus to the growth of the industry (Hill 1992). In 1968, some 481,780 ring spindles were in use in the spinning industry; 10 years later this number had multiplied by more than three times. In the 1980s growth was even faster, and in a few years time the total number of spinning units will probably be 20 times as high as in 1968.

The first data on open-end systems in Indonesia go back to 1976. At that time, some 10,216 open-end rotors were involved in the production process, while 1.24 million ring spindles were being used. Growth in the numbers of open-end units has never exceeded ring spindle growth; in fact, the total number of open-end systems exceeded 1% of the total number of ring spindles only in 1984. It will be a long time yet before the technologically superior open-end rotor overtakes the ring spinning unit in popularity. Its deteriorating performance with increasing yarn count and its subsequent specialisation in thicker yarn, together with its larger investment costs, are important reasons for the modest role played by open-end systems in Indonesia. In any case, ring spinning is by no means an obsolete technology: continued improvements in speed and in the various side technologies used, together with its 'universal' applicability, are important factors explaining its continued popularity.

Weaving

Weaving refers to the process by which yarn is transformed into woven cloth. The simplest form of weaving is achieved by raising alternate warp yarns and inserting a pick (that also carries yarn). The raising of the ends is normally done by heddles that are controlled by a loom driving mechanism. Yarn used in the weaving process goes through a number of preparatory stages. Technically, *twisting and winding* are among these stages. In *warping and sizing,* the threads are arranged in long lengths parallel to one another preparatory to further processing. During *looming,* the loom warps are prepared for the actual weaving process. The looming process has long been an expensive and time consuming activity, although recent developments have made it somewhat faster. For the *weaving* process itself, several technologies are available. In Indonesia, three main generations of technology have been used over the years.

More so than with spinning, developments in weaving have focused on the core weaving technology; this was where major improvements could best be achieved given the limited number of processes in the weaving cycle. According to Hill (1983), developments have been rapid enough to justify the term 'technological revolution'.

Mechanised Handlooms

As early as the 1920s, a textile education and research institute in Bandung (now called STTT Bandung) played a key role in the upgrading of traditional handlooms. By mechanising part of the shuttle changing of the loom, a faster machine was developed, though it was still operated by hand. In spite of its reliance on manual power and its high employer:loom ratio (1:1), the partly mechanised handloom (*alat tenun bukan mesin,* ATBM) was the first piece of weaving equipment used for industrial weaving activity on a broad scale.

Shuttlelooms

Shuttlelooms, or power looms, have been standard technology in many countries since their development in the early 1900s. The use of a shuttle has always been the most important limiting factor in loom speed, since the kinetic energy of the loom equals one-half of its mass multiplied by its squared velocity, and doubling the speed therefore requires quadrupling the energy transferred to each shuttle move. This is one of the most common technical problems of the shuttlelooms. Shuttlelooms have existed for some 65 years, with various types being developed during that time.

Shuttleless Looms

Because of the limitations of shuttlelooms in speed and thus productivity, producers of weaving equipment focused on the development of a loom that was not hampered by shuttle use. This resulted in the development of machines using a variety of shuttle replacing techniques, mostly in the 1950s. *Rapier looms* use a rigid or flexible rapier which reciprocates in and out of the shed. Apart from this the weaving principle remains intact. *Jet looms* are of two basic types: *air-jet looms* and *water-jet looms*. The principle of this technology is the ejection of a metered length of weft from a nozzle, with the nozzle ejection being performed with jets of air or water.

Progress has been made in the maximum speed and yarn insert of shuttleless looms, but a comparison is difficult. Sources specify differences of up to 30%, but it is unclear whether they are referring to exactly the same type of shuttleless loom.[121]

Comparing Weaving Technologies

In our comparison of weaving technologies (Table 15.2), the focus is on speed, yarn insert, output, labour requirements, productivity, investment costs, lifespan and product specialisation. Shuttlelooms, the prevalent technology in Indonesia for many years, are considered the standard and hence take a value of 100 in the indices presented in Table 15.2. In the final equations for productivity and technology indices, the shuttleless looms will be taken as one technology, despite differences in technique and achievements. This is done because of lack of detailed information on the various types of shuttleless looms, because their output figures are in the same range, and because this simplification has not been uncommon in the literature.

Weaving Systems in Indonesia, Past and Present

Weaving technologies in Indonesia provide an excellent example of the transition of different generations of technologies, as illustrated in Figure 15.1. The partly mechanised handlooms, the ATBMs, were the dominant technology from 1930 to 1970. They were then quickly

---

[121]For instance, the maximum speed of air-jet looms is 600 picks per minute (ppm) according to Ishida (1991), but 500 ppm according to Departemen Perindustrian–Bank Indonesia (1989).

TABLE 15.2 Comparison of Weaving Technologies

	Mechanised Handlooms	Shuttle-looms	Shuttleless Looms
Maximum speed (picks per minute)	25	185	576
Yarn insert (metres/minute)	36	265	1,599
Maximum output (M2 per loom, index)	14	100	603
Labour requirements (operators per loom, index)	1,450	100	59
Productivity (maximum output per employee, index)	1.0	100	1022
Investment costs (1993 $)	100	2,500	41,250
Lifespan (years)	40	30	10 (excluding scrap value)
Product specialisation	–	–	Water-jet only for synthetic fibres

Sources:
Speed, yarn insert and output per loom: Departemen Perindustrian–Bank Indonesia (1988).
Labour requirements: Hill (1983); interviews with experts.
Lifespan and product specialisation: interviews with experts, Ishida (1991).

replaced by power looms, such that by around 1980 they contributed only marginally to industrial production. However, with increasing exports of textile products and pressure from international markets starting in the mid 1980s, the Indonesian weaving sector began to invest in the latest generation of weaving technologies, shuttleless looms. These looms have already surpassed shuttlelooms in terms of production capacity, and the replacement of the older shuttlelooms by high-tech shuttleless looms is expected to continue at a rapid pace.

Constructing a Technology Index

Deriving one time series containing all technological developments in a particular industrial sector necessarily requires making

FIGURE 15.1 Weaving Systems in Indonesia, 1930–95[a]
(no. of machines)

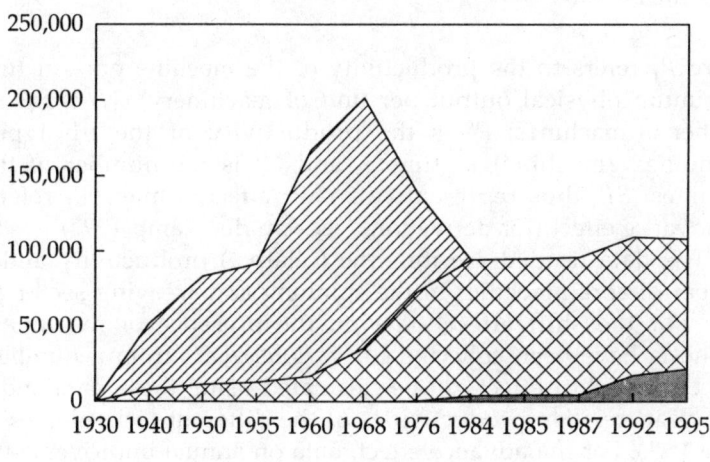

ATBMs (mechanised handlooms)

Shuttlelooms

Shuttleless looms

[a]Note that the horizontal axis employs a variable scale.

Source: Media Tekstil (1988); PT Capricorn Indonesia Consult Inc. (1995), 'Perkembangan dan Prospek Industri Pemintalan di Indonesia' [Development of and Prospects for the Spinning Industry in Indonesia], *Laporan Bisnis: Indotextile*, No. 12, August; Departemen Perindustrian–Bank Indonesia (1987).

assumptions about various technology-related aspects. To summarise, we have assumed that developments in the core technologies of the 'machinery and equipment' segment of capital goods in the Indonesian spinning and weaving sector are representative for all technological developments in the sector. The correct name for the index that we will derive would therefore probably be 'index for the maximum productivity of the machine park', which we will refer to as the technology index. It can be shown that our technology index can be represented by:

$$I_{Tech,t} = \frac{P_t}{P_0} = \left[\frac{N_0}{N_t} \frac{\sum P_0^i N_t^i}{\sum P_0^i N_0^i} \right] + \left[\frac{N_0}{N_t} \frac{\sum N_t^i (P_t^i - P_0^i)}{\sum P_0^i N_0^i} \right]$$

or

$$I_{Tech,t} = SE_t + AE_t$$

where P_t refers to the productivity of the machine park at time t (maximum physical output per unit of machinery), N_t is the total number of machines, P_t^i is the productivity of the ith type of technology (machine) at time t, and N_t^i is the number of those machines. SE_t thus represents the shift effect, while AE_t refers to the advance effect (for derivations, see van der Kamp 1997).

The data needed are thus the (relative) productivity and the number of core machines in the spinning and weaving sector over time. For the shift effect, data on the presence of machines in Indonesia have been collected for various years from a number of sources. The shift effect can be calculated using these data and the 'base year'·relative productivities of the different technologies (see Table 15.1). For the advance effect, data on annual improvements in the technologies are also needed. Since the information available was not sufficient to provide a good annual estimate, we have chosen to estimate the 'rate of advance' as a fixed annual percentage, based on the empirical data available. This can be expressed as:

$$P_t^i = P_0^i * \left(1 + \frac{A}{100}\right)^{t-t_0}$$

where A represents the (sectoral) advance effect as a fixed percentage, and t_0 is the base year.

Since technological developments in the spinning sector have, to a large extent, concentrated on increasing the speed of the spinning frames, and on the side technologies of the production process, its advance effect is estimated to be higher than that of the weaving sector. In our analysis, we have used annual estimates of 3% for the spinning sector and 1% for the weaving sector. These estimates are based on changes in spindle speed, interviews with experts, changes in productivity of non-core machinery in weaving, and the literature (for more detail, see van der Kamp 1997). The estimates of the advance effect are of necessity rather crude, but they are the most plausible set of figures available in the circumstances. Our data are presented in such a fashion that the reader may exclude the advance effect estimates in considering the results suggested by the technology index. Table 15.3 presents the resulting indices, taking 1976 as a base year.

TABLE 15.3 Technology Index for the Spinning and Weaving Industry, 1940–95
(1976 = 100)

	Spinning Index			Weaving Index		
	Contri-bution of Shift Effect	Contri-bution of Advance Effect	Contri-bution of Shift Effect and Advance Effect	Contri-bution of Shift Effect	Contri-bution of Advance Effect	Contri-bution of Shift Effect and Advance Effect
1940				45	−14	31
1950				42	−10	33
1960				38	−6	32
1968	97	−21	76	48	−4	44
1976	100	0	100	100	0	100
1984	101	27	128	201	17	218
1987				205	24	229
1989	99	46	145			
1992				298	53	351
1993	100	65	165			
1994	100	70	170			
1995				332	71	403

Sources:
Productivity data: see previous tables.
Advance effect: see text.
Number of core machines:
Spinning: Media Tekstil (1995), *Spinners Guide 1995*, Jakarta; Media Tekstil (1988), *Industri Serat Sintetis Indonesia*, Jakarta; Media Tekstil (1991), *Spinners Guide 1991*, Jakarta; PT Capricorn Indonesia Consult Inc. (1995), 'Perkembangan dan Prospek Industri Pemintalan di Indonesia' [Development of and Prospects for the Spinning Industry in Indonesia], *Laporan Bisnis: Indotextile*, No. 12, August.
Weaving: As for Figure 15.1.

The table shows that there are enormous differences between the spinning and the weaving industry with respect to the nature and rate of technological change. The data on textile technologies reveal that, while a mild 'technological revolution' has taken place in weaving, progress in the spinning industry has not been embodied in the replacement of older generations of technology (shift effect). The standard technology in spinning, the ring spinning frame, is indeed still overwhelmingly present in Indonesia, and this is not expected to change in the short to medium term. The estimated advance effect for spinning of 3% is in fact the only important source of technological progress. Technological progress in spinning is now some 94% in 26 years, whereas it is only 3% without the advance effect. For weaving, the equation used to calculate the role of the advance effect draws attention to the rapidly increasing importance of the advance effect as the shift effect accelerates. The reason for this is that when advance effects are embodied in more productive capital goods, the increase in productive capacity will be larger than when these effects are embodied in inferior capital goods.

HUMAN CAPITAL AND HUMAN RESOURCE DEVELOPMENT

In this section we take a look at some of the important factors that determine the nature and direction of technological progress. In the economics literature, much attention has been devoted to the role of education, on-the-job training and other aspects related to the concept of human capital. Traditionally, human capital (or human resources) is defined as the stock of skills and knowledge embedded in an individual, or labour force, that supplies a productive input into the economic process. It can be understood intuitively that the level of human capital plays an important role in the successful adoption of new technologies. Starting with Denison (1967), many scholars have investigated the role of human capital in economic growth, usually taking educational attainment as an indicator for this level. In our analysis we take the average number of years of education per person employed in total manufacturing as a proxy for human capital in the textile labour force. From 1975 to 1995, the average rose from 3.21 to 6.81 years, an increase of some 110%. Weighted with the income differentials of the various educational attainment groups, this resulted in an index growth of 57 percentage points (see van der Kamp 1997, Table IV.3).

The managers and experts interviewed claimed that the role of education in improving the level of human capital was of increasing importance. Firms have started hiring better educated personnel in the wake of technological modernisation.

Human Resource Development

Whereas statistical data on (national) levels of education are often available, information on HRD or on-the-job training efforts in economic sectors is frequently hard to find. Our findings on this topic have been derived from field visits to 10 Indonesian textile companies, five of which completed our questionnaire.[122]

One of the first observations that can be made, and one that is relevant to the issues discussed by Manning in Chapter 10, is that training of newly recruited personnel and instruction in the use of new equipment and technologies are already very much part of current practice in the Indonesian textile industry. All the companies we visited provided training programs of this type. In addition, many offered training aimed at achieving general improvements in efficiency and productivity. All the managers and textile experts we interviewed agreed on the importance of training in improving the productivity of their firms. The level of human resources was, moreover, deemed relevant to decisions on new technology, and some firms had started to hire better educated personnel in recent years to complement their technological modernisation. High rates of labour turnover, often viewed by neoclassical authors as an obstacle to the provision of on-the-job training, and a typical feature of Indonesian textile firms, were not mentioned as a depressing factor in HRD activities. It seems that improving the level of human resources is seen as a prerequisite for technological progress, a *sine qua non* for modernisation of the machinery used. Furthermore, textile managers are noticing that their investments in human resources are paying off in improved firm productivity.

However, there appears to be a general consensus among researchers, experts and textile managers that the level of human resources and the provision of on-the-job training are still lagging behind technological developments in the sector. Average expenditure on HRD programs in the firms we visited was about Rp 34,000 per worker annually, or some 2% of annual wages. Many personnel managers claim that senior management still considers HRD programs a cost rather than an investment, and this may obstruct future HRD efforts. STTT Bandung (forthcoming), which will present results for the spinning industry, uses the technometric methodology developed by the UN to show that the 'humanware'

---

[122]The textile companies were affiliated with the West Java Skill Development Project in Bandung.

component of the spinning sector is still considerably (2.5 times) lower than the hardware component. Although no definite conclusions can be drawn from these results, they are broadly indicative of trends in the industry. Also, there is general agreement that steps need to be taken to increase HRD efforts in the textile industry, in order to improve the efficiency and productivity of firms.

OTHER DRIVING FORCES BEHIND TECHNOLOGICAL PROGRESS: FOREIGN COMPETITION

In his dissertation, Koesmawan (1996) observes a correlation between technological capability and export orientation towards the so-called affluent countries. Moreover, empirical data show that countries with high wage levels in textiles, such as Germany, Italy and Japan, are still heavily involved in textile production and exports (van der Kamp 1997, Tables 1.3 and 1.5). These observations give rise to an interesting hypothesis: that the textile product market can be divided into a number of segments, with varying degrees of technological complexity and different levels of human resources. Koesmawan distinguishes between two such segments: the standard-quality market, where the focus is on low production costs; and the high-quality market, where the focus is more on product quality and diversity in design. Customers in affluent markets are apparently willing to pay more for exclusive, high-quality textiles.

With the shift towards export orientation in the mid 1980s, Indonesian textile producers have faced increasing foreign competition, most notably from other low-income countries such as Vietnam, China, Bangladesh and Pakistan. In these countries, wages are even lower than in Indonesia, giving their textile producers a competitive advantage on world markets. With the prospect of continued growth in Indonesia in the near future, the level of wages is expected to rise further. This means that foreign competition from low-income countries will be a growing problem for Indonesian manufacturers wishing to produce for the standard-quality segment of the textile market, and could eventually drive them out of business. It is therefore often argued that Indonesian textile producers should shift their focus to the higher quality segment of the market where they still hold a sizeable wage-related advantage. Malaysia, Korea and even Japan are important producers in this market.

This shift in focus would necessarily be accompanied by an increasingly rapid replacement of older with more modern machinery. This subject has received much attention in the Indonesian literature on the textile sector (see, for example, Safioen 1990; Sudradjat 1996). Inherently, the level of human resources would need to be improved as well. In their initial reaction to these problems, the Indonesian government, as indicated in the Sixth Five-Year Development Plan (Repelita VI, 1994–99), has targeted technology and human resource levels. The pressure on the standard-quality segment of the textile market will most likely continue to grow. With increasing globalisation, foreign competition will continue to play an important role in the nature and speed of technological progress in Indonesia.

ESTIMATING THE ROLE OF TECHNOLOGY AND HUMAN CAPITAL IN ECONOMIC PERFORMANCE

In assessing the role of technology in economic performance, we are faced with a number of theoretical difficulties. The first, measurement of technological progress, has already been dealt with, and we have obtained an index that can be deemed representative of technological developments in the spinning and weaving sectors in Indonesia.

Second, an economic framework is needed to fit technological progress into an economic growth or performance model. In our analysis we have used a growth accounting model, which relates growth in value added to growth in the factor inputs labour and capital, and to growth in the overall level of outputs per unit of inputs, or total factor productivity (TFP). We use a standard Cobb–Douglas production function, which allows us to use the shares of labour and capital in value added as the coefficients for labour and capital.

Finally, and most importantly, we have to find a manner in which technological progress can be integrated into a growth accounting framework. An elegant way of doing this, albeit one that is not free of theoretical difficulties, can be found by interpreting technology, as defined in our analysis, as referring to the quality of capital goods. If we consequently define the quantity of capital goods as the total number of core machines present at a certain time, we can construct a series for capital services by multiplying the quantity and quality series for capital goods. In other words, we can use the technology index created in the previous section, together with information on the total number of machines in the spinning and weaving sectors, to arrive at an estimate for total capital services, which thus includes embodied technological progress.

Although not new, this methodology for deriving capital services estimates has not been widely used.[123] It differs from the more widely used perpetual inventory method (PIM), which has the advantage of including all capital goods in a sector, and is therefore theoretically a better representation of the capital stock. This methodology requires us to focus on one link in the production process, in order to compare the productivities of different types of technology without having to resort to price-based estimates. However, the exact data needed to derive a PIM estimate are rarely available. In particular, exact price indices for capital goods in a certain sector are very hard to obtain, and in fact quality improvements in capital goods can be incorporated fully only if hedonic price indices are used. In addition, the method relies critically on the use of prices of capital goods as indicators for their productivity, assuming perfect competition conditions in the capital goods market. We have seen earlier that for spinning frames and weaving looms, at least at some points in time, the ratio between price and productivity was not a constant between machines.

In our growth accounting analysis, we have used the method described above to estimate capital services. Under the assumptions of perfect competition, profit maximisation and constant returns to scale, we have used the shares of labour and capital in value added for the period 1975–93 to assign economic growth to the factor inputs and TFP.[124] Within this interval, according to Wibisono (1987, as quoted in Koesmawan 1996), two time periods can be distinguished: 1975–82, which was the period of import substitution; and 1983–93, the period of export promotion and liberalisation of textile regulations. A correction for the quality of labour has been carried out in the tradition of Denison (1967), using education as an indicator for the quality of labour inputs. The average hours worked has been

---

[123]One notable exception is Otsuka, Ranis and Saxonhouse (1988). In their analysis of the Japanese cotton textile industry, the number of spindles and looms is used as an indicator for the physical quantity of capital goods, and in a growth accepting exercise to determine growth rates in TFP.

[124]Using these assumptions, we have estimated the partial elasticity of value added with respect to labour inputs as the share of employment costs in total value added, and the partial elasticity with respect to capital inputs as the remaining share.

**TABLE 15.4 Growth Accounting for the Spinning
and Weaving Industry, 1975–93
(%)**

	Spinning			Weaving		
	1975–93	1975–82	1983–93	1975–93	1975–82	1983–93
Value added (total growth, constant Rp million)	1,118	110	1,008	1,425	165	1,259
Average growth rate	21.2	22.1	19.6	14.2	11.5	17.4
Labour inputs	14.7	14.7	14.7	13.8	–4.0	16.0
Employment	9.5	14.9	8.9	6.9	–4.5	8.4
Average hours worked	–0.3	–2.7	–0.1	–0.3	–4.5	0.3
Educational attainment	5.5	2.6	5.8	7.1	5.0	7.4
Capital inputs	68.6	73.3	68.0	34.4	27.6	35.3
Number of machines	36.7	50.5	35.2	–0.7	–24.0	2.4
Technological progress	31.9	22.8	32.9	35.1	51.6	32.9
Shift effect	–0.2	0.6	–0.3	25.7	44.5	23.2
Advance effect	32.1	22.2	33.1	9.4	7.1	9.7
TFP	16.7	12.0	17.3	51.8	76.4	48.6

Sources:
Value added, employment: BPS (various years), *Statistik Industri*, Jakarta.
Average hours worked, educational attainment: BPS (various years), *Keadaan Angkatan Kerja di Indonesia*.
Number of machines, shift effect, advance effect: see previous tables in this chapter.

used as a correction for the quantity of labour inputs, using the procedure followed by Szirmai (1993). The results for the spinning and the weaving industry are presented in Table 15.4.

In spinning, most growth is explained by capital inputs adjusted for technological change (68.6%). Both the expansion of the machine park (capital widening) and technological change have been important. Technological change works via the advance effect (more effective vintages of the same type of machinery). The shift from one type of machinery to another is unimportant. It is interesting that in the second subperiod, 1983–93, capital widening is less important relative to the advance effect than in the first subperiod. In other words, embodied technological change becomes more important. Growth in augmented labour inputs explains about 15% of growth in value added. There is a negative effect due to a decline in average hours worked. Increased educational levels make a positive contribution (5.5%). As in the case of technological change, the role of human capital is more important in the second subperiod than in the first.

An important difference between weaving and spinning is that in the former sector the contribution of capital is due primarily to technological change and not to an increase in the number of machines—in fact, the number of machines even declines in the first subperiod, 1975–82. The main positive effect on growth is the shift to more modern types of weaving machinery, especially in the first subperiod. The advance effect contributes 9.4% to growth over the whole period, much less than in the case of spinning where, in contrast, the shift effect is negligible. A decline in employment in the first subperiod has negative effects on growth, as does a decline in hours worked. In the second subperiod employment expands again. Over the whole period, increases in human capital have a positive effect on growth, more so in the second period.

In spinning, TFP growth accounts for some 17% of the expansion, and so one may conclude that most growth is explained by that occurring in augmented factor inputs. In the case of weaving, TFP growth accounts for 52% of the expansion. The difference has much to do with our estimate of the advance effect, which is much higher for spinning than for weaving.

TFP growth in weaving is much higher than usually found in growth accounting studies for developing countries. This may be attributable to very substantial increases in efficiency due to economies of scale, capacity utilisation and better resource allocation, but it may also have to do with our measurement techniques. Alternatively, our indexes may be too modest in representing the growth of inputs. For example, our index of capital augmented for technological change based on core machinery may underrepresent the increasing capital intensity of production. Further comparison between direct measurement of machinery and perpetual inventory approaches is required. Quantifying our findings in the form of indices and placing them in a growth accounting framework should be seen as a challenge for further research in this area.

Changes in HRD have not been included in the growth accounting scheme. Apart from measurement problems, human resource efforts are too recent to have much impact on the results. However, a number of points are worth noting. First, human capital becomes more important from the first period to the second. Second, technological change is an important force. More advanced technologies have a necessary complement in HRD, which takes the form of on-the-job training. The managers and experts we interviewed were increasingly convinced of the importance of investments in HRD for export performance and future competitiveness. This conclusion is supported by recent literature on the Indonesian textile industry.

CONCLUSIONS

In this chapter we have attempted to measure technological progress in the Indonesian textile industry. Using a narrow definition of technology, referring essentially to capital-embodied technology, it was possible to measure technological progress by looking at technical data on the various types of core technologies used. In this way a technology index was constructed for the period 1940–95 (1968–94 for spinning). Technological progress in the weaving sector has been very rapid, with the technology index for the sector increasing over 13 times since 1940. Most of this progress has been due to the replacement of older generations of technology with more modern machinery (shift effect). In spinning, technological progress has been less rapid, with the technology index increasing some 70% in the last 26 years. All of this progress was caused by improvements in existing technologies for the spinning process rather than the introduction of new technologies. In addition, improvements have been made in several side technologies, such as drawing and carding; these have not been included in our analysis. Together with other factors such as increasing capital intensity and higher levels of human capital, technological change has resulted in a 5.5-fold increase in labour productivity from 1975 to 1995.

The level of human capital (or human resources) is an important determinant of the speed and nature of technological progress. Without sufficient levels of human resources, textile firms are unable to realise the increased productive potential associated with more modern technologies. Levels of human capital have been increasing, but efforts in the HRD field within firms have been limited. In recent years, with the increasing rate of technological progress, on-the-job training has become popular among managers as a way to improve the productive performance of their firms. Also, some firms have begun to employ better educated personnel as they modernise their technology.

Another factor that is considered a driving force behind technological development is increased foreign competition, which has in recent years forced Indonesian textile producers to shift their focus to the high-quality segment of the textile market.

Using a growth accounting framework, we found that the contribution of technological progress to the growth of the Indonesian textile sector was 31.9% for spinning and 35.1% for weaving. Although these percentages are similar, the nature of technological progress in the two sectors has been different, with advance effects predominating in spinning and shift effects predominating in weaving. Increases in the level of education were found to contribute some 5.5% of growth in spinning and 7.1% in weaving, with a growing importance in the later period (1983–93).

The main recommendation to be derived from our analysis is that a continued shift in focus towards the high-quality segment of the textile market should be a priority for Indonesian textile manufacturing. This shift requires a continuing rapid replacement of older machines with more modern ones, together with increasing efforts in the human resource field. This will help prevent the Indonesian textile sector from being 'outcompeted' internationally by producers in low-income countries, and will constitute part of Indonesia's progression from a low to a middle-income country.

PART V

TECHNOLOGY: INTERNATIONAL DIMENSIONS

16

EMERGING PATTERNS OF TECHNOLOGY FLOWS IN THE ASIA–PACIFIC REGION: THE RELEVANCE TO INDONESIA

Hadi Soesastro

INTRODUCTION

Indonesia needs technology to sustain its economic growth. Like other East Asian developing economies, it is a net importer of technology. Inward technology flows into Indonesia, and into other East Asian developing economies, have continued to rise. Increased imports of technology are observed even in advanced industrialised countries where most technologies are being generated. This suggests that technology flows have intensified, globally as well as in the Asia–Pacific region. As with trade, it is probable that global technology flows have been increasing more rapidly than growth in world output.

Technology flows take many forms, and cannot be measured easily. To use the definition proposed by Rosenberg and Frischtak (1985), technology is a quantum of knowledge resulting from accumulated experience in design, production and investment activities that is retained by individual teams of specialised personnel. This knowledge is mostly tacit and often is not made explicit in blueprints or manuals. Krugman (1991) has argued that anything at all can be assumed about flows of knowledge as they leave no paper trail by which they can be measured or tracked. Despite this difficulty, it is hard to dispute that something does flow, and that it contributes to higher growth, increased productivity and greater technological capabilities on the part of the recipient.

The literature on technology transfer regards the 'flow' of technology as technology that is actually 'transferred', namely the knowledge that is being absorbed by the recipient. Thus the flow of technology is not simply a matter of the purchase of equipment or the acquisition of blueprints. Cross-border flows of technology, or the international transfer of technology, are defined by Robinson (1988) as the development by one country of the capacity to use, adopt, replicate, modify or further expand the knowledge and skills developed in another country, with the ensuing product or service often becoming associated with either a different manner of consumption or product use, or a different method of manufacture or performance.

The flow of technology can also be viewed in terms of the diffusion of innovation, which involves active adoption, adaptation and change, and not simply the passive implementation of *new* technologies (OECD 1996a). In its *disembodied* form, such diffusion can occur through an organised process in which the rights to a patent are sold or an innovation is licensed, but it could also occur simply as a consequence of the innovation activity itself, namely in the form of research spillovers. *Embodied* technology flows take place as a result of the introduction into production processes of machinery, equipment and components that incorporate the new technology. It should be noted, however, that new technology or innovation can be understood to involve any product or process that is *new to the recipient* (Hobday 1997), and does not necessarily mean high or frontier technologies.

As flows of knowledge cannot be measured directly, studies of cross-border technology flows have focused on the different ways or vehicles by which technology moves across national borders. A number of taxonomies and models have been developed to describe these various transfer mechanisms (see, for example, Robinson 1988; Simon 1991). Such mechanisms could involve flows through the international technology market or through education, training or conferences; they could be government-driven flows or result from commercial arrangements among private firms. Technology can be transferred as part of a foreign direct investment (FDI) package or by means of a variety of 'unbundled' contractual arrangements; intrafirm technology flows take place between parent multinational corporations (MNCs) and their subsidiaries abroad.

The flow of technology in the broadest sense has been associated with a host of mechanisms: trade in capital goods; FDI; licensing; technical agreements; consultancies; turnkey plants and project contracts; flows of public technological information through patent descriptions and scientific and technological journals; reverse engineering; visits to production facilities; and the transfer of person or institution-embodied knowhow. Most of these modes of transfer are hard to detect and therefore difficult to measure. For instance, it would not be easy to separate out the value of embodied technology in the trade of capital goods. The monetary value of licensing agreements can, in principle, capture flows of knowhow, but equally if not more significant flows through academic institutions or public technological information remain unrecorded (Pavitt 1985). Changes in the magnitude of FDI flows may be used as an indicator of whether the flows of technology through this channel increase or decrease. However, total FDI flows can be a misleading indicator, since the technology content of a dollar investment varies by sector, and within the same sector may differ according to the scale of the project. It may well be that the stock, rather than flow, of FDI provides a better indicator of the actual flow of technology.

These difficulties notwithstanding, it should be possible to identify the broad patterns of technology flows in the Asia–Pacific region. This region is of immediate interest to Indonesia for a number of reasons. As more than 75% of its trade is with the Asia–Pacific, and as a large proportion of inward FDI flows also originate in the region, Indonesia's main sources of technology are likely to be Asia–Pacific countries. Thus changes in the pattern of technology flows in the region may have an effect on the structure of inward technology flows to Indonesia. It appears that until now FDI has been the main channel for inward technology flows to Indonesia, as is the case with many other East Asian economies. Would countries with similar patterns of inward technology flows exhibit similar technological performances? If not, what would explain the difference? These are some of the interesting questions that arise from a study of technology flows in the region. More specifically, a comparison of the structures of inward technology flows between Indonesia and other East Asian developing economies may shed some light on the opportunities and challenges for Indonesia's strategy of technology acquisition *and* accumulation.

Figure 16.1 provides a schematic presentation of the possible influences that changes in the global and regional technology landscape can have on the structure of inward technology flows to

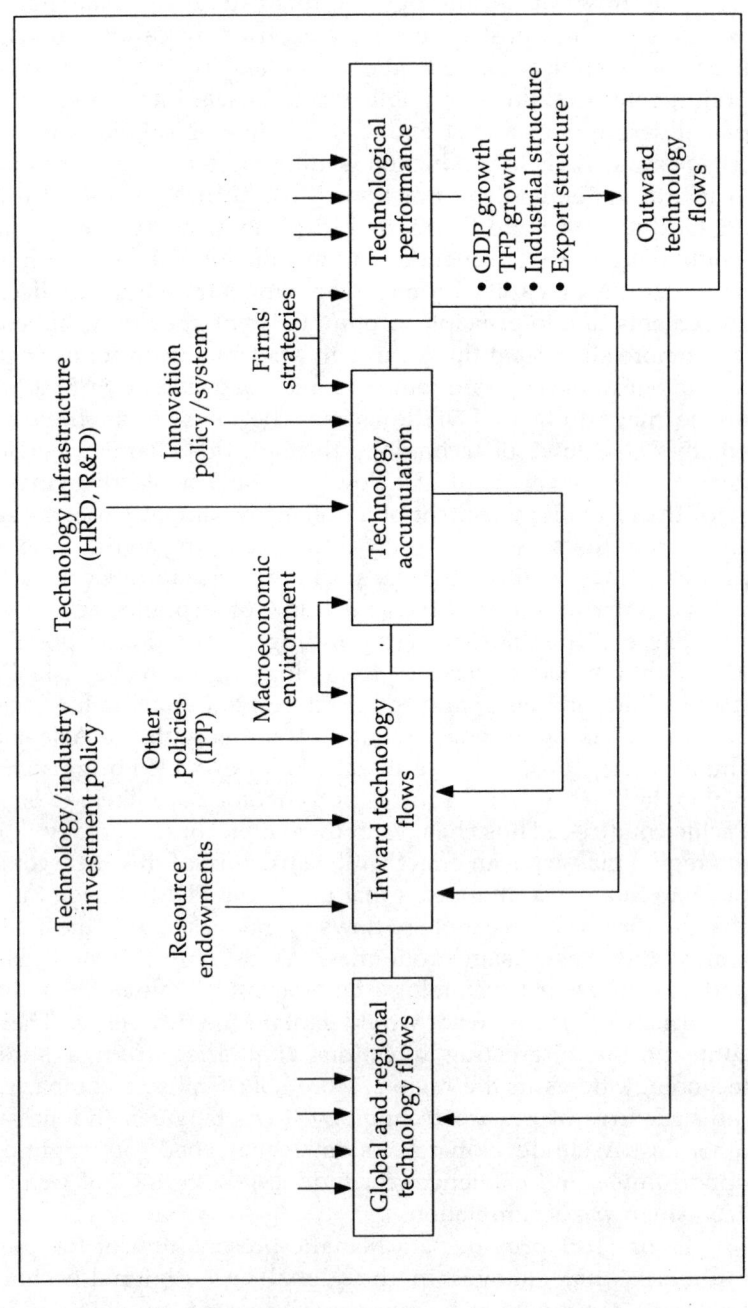

FIGURE 16.1 Technology Flows, Accumulation and Performance

countries such as Indonesia. Technology flows in the Asia–Pacific region, particularly those that are of relevance to the East Asian developing economies, will be examined in the first section of this chapter. The literature has identified a number of factors in host countries that can have a significant effect on inward technology flows (Enos, Lall and Yun 1997). These include the country's overall macroeconomic environment; market size; the set of technology, industry and investment policies; and the strength of specific policies on, for example, intellectual property protection.

Given the structure of its inward technology flows, a country's immediate challenge is to find effective ways to promote the process of technology accumulation within the economy. The degree of, and progress in, technology accumulation may themselves affect the country's pattern of technology inflows. With an increase in the level of technology accumulation, more diverse channels for technology acquisition will be open to a recipient country. Also, that country may have greater access to more sophisticated technologies. Technology accumulation is a result of specific and systematic efforts on the part of the country (through government policies) and of individual firms in that country, as manifested in the strategies they adopt. Government policy can help improve the country's technology infrastructure through investment in education, information and communication networks, or a set of R&D support infrastructures. The overall macroeconomic environment, including a healthy environment for competition, can create the necessary incentives for firms to invest in technology development efforts. Policy and institutional frameworks and the combined efforts of government, the business sector and academic institutions form a country's system of technology accumulation efforts, known as the national innovation system.

Ultimately, the degree of and progress in technology accumulation will be reflected in the country's technological performance. Indicators of this performance could include broad ones such as sustained growth in GDP and total factor productivity (TFP), or changing industrial and export structures towards more skill and technology-intensive products. They may also include more specific indicators such as greater innovation capabilities, as demonstrated by increased applications of patents (domestically or abroad) or increased exports of disembodied or embodied technologies. However, other factors do come into play. A country which has a higher level of technology accumulation but which adopts inward-oriented policies may not have a better technological performance than an export-oriented country with a lower level of technological accumulation. A comparison between India's performance on the one hand, and Taiwan's on the other, clearly illustrates this point. Also, increased resources for R&D may not necessarily improve a country's technological performance.

PATTERNS OF CROSS-BORDER TECHNOLOGY FLOWS

A number of developments have influenced the patterns of cross-border technology flows in the world today. Ostry and Nelson (1995) have noted that the rise of the MNC and the surge in FDI, particularly since the mid 1980s, have resulted in a rapid inter-nationalisation of trade and business which, in turn, has eroded national barriers to the flow of technology. In their effort to exploit their technology globally, MNCs are diffusing their technology. In addition, they are continuously strengthening their access to new technology through the worldwide diffusion of R&D activities, because such activities allow them to tap a host country's tech-nological resources. They are also increasingly engaged in various forms of collaboration and technology-based alliances as a way to broaden their access to new technologies.

The role of the corporate sector in the generation and diffusion of technology has definitely increased. In OECD countries, where most technologies are being generated, corporate R&D expenditures now account for an increasingly large share of national R&D expenditures, and civilian-based—rather than defence-related— R&D is now defining the frontiers of technology. Technology policies of the private sector have become increasingly transnational in scope, leading to the emergence of *techno-globalism*.

In their survey of the literature, Enos, Lall and Yun (1997) raised the question as to whether the issue of technology transfer may have to be removed from the public arena because it is now largely being decided privately by firms rather than publicly through the intervention of government agencies. The immediate question of interest to developing countries in the region, including Indonesia, is whether this techno-globalism means that techno-logies will become more accessible as they are diffused more widely. Or, on the contrary, will it become more difficult to access new technologies which, although becoming internationalised, are almost exclusively in private hands? The issue of techno-globalism has also attracted the attention of governments in the advanced industrialised countries. In the US, for instance, concerns have been expressed about the loss of the country's critical technological assets as a result of the globalisation of technology. Increased attention has been given to trade in high-technology products, with several attempts being made to control it. This suggests that the technology policies of governments are still being dictated in part by *techno-nationalism* (Simon 1997). It can be concluded that the global, and perhaps the regional, technology landscape is influenced by these two conflicting trends: techno-globalism and techno-nationalism.

Archibugi and Michie (1995) have argued that the globalisation of technology is taking place only partially. In terms of the global *exploitation* of technology, this globalisation process has clearly accelerated as more firms are involved and as foreign investment regimes everywhere in the world are progressively being liberalised. Global *technological collaboration* tends to be more limited in scope. This process takes place only in fast growing technological fields such as biotechnology, information technology and new materials, and involves almost exclusively firms from advanced industrial countries possessing complementary technological assets. The global *generation* of technology applies mainly to MNCs, which are increasingly decentralising their R&D activities by creating a network of globally connected subsidiaries engaged in R&D. This process is not yet a truly globalised phenomenon, and appears to be mainly confined to European countries.

With this qualification in mind, it appears that the broad trend towards freer flows of technology globally is undeniable. A brief examination of this development in a number of important channels of technology flows will confirm this trend.

Trade in Disembodied Technology

Trade in disembodied technology, in the form of patents, knowhow or services with a technology content (engineering studies, R&D services), is being recorded in a country's 'technology balance of payments'. In 1992 the OECD area as a whole (including intra-OECD trade) recorded receipts from technology exports of more than $47 billion and payments for technology imports of about $40 billion. This trade in disembodied technology takes place almost exclusively between advanced industrialised countries (OECD 1995). More than 90% of registered payments worldwide in patents and licences involve about 10 major OECD countries.

In the Asia–Pacific region, the US is the only country that has a large and increasing surplus in its technology balance of payments, of about $15 billion in 1992 and $20 billion in 1995 (OECD 1996b; IMF 1996). The ratio of receipts to payments has declined for the US, from 5.3 in 1990 to about 4 in 1995, suggesting that technology imports by the US have increased much faster than technology exports. Conversely, other OECD countries in the region have exhibited increased ratios, which shows that they have become more important technology exporters. This suggests another important development, namely a growing convergence in the technological capabilities of OECD countries, away from the earlier situation in which the US was the predominant supplier of technology.

OECD statistics show that since 1993 Japan has experienced a surplus in its technology balance of payments. However, Japan has always been an important source of technology for Asian developing countries. In 1980 Asia accounted for about 34% of Japan's technology exports, with such industry-specific technologies as electrical machinery, chemicals and transport amounting to over half of the exports (Lynn 1985). In 1995 East Asia's share of Japan's technology exports increased to 46% (OECD 1995). Imports of disembodied technology by developing East Asian economies have been on the rise. Korea's imports increased from $136 million in 1990 to $2.4 billion in 1995. Over the same period, Thailand's imports increased from $170 million to $630 million, while those of the Philippines increased from less than $40 million to nearly $100 million (IMF 1996). The share of disembodied technology in the value of China's total technology import contracts increased from 4% in the 1970s to 15.8% in 1991 and to 28.7% in 1992, when they amounted to close to $2 billion (Hu 1994; Zhao 1995). In 1992, almost 80% of the contracts for disembodied technology imports were in the form of licensing. Indonesian data for 1990 recorded imports (patents and licence fees) of only $11 million (Garrett-Jones 1996), but this appears to have been underreported; royalty payments and licence fees by US subsidiaries in Indonesia to their parents alone amounted to $24 million in 1989 (US Department of Commerce 1992).

International Patent Flows

International patent flows are increasingly used as an indicator of technology flows, particularly among advanced industrialised countries. Studies using patent data have shown that intra-OECD flows of technology have accelerated (Schmoch 1996; Eaton and Kortum 1996). Patent applications with the United States Patent and Trademark Office reveal a significant rise in patents granted to Asia–Pacific countries, whose share doubled during 1980–93 from 12% to 24%. Over the same period, the share of East Asian developing countries increased ten-fold, from 0.2 to 2.1%, exhibiting the highest growth rate (OECD 1995). The data on US patenting by industrial product group show that electrical/electronics industries grew fastest, followed by scientific instruments and chemicals. It is interesting to note that by 1993, the share of US patents taken out by Asia–Pacific countries in the electrical/electronics group was 2.6 times that of Europe. In the scientific instruments group as well, Europe has been outdistanced by the Asia–Pacific region. The same trend, albeit less dramatic, can be discerned in transport equipment and machinery.

Foreign Direct Investment

There is no doubt that FDI plays an important role in cross-border flows, transfers and the diffusion of technology. The story of technology flows in the Asia–Pacific region has centred on the dramatic surge in FDI, particularly in the East Asian developing economies. As a percentage of the world's total inward FDI flows, the share of developing East Asia has risen rapidly. By the end of 1995, the stock of inward FDI in developing East Asia was about $350 billion (Table 16.1), or about 13% of the world's total stock of inward FDI. In 1980 it stood at about 6%.

During 1991–95, about 84% of cumulative FDI flows to developing East Asia were accounted for by flows to the four largest recipients: China, Singapore, Indonesia and Malaysia. Of the total stock of inward FDI to China, nearly 90% was of more recent vintage (1991–95). This was also the case with Malaysia (63%) and Thailand (56%), but only about 20% of the stock of inward FDI to Indonesia was of more recent vintage. It is widely believed that the later waves of FDI involved more advanced technologies. Technology transfer mechanisms through FDI have yet to be studied in greater depth.

It is generally believed that FDI brings in more advanced technologies than alternative channels. This is particularly the case with MNCs, because they play a dominant role in the generation of technology and are usually associated with new or

HADI SOESASTRO

TABLE 16.1 FDI into East Asia, 1991–95
($ billion)

	Total Flows, 1991–95	Stock, End of 1995
China	112.7	129.0
Malaysia	24.3	38.5
Singapore	24.2	55.5
Indonesia	11.7	50.8
Thailand	9.4	16.8
Hong Kong	8.4	21.8
Taiwan	5.9	15.6
Korea	5.1	14.0
Philippines	4.7	6.9

UN (1996), *World Investment Report 1996*, New York.

technologically complex products. International technology diffusion through MNCs takes the form of parent–subsidiary flows of technology. The challenge to the host country is to create an environment in which the subsidiaries of MNCs will be encouraged to upgrade their technological capabilities continuously and maximise technology spillovers of inflowing advanced technologies to the host country.

The transfer of a parent MNC's proprietary technology can take various forms, including technical documentation; education and training of the subsidiary's labour force; exchanges of technical personnel; shipment of machinery and equipment; and continuing communication to solve problems that occur in the production process. Such parent–subsidiary flows of technology are usually not recorded (Kokko and Blomstrom 1995). US benchmark surveys on FDI have, however, collected data on payments of royalties and licence fees by American overseas subsidiaries. They have also recorded exports of capital goods to subsidiaries by US parents. A study by Kokko and Blomstrom (1995) using the 1982 Benchmark Survey data concluded that: (1) for subsidiaries in developing countries, the main mode of technology transfer was imports of capital equipment; and

(2) subsidiaries in developed countries rely more on imports of disembodied technologies. This conclusion no longer held for the East Asian developing economies in 1989: US subsidiaries paid about $1.3 billion for licences compared to $640 million for capital goods imports.

Increased attention has been given to the internationalisation of MNCs' R&D activities. The spillover effect of such activity on the host country is believed to be quite significant. The 1989 Benchmark Survey data show that subsidiaries of US MNCs spent nearly $8 billion on R&D. Preliminary results of the 1994 Benchmark Survey indicate that this amount has increased to $12 billion, of which about 75% was spent in Europe. Data compiled by Japan's Ministry of Trade and Industry (MITI) on overseas R&D expenditure by Japanese corporations reveal that about 94% was spent in the US and Europe (Kumar 1996).

In an empirical study on the determinants of overseas R&D, Kumar (1996) put forward the hypothesis that the location of overseas R&D could be affected by three sets of factors: (1) the scale and nature of local production; (2) host country resources; and (3) the policy environment of the host country. Thirteen variables were constructed to capture these factors, and regressions were undertaken, with R&D intensity of US subsidiaries as the dependent variable, for a full sample of 54 countries, a subsample of advanced industrial countries and a subsample of developing countries. The results showed that size of the host country market and technological resources in the host country are the most important factors. Intellectual property protection is a significant explanatory variable in the subsample of industrial countries, but not in the subsample of developing countries, indicating perhaps the different types of R&D being undertaken in the two groups of countries.

The 1989 and 1994 Benchmark Surveys show that R&D expenditures by US subsidiaries in East Asia have been rising. Singapore ranked first, with R&D expenditures of $238 million in 1994, about the size of R&D expenditures by US subsidiaries in Australia. This clearly suggests the importance of the technology resources of the host country. Taiwan ranked second, followed by Hong Kong and Malaysia. Among developing East Asian economies, such expenditures have risen most rapidly in Malaysia, suggesting that Malaysia's technological resources may also have increased rapidly. In 1994, US subsidiaries' R&D expenditures in Indonesia were only about $6 million.

Technology-Based International Strategic Alliances

Technology-based international strategic alliances (TISAs) among firms from different countries have become another important vehicle for cross-border flows of technology. This mode, however, mostly involves firms from the advanced industrialised countries. There are a number of reasons for the emergence of such alliances. The greater convergence of technological capabilities among advanced industrialised countries may be one important factor. Another is the increased cost and risk of R&D to bring a new product or process to the market, particularly for high-technology industries. The shortening of product lifecycles has also increased the need for rapid and simultaneous penetration of global markets. A related factor is the growing importance of de facto technical standards or dominant designs as a source of first-mover advantage, adding to the urgency to shorten development times.

An analysis by Freeman and Hagedoorn (1994), based on the MERIT-CATI databank at the University of Limburg in the Netherlands, shows that over 95% of a total of about 4,200 TISAs established during the 1980s involve companies from advanced industrialised countries. Of the total number, around 20% were in biotechnology and 10% each in new materials and chemicals. These were followed by microelectronics, telecommunications and software. Only about 2.3% of TISAs involved firms from East Asian developing countries. The most important fields of technology in which such TISAs were engaged were automotive and food and beverages (10% each), information technology, chemicals and electronics. In a study of the nature of some recent TISAs involving firms from East Asia, Wong (1997) concluded that most evolved from an earlier stage of an original equipment manufacturing (OEM) subcontract performed by East Asian firms for MNCs. As some firms from East Asia have become technologically more advanced, the relationship has developed into alliances of a joint product development type.

Trade in High-Technology Products

Trade in high-technology products is believed to be a major channel for technology diffusion. The loss of technologies critical to international competitiveness has resulted in growing concern on the part of some governments. Ostry and Nelson (1995) have argued that the sharp increase in international trade in high-technology products has made them available even in countries that do not have firms to produce them. In countries with a high level of scientific and technical sophistication, access to and familiarity with such products can generate considerable understanding of the technology incorporated in them.

Initial approaches have used R&D intensity as the basis for defining high technology. High-technology industries are leading-edge growth sectors that rely heavily on the application of new science-based technologies, and as such are characterised by large investment requirements in R&D (Green 1996). Over the years a number of modifications have been made in defining high technology. In addition to the use of such 'objective' indicators as R&D expenditure per unit of production, some 'subjective' judgement by industry analysts is also used to evaluate and determine whether a product can be classified as involving high technology.

A classification system that combines the objective and subjective approaches has been constructed by Guerrieri and Milana (1995), who have produced a list of high-technology products at the five-digit SITC level. In broad outline the list encompasses five groups of products: (1) chemicals and drugs (synthetic organic dyestuff, radioactive and associated materials, polymerisation and copolymerisation products, antibiotics); (2) mechanicals (turbines, electric motors, electric power machinery and apparatus, internal combustion piston engines); (3) electronics (automatic data processing machines and units, telecommunications equipment, semiconductor devices, electronic microcircuits); (4) scientific and professional instruments (electronic measuring instruments, medical instruments, optical instruments and appliances, photographic apparatus and equipment); and (5) aerospace (aircraft and associated equipment, spacecraft).

Table 16.2, based on the above classification, shows the market share and trade balance, in each of the five high-technology product groups, of the US, Japan and the Asian NIEs (Korea, Taiwan, Hong Kong and Singapore). Both market share and trade balance have been declining in the US in all groups except aerospace. According to Guerrieri and Milana, the decline has been caused mainly by a loss in competitiveness, and not so much by so-called structural effects, which refer to changes in the export composition of commodities and markets. Japan's market share has been increasing except in chemicals. Japan's gain resulted from both positive competitiveness and structural effects. The performance of the Asian NIEs has been remarkable, with their market share increasing in all

**TABLE 16.2 Market Share and Trade Balance in
High-Technology Product Groups, 1981–90**

Product	Market Share[a]		Trade Balance[b]	
	1981–84	1987–90	1981–84	1987–90
Electronics				
US	25.1	17.8	2.5	−4.7
Japan	20.0	23.0	16.6	19.4
East Asia	10.3	16.1	0.5	4.3
Chemicals				
US	16.9	13.0	8.4	5.5
Japan	5.7	5.4	0.1	0.3
East Asia	2.6	4.4	−2.4	−2.8
Mechanicals				
US	20.0	15.7	4.4	−4.1
Japan	14.8	16.8	12.6	14.3
East Asia	3.9	5.8	−2.5	−2.9
Aerospace				
US	42.0	43.3	29.7	28.9
Japan	0.5	0.8	−4.1	−4.0
East Asia	1.4	1.4	−2.9	−3.8
Instruments				
US	29.7	22.3	19.8	9.8
Japan	9.0	11.9	4.0	6.1
East Asia	2.0	3.7	−3.4	−4.7

[a]Ratio of a country or area's exports to total world exports in the respective
product group.
[b]As a percentage of total world trade in the respective product group.
Source: Guerrieri and Milana (1995).

product groups except aerospace. Guerrieri and Milana show that
this increase resulted mainly from positive structural effects,
namely shifts in export composition towards high-growth markets
and commodities. However, the trade balances of the Asian NIEs in
all product groups except electronics worsened during the 1980s and
were negative. These trends deserve further analysis, but the above
examination shows that the involvement of the East Asian

economies in high-technology production and trade is rather narrowly based, and is mainly in electronics.

Technology Exports by Developing Countries

The above examination suggests that technology flows are mainly two-way among advanced industrialised countries, and one-way from advanced industrialised countries to developing countries. Technology exports by developing countries are not a new phenomenon, but they may have increased. These exports are mainly to other developing countries, and appear to be in the form of embodied technology plus the technical services that come with it.

China's technology exports show an increase, from less than $500,000 in 1981 to $1.5 billion in 1992 (*Almanac of China's Foreign Economic Relations and Trade*, 1990; Hu 1994). Most of the technology exports in 1992 were in the form of turnkey plants, valued at $1.37 billion. These included power generation plants, a clinker cement production line, a floating glass production line, and a complete plant for ion film caustic soda production. The remaining flows, of just $140 million, took the form of disembodied technology exports. As would be expected, China's chief export markets for its technology are developing countries. The government of China has been actively promoting technology exports with the primary objective of earning foreign exchange.

The Korean government has also encouraged technology exports. Korea's exports of disembodied technology have increased rapidly during the 1990s. IMF statistics show that receipts from royalty payments and licence fees increased about eight-fold between 1990 and 1995; by 1995 Korea's technology exports were worth about 20% more than Australia's. In the early phase, disembodied technology accounted for only about 1% of Korea's total technology exports cumulatively to the end of 1981 (Westphal, Kim and Dahlman 1985). Up to the early 1980s, the single most important form of technology exports from Korea was overseas construction. Competence in this activity was gained through learning-by-doing under US military procurement contracts, including outside Korea. Plant exports began in connection with overseas construction activity. Today, Korea is a significant exporter of capital goods.

Another developing East Asian economy, Singapore, has also developed a capacity for technology exports, capitalising on specific local skills and advantages in such areas as urban planning and housing development, and tropical agricultural research. As described by Hill and Pang (1991), Singaporean firms have acquired the capability to facilitate 'two-stage technology transfer', to act as an intermediate step between the technology employed in large, high-wage economies and that used in smaller, low-wage economies in the region.

The Flow of People

Finally, the flow of people has become an increasingly important vehicle for the flow of technology, particularly the transfer of tacit knowledge. Technical assistance, consultancies and supplier–buyer trading relations are examples of the mechanisms by which technology can flow through people-to-people contacts. There are no databases detailing these types of activities, but it appears that they have intensified in the Asia–Pacific region.

Some General Observations

Tran and Urata (1995) have argued that intensive flows of technology among Asia–Pacific countries are perhaps the most important development in the region since the mid 1980s. The above examination of the various channels of technology flows shows that cross-border flows, globally as well as in the Asia–Pacific region, are clearly on the rise. This trend has been accompanied by a diversification in the channels for technology flows. The expansion and diversification of such channels in the Asia–Pacific region have resulted from favourable developments on both the demand and the supply side.

On the demand side, the important factors include: (1) expanding economies and markets resulting from sustained long-term growth, in turn leading to increased needs for more complex technologies; (2) the steady accumulation of technological capabilities at the country level and by individual firms; (3) increasingly open trade and investment regimes, resulting in greater competition among firms and gradual improvement in non-cross-border policy areas such as intellectual property protection; and perhaps (4) increasing confidence on the part of latecomers about their ability to catch up technologically.

On the supply side, the push factors that are at work mirror these pull factors. They include: (1) greater readiness on the part of owners of technology to exploit their technology globally, particularly in expanding markets; (2) the imperative of international competition, which has led companies to relocate their activities and engage in continuous efforts to develop local technological capabilities in order to remain internationally competitive; (3) reductions in impediments to trade and investment, resulting in greater technology flows through the various channels; and (4) greater readiness to engage in mutually beneficial interfirm technology collaborations as the technological capabilities of larger corporations, including those from developing countries, have converged.

The important shift in the patterns of technology flows in the Asia–Pacific region is that the flow of technology through one particular channel is no longer seen as a substitute for the flow of technology through another. In an earlier study of technology flows in the Asia–Pacific region, Mowery and Oxley (1995) proposed that the relationship between FDI and licensing in the Asia–Pacific region appeared to be one of substitution, rather than complementarity as is the case in Western Europe. Other conclusions of interest from the study were that the importation of capital goods occasionally complements the flow of technology through licensing; that different channels of inward technology flows convey different vintages of technology, with FDI bringing in more advanced technologies than alternative channels; and that the pattern of inward technology flows appears to be changing with the stages of development of the respective country.

THE RELEVANCE TO INDONESIA

The pattern of inward technology flows for Indonesia seems to be dominated by the use of FDI as the main channel for technology acquisition. In some sense this has been the country's implicit 'technology policy', and the favourable attitude of the government towards FDI has been based to a large extent on the promise of technology that will be brought in as part of the investment package. The government has attempted to use some performance requirements in its foreign investment regulations to effect more rapid transfers of technology. The regulations have been weak or have not been enforced, and no specific incentives have been given to encourage FDI that will upgrade local technological capabilities.

Indonesia's more explicit technology policy rests on the development of increasingly advanced technologies within the framework of the promotion of 'strategic industries'. Efforts are being made to develop indigenous technological capabilities, complemented by the use of licensing and technology collaboration arrangements with foreign producers. This effort is government-led, with the implementation of the policy involving state enterprises and government research institutions.

No attempt has been made to integrate Indonesia's contrasting technology policies, with the result that the pattern of inward technology flows is being influenced by two separate policies. Indonesia is thus unable to maximise the synergies that could be had from the diverse channels of technology flows.

The experiences of various East Asian countries provide models of technology acquisition and technology accumulation that are of relevance to Indonesia. Korea is an example of a country which has historically placed little reliance on FDI, as a channel of technology inflow. In the 1960s and 1970s the Korean government explicitly restricted the use of FDI and foreign licensing, emphasising instead capital goods imports as a means of acquiring foreign technologies. As described by Kim (1997), this policy was adopted with the aim of protecting the local market from exploitation by MNCs and developing an independent approach to the assimilation of imported technology. In turn, the policy was designed to strengthen Korea's bargaining power in negotiations on the transfer of more complex technologies. Korean companies, the *chaebol*, which were given an important role in technology accumulation, were forced to rely on reverse engineering. They were also forced to accelerate their technological learning by being exposed to greater international competition.

Korea's experience has produced mixed results. There is no doubt that the country has been successful in accumulating and upgrading its technological capabilities through its 'export-led technological learning'. R&D investments have increased rapidly, as have patent applications by Korean inventors. However, the Korean approach, which could be called following the 'hard path', exhibits a number of weaknesses. Korea's strategy has been costly. It has made heavy capital investments, but the bias in favour of the *chaebol* has resulted in an underdeveloped small and medium-sized enterprise (SME) sector. This to a large extent explains Korea's continued dependence on Japan for high-technology parts and components.

Korea's policy for achieving technological independence may have been misguided. The government has now reversed its previous policy, to actively promote inward FDI in high-technology sectors.

It also undertakes outward FDI, and engages in strategic alliances as a means of accessing new and more advanced technologies. Despite limited use of FDI in the past, the development of Korea's electronics industry was spearheaded by FDI. However, this industry is highly concentrated; it is dominated by three conglomerates, Samsung, LGE and Hyundai. The industry is characterised by a relatively shallow indigenous innovation capability and support structure; large component and material imports; and high levels of royalty payments (Mathews 1996). It also continues to depend heavily on OEM. About 70–80% of Korea's electronics exports—and 60% of the electronics exports of the three *chaebol*—are based on such arrangements (Hobday 1997),

In contrast to Korea, Taiwan has followed the 'soft path'. It has made some use of FDI, but has relied extensively on OEM as the main technology acquisition channel. Its more broad-based industrial structure, which involves strong SMEs, has created clusters of SMEs around MNCs through a network of subcontracting/OEM arrangements and extensive backward and forward linkages. Over the years, a number of local companies have accumulated technological capabilities that allow them to move into original design manufacturing (ODM), and even into own brand manufacturing (OBM), although this process continues to be a difficult struggle. The authorities have introduced a number of policies to influence the pattern of inward technology flows and to encourage the upgrading of technological capabilities. Tax holidays were introduced to attract high-technology industries, and at other times performance requirements were set to promote technology transfers by FDI. These policies proved ineffective. Current policies focus on strengthening the country's technological and public infrastructure, such as through the development of the Hsinchu science-based industry park. Technology diffusion seems to be the key to Taiwan's current technology policy.

Singapore provides an example of 'extreme use' of FDI. About 75% of Singapore's industrial output and 95% of its exports are accounted for by MNCs. Hence, technology flows take place largely within MNCs. The government has emphasised the importance of public infrastructure (education, training and physical infrastructure) to encourage subsidiaries of MNCs to upgrade their technological capabilities. These investments appear to have paid off; Singapore has gradually been able to move up the technological ladder. Subsidiaries of MNCs in Singapore have also developed into regional hubs that undertake OEM arrangements with firms throughout Southeast Asia. This could result in region-wide diffusion of technology.

The above brief examination of the different experiences in, and approaches toward, technology acquisition and accumulation does not suggest that there is a clear-cut model to be emulated by other countries, including Indonesia. In the final analysis, a model works best for a country if it produces an outcome that meets specified objectives, standards or targets. As proposed earlier (Figure 16.1), it is therefore important to examine the technological performance that results from the country's technological efforts. Again, no single measure would suffice to describe technological performance. However, a closer examination of the export structures of the above countries could shed light on the strengths and weaknesses of the different approaches.

Figure 16.2 exhibits the structure of exports by technology intensity of a number of developing East Asian economies, as well as that of the US and Japan as a reference. A few general observations can be made from a cursory examination of these different structures. Developing East Asian economies, with the exception of China and Indonesia, have an export structure which suggests that their comparative advantage in medium-technology products is weaker than their comparative advantage in high-technology products. This is in contrast to the export structures of more advanced industrialised countries such as the US and Japan, which show a stronger comparative advantage in medium-technology products. On the basis of this comparison, it might be argued that a country's industrial strength and its broad technological base are reflected in its comparative advantage in medium-technology products relative to that in the other technology categories (high and low).

If this is the case, Korea and Taiwan have a stronger technological base than Singapore, Malaysia or the Philippines. The share of high-technology products in these countries' exports is highest in the Asia–Pacific region. A closer examination reveals that their high-technology exports comprise mainly electronics. The value of Singapore's electronics exports is much larger than that of Korea or Taiwan, and Singapore is therefore often seen as having a better technological performance than Korea. However, this is the case only if one is interested in just this one product group. In theory, the development of a country's technological capabilities should involve the dynamic creation of comparative advantages to achieve *industrial sustainability* (Mathews 1996). A heavy dependence on a

FIGURE 16.2 Structure of Exports by Technology Intensity, 1993[a]

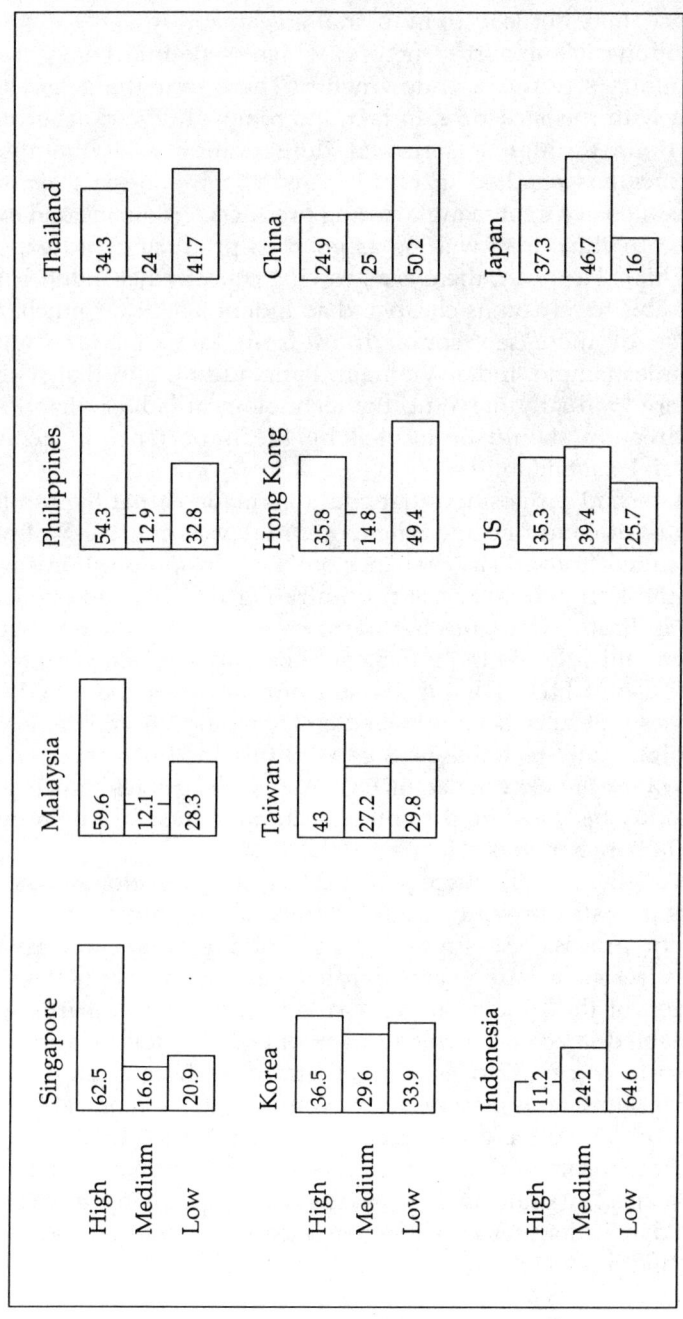

[a]'High', 'medium' and 'low' technology groups are based on the classification by OECD (1996a).

particular industry cluster, albeit one located in a high-technology category, may not lead to industrial sustainability.

Indonesia's export structure, which is bottom heavy, reflects the country's resource endowments. There is nothing necessarily wrong with this structure. In fact, the removal of various economic distortions through a series of deregulation and liberalisation initiatives has resulted in rapid growth in Indonesia's exports of labour-intensive light manufacturing products. Producing and exporting potato chips may well be as good as producing and exporting microchips. However, there is a growing concern that Indonesia may not be able to rely on its cheap and abundant labour for much longer because of increased competition from large labour-abundant economies (China, India, Vietnam, Bangladesh), and that it should therefore gradually move up the technological ladder. Its efforts in this direction should be guided by the importance of achieving industrial sustainability.

A second proposition that could be made is that the top-heavy structure of countries like Singapore, Malaysia and the Philippines has resulted from a heavy reliance on FDI. This proposition suggests that MNCs in particular are concentrating their technology exploitation in East Asia in a rather narrow sector of production activity. Knowing this, should Indonesia continue to place such great reliance on the use of FDI? The jury is still out, however, on whether the successes that have been achieved so far would not motivate MNCs to exploit their technological capabilities in other sectors. Some believe that the experience in East Asia's electronics industry may not readily be transferred to other sectors because of the industry's special characteristics (Hobday 1995).

Nonetheless the success of East Asia in developing the electronics industry provides useful lessons for countries that are in a catch-up process. As stipulated by Hobday (1997), the common success factors include: (1) the gradual learning of technology; (2) the adoption of the strategy of 'export-led technological learning' that has enabled latecomer firms to overcome technological and market barriers to entry; (3) investments that are made to catch up (not to leapfrog) and to assimilate technology; (4) technological progress acquired cumulatively through time; and (5) a strategy of technology acquisition and accumulation which begins with mature, standardised manufacturing processes and gradually moves on to more advanced stages of process engineering, product–process interfacing and product design.

Mathews (1996) stressed the importance of avoiding the conventional R&D-led innovation strategy as the mechanism for technology acquisition and accumulation. This is the mistake that is often made by countries that seek to achieve technological self-reliance. Indonesia has always been tempted to embark on this course.

17

THE TRANSFER OF JAPANESE MANAGEMENT TECHNOLOGY TO INDONESIA

Yuri Sato

WHAT IS MANAGEMENT TECHNOLOGY?

With the waves of foreign direct investment (FDI) into Southeast Asia, particularly from Japan, technology transfer has become a very important and much discussed issue. So far there seems to be no consensus on the effects of technology transfer on the host country. Some observers judge these effects to be positive, while others argue that they have been very limited. In some studies, the degree of technology transfer has been measured quantitatively by indicators such as the decline in the number of foreign personnel over time, the number of technical licences granted by Japanese firms, royalty fees,[125] the growth of R&D expenditures in the Japanese affiliates, and growth in technology-intensive exports. Other studies (for example, Koike and Inoki 1987) have attempted to observe technology transfer qualitatively in the form of skill formation within some sample Japanese-controlled companies. These studies focus primarily on the transfer of production technology, embodied either in production facilities or in persons.

---

[125]Statistics on technology exports from Japan compiled by the General Affairs Agency (Somu-cho) of the Japanese government are useful in this regard. There are practically no statistics on technology imports by ASEAN countries.

This chapter deals with management technology, as distinct from production technology or production technique. If technology is defined in a wide sense as 'all the knowledge and information accumulated in any form, either in machinery or in persons, necessary for the production and supply of goods and services', the scope of technology can be widened beyond production technology to cover management technology as well. Management technology here simply means 'all the knowledge to manage production and supply activities in various aspects'. It may be regarded as an infrastructure enabling the various production technologies to work effectively.

A major difference between management technology and production technology is that whereas the former is always embodied in persons, the latter can be embodied in either machinery or in persons. Thus management technology is often hard to recognise visibly and to measure quantitatively. For this reason, there has been little research as yet on the transfer of management technology, which can only be studied by undertaking qualitative analysis based on close field observations.

Quality control (QC) is a major aspect of management technology. QC was originally developed in the prewar period in the US as pure manufacturing technology. Its purpose was to keep the measurement of mass produced products within a certain error range. The technique was then extended to the control of entire production processes (process control), for the purpose of stabilising fluctuations in measurements statistically. 'Statistical QC', as it was called, was introduced to postwar Japan by American experts, where it developed into the unique teamwork-based activity known as the 'QC circle'. This method of managing and improving the quality of various aspects of the production process can be said to be a purely Japanese innovation of management technology. QC has thus been transformed from production technology in the era of mass production into management technology that can be adapted even to today's small-lot, large-variety production environment.

Total quality control (TQC) is another example of management technology. TQC can be considered as a system to integrate all the quality development, quality maintenance and quality improvement measures made by the groups within an organisation, including non-production divisions such as marketing, financing and accounting, planning and R&D. For this reason TQC is also called company-wide QC. The concept of TQC originated in the US (Feigenbaum 1961), but was further developed in Japan in the 1970s.

In this chapter I examine the nature of the transfer of management technology from Japan, based on a case study of a leading group of Indonesian companies. I give a brief overview of the transfer of Japanese management technology to Indonesia before providing a case study of the Astra group. I then discuss the implications of this case study and summarise the major findings.

JAPANESE MANAGEMENT TECHNOLOGY AND ITS TRANSFER TO INDONESIA

Japanese firms regards 'process control' as a key element in managing the quality of production activities. This is a way of controlling the interrelations between different production processes in the division of work—for example, timing, quantity, distance and sometimes prices—in order to optimise efficiency. A *kanban* [just-in-time delivery] system, with no semi-processed stock in any production process, is an extreme example of the optimisation of process control. This innovative technology, introduced by the Toyota Motor Corporation, encompasses all processes, not only at the assembly level, but even extending to subcontractors. Process control usually starts with improvement of the working environment. Examples include cleaning and tidying the production site, classifying components and semi-processed products, rearranging the layout of production flows, and announcing work targets and results.

Another key to the management of production activities has been to let workers themselves undertake process control, thus tending to produce a sense of responsibility and a will to work among employees. Learning, maintaining and improving methods of process control also fosters teamwork, while cultivating the problem finding and problem solving capabilities of members. Conversely, process control is often carried out effectively by teamwork between related processes. Prerequisites for process control conducted by the workers themselves are, first, harmonious human relations among team members through the sharing of information, and second, effective leadership, usually by a team leader, group leader or supervisor. Such teamwork naturally leads to the emergence of QC and TQC activity.

Process control as described above reveals a major characteristic of Japanese management technology, namely that information and knowledge tend to be mediated through human channels, rather than physical channels such as machinery, manuals, specifications and blueprints. For this reason, Japanese management technology is also referred to as technology to develop human resources.

Such characteristics of Japanese management technology have long been regarded as peculiar to Japan with its particular socio-cultural background. However, the technology may contain universal values if adopted effectively in other countries. Indonesia may provide a good 'laboratory' for examining whether Japanese management technology can be adopted in another country, and if it can, how and why firms should do so.

Over the past three decades, Japan has been the dominant foreign investor in Indonesia. According to Investment Coordinating Board (BKPM) approvals data, from 1967 to 1996 there were 4,843 foreign investment projects (PMA) totalling $173.6 billion.[126] Of this, Japan accounted for 919 projects (19% of the total) valued at $27.8 billion (16% of the total). This far exceeded the next largest investors, Singapore (595 projects valued at $13.3 billion) and the UK (217 projects valued at $20,809 million). Since some Japanese firms have invested through, or jointly with, their Singapore subsidiaries, the actual number of Japanese investments is estimated to be a good deal larger still, and probably exceeds 1,000 projects.

Due to this accumulation of Japanese direct investment, Indonesia has a number of suitable candidates for the study of the transfer of Japanese management technology (Thee 1990). These include Japanese joint venture companies with more than a decade of experience in Indonesia, particularly in assembly-type manu-facturing industries. Compared with single-line processing industries (thread to clothes, or logs to plywood), assembly operations (manufacture of automobiles, motorcycles, electronics, machinery, components) require a far more precise process control to handle multiple components and multiple manufacturing processes simultaneously. Such industries are therefore appropriate for the development of Japanese management technology. The following case study is chosen from several candidates satisfying the above conditions.

---

[126]The data compiled by BKPM do not cover the oil, gas and finance sectors. These figures refer to aggregate approvals less projects withdrawn after approval. There are no accurate data on realised investment, which is thought to total 60–70% of the approvals figures.

THE CASE OF THE ASTRA GROUP

The Astra group was chosen as a case study of the transfer of Japanese management technology because of its multiple joint ventures with Japanese companies over almost three decades, and because of its business experience in assembly-type machinery industries.

Main Features of the Astra Group[127]

The Astra group is the third largest business group in Indonesia, after the Salim group and the Sinar Mas group. The group's holding company cum headquarters, PT Astra International, has the largest net revenues of over 250 companies listed on the Jakarta Stock Exchange, amounting before the recent financial crisis to Rp 12,284 billion (approximately \$5.2 billion at the June 1997 exchange rate). It had 123,000 employees in 1996 (Astra International 1997, annual).

The Astra group is Indonesia's largest auto maker, with more than 50% of the domestic four-wheel and motorcycle markets. While it has been diversifying into the agri-business and infrastructure sectors, the group's business base still lies in the machinery industry, which constitutes its major activity. Its major products include automobiles (with Toyota, Daihatsu, Isuzu, Nissan Diesel, BMW and Peugeot); motorcycles (Honda); heavy equipment (Komatsu); and office equipment and electrical appliances (Fuji-Xerox and Goldstar).

The group has taken part in an abundance of joint ventures and technical cooperations with foreign partners. Of around 120 affiliated companies under Astra's umbrella, at least 31 are joint ventures with foreign companies; as of 1997, 24 of these were with Japanese partners. Its business ties with Toyota date back to 1967, with Honda to 1970, and with Komatsu and Daihatsu to 1973. This group is one of the most prominent cases in Indonesia of long and close tie-ups with Japanese manufacturers.

In the course of the Astra group's development, there has been some concrete evidence of transfer of Japanese management technology. Two such instances are examined in this chapter: the production of financial statements, and the development of QC and TQC activities.

---

[127]See Sato (1996) for a broader and more detailed study of the Astra group.

Diffusion of Financial Management Technology

PT Astra International is probably the first domestic private company in Indonesia to have produced audited financial statements. This occurred in 1972–73, when the value of audited financial statements was hardly recognised in Indonesian business. Though certification of the first issue was rejected in 1972 by an Indonesian auditor (Drs. Utomo Mulia & Co.), in the following year, with the guidance of a Dutch financial expert, the second issue was unconditionally certified by the same auditor. This constituted the first step in Astra's steady course of modernising its financial management.

What should be noted here is that both financial statements adopted exactly the same format as those issued by Japanese-affiliated companies in the Astra group. These affiliates were obliged by their principals in Japan to prepare financial reports from the first year of their operations, with the task being carried out by the firms' Japanese financial managers using a standardised format. It was this format that Astra International adopted, from the classification of assets/liabilities and income/expenditure to the colour of the cover sheets, despite the fact that the Indonesian accounting system basically followed Dutch methods. This can be regarded as a diffusion of financial management technology from a Japanese joint venture. But it was in fact instigated at the initiative of Astra International—in this case, probably the founder and CEO of the group, William Soeryadjaya—rather than by the Japanese joint venture partner.

Development of Astra's QC and TQC activities[128]

The second aspect of the transfer of Japanese management technology—QC and TQC activities—is, unlike the first case, an example of deliberate transfer by Japanese partners. According to my own observations, QC and TQC activities in the Astra group have progressed through the following five stages.

---

[128]This section is based on information given in *Astra* (various issues) and interviews conducted by the author with the public relations division of Astra International.

Transplantation of QC

QC circle activities were first introduced into the Astra group in the mid 1970s. It was at this time that the component localisation program started, with Honda (motorcycles) and Toyota (commercial cars) shifting from completely knock-down (CKD) assembly to body and part pressing. The Japanese partners initiated transplantation of QC circle activities in the new body pressing joint venture companies, PT Honda Federal and PT Toyota Mobilindo, and then in the respective locally owned assemblers, PT Federal Motor and PT Multi Astra.

These QC circles were quite different from their Japanese counterparts, consisting of courses run by Japanese experts for the benefit of Indonesian upper level personnel (general managers, managers and engineers), rather than teamwork activities by workers on the shop floor. The goal was not to teach actual process control to the workers themselves, but simply to introduce basic ideas on quality control, process control and the importance of maintaining the working environment in good order. Thus what was transplanted was the Japanese *concept* of QC rather than QC activities as management technology.

Diffusion of QC

The introductory courses on Japanese QC circles gradually permeated among supervisors, group leaders and other upper level personnel. A series of courses was held to generate harmonious and information sharing relations among the participants. Some personnel, mainly supervisors, were then put in charge of QC (or *kaizen*, which means 'improvement' in Japanese), to lead teamwork activities in the plants. They began by focusing on the cleanliness of the work site, and by seeking proposals to improve work processes and environments. In this manner, QC permeated the workplace, and progressed from concepts to activities.

The diffusion of QC occurred not only within one company but also across company boundaries. It spread to automobile-related, non-joint venture companies within the group, including PT Gaya Motor, the assembler of Daihatsu, Peugeot and Renault automobiles. In this second stage of QC development, the lecturers of introductory courses were no longer Japanese experts, but Indonesian personnel seconded from plants where QC activities had first been introduced.

In the new participant companies as well, QC permeated from engineer level to production sites, progressing from guidance courses to teamwork activities led by Indonesian core leaders.

Transplantation of TQC

Around 1980, when QC began to be adapted to the management of non-production sections, it was broadened to include TQC. It is noteworthy that this broadening first occurred in four companies, three of which (PT Gaya Motor, PT United Tractor and PT Multi Astra) were not Japanese-affiliated companies. While the concept of TQC was initially transplanted to the group by Japanese experts through seminars and guidance courses, the initiative for putting it into practice was clearly taken by the Indonesian side. This is perhaps to be expected: Japanese-affiliated companies were managed by a limited number of Japanese personnel and so had no pressing need to implement TQC. However, the management of non-Japanese-affiliated companies was, with the exception of technical matters, in the hands of Indonesian managers, and there was no room for the Japanese to intrude. Whether TQC would be adopted, and if so, how, was therefore left entirely to the Indonesian side. The most prominent example of transplantation and adaptation of TQC was PT Gaya Motor. This company, under the leadership of some managers/core leaders who had participated in QC and TQC courses, actively applied TQC to the financial, planning and stock management of the company. The company later became a leading promoter of TQC within the Astra group, dispatching lecturers and experts to affiliated companies.

Diffusion of TQC

TQC spread from a few advanced companies to other affiliates. At this stage, important diffusion vehicles included TQC training courses and contests throughout the Astra group. These devices were planned and sponsored by Astra International headquarters, and were carried out with the participation of core leaders from the advanced companies. The group's top management, including managers from company headquarters and each operating company, were thus involved in these programs.

From these activities, the designation 'ATQC', or Astra Total Quality Control, emerged in 1983. The Indonesian government,

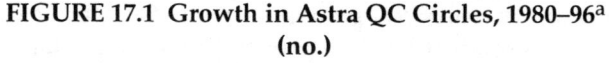

FIGURE 17.1 Growth in Astra QC Circles, 1980–96[a]
(no.)

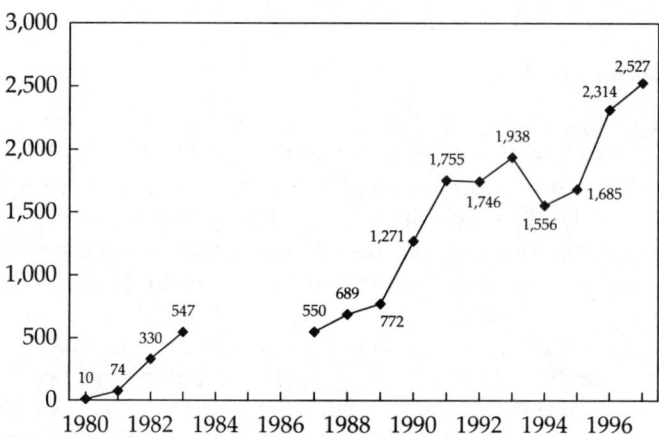

[a]Data for 1984–86 are not available. Data for 1997 are through to June.

Source: Information supplied by the Public Relations Division of PT Astra International.

through a statement by the then Minister of Industry, praised the Astra group for pioneering TQC in Indonesia. However, the real state of ATQC at this stage was that it was no more than a cluster of firms, each conducting TQC activities.

Creation of ATQC

After 1983, Astra headquarters adapted TQC to encompass the entire group. This form of ATQC was a new creation of the Astra group's own management system, and can be regarded as a genuine innovation. The group-wide ATQC constituted a unified system, characterised by regular managerial meetings, business planning methods, and recruitment and training initiatives. Figure 17.1 shows the rapid growth of these Astra circles, from a mere handful in 1980 to several hundred by the mid 1980s, and to over 2,500 in 1996.

Managerial meetings include weekly meetings of the Board of Directors at headquarters, monthly meetings of division managers, and semi-annual (now annual) group-wide 'leaders meetings' (*rapat pemimpin*) attended by managers of affiliated companies. At Astra

International, the previous year's business results from each division and the business plans for the next two years (now the next five years) are presented in a standard format and used as reference materials at the regular managerial meetings. A unified recruitment system has been implemented, with a batch of new university graduates being hired at one time for the entire group. Newly hired employees are trained at the Astra Training and Education Centre under a unified training curriculum. Leaders of the Astra group now describe all these managerial devices generically as 'Astra TQC'. According to the group's calculation, its productivity increased by about 25% after ATQC was introduced; the amount saved by ATQC was estimated to reach Rp 500 million annually by 1997.

This most recent stage of ATQC had no existing model to follow, in Indonesia or in Japan, and does not appear to have been influenced by the Japanese practice of QC and TQC. ATQC is very much Astra's original management technology, and may illustrate the final stage of technology transfer.

FINDINGS AND IMPLICATIONS OF THE CASE STUDY

One case study is, needless to say, hardly sufficient to draw general conclusions about the transfer of management technology to Indonesia. Nonetheless, Astra's experience provides some important clues to the nature of the technology transfer process, and some pointers to successful practice. At the very least, this case study demonstrates that Japanese management technology can be transferred to Indonesia. Findings from the Astra case can also shed light on other studies of technology transfer.

Summary of Findings

- Japanese management technology can be transferred to Indonesia, even though the technology is allegedly Japan-specific, and even though it is sometimes argued that Japanese partners are reluctant to transfer it.
- The management technology was transferred not only within one Japanese joint venture company, but also to a group of affiliated firms, including locally owned companies.
- QC/TQC concepts, not activities, were transplanted first.

- For the transplantation of QC concepts, formal guidance courses for Indonesian personnel at all levels needed to—and did—function effectively.

- Concepts were conveyed by the Japanese partners. The transmitters of information were Japanese and then Indonesian in the case of QC, and Indonesian from the early stage in the case of TQC.

- Prior to start-up, communicative human relations, featuring shared common information, were expanded and deepened within the group.

- On the Indonesian side, there emerged not only transmitters who had absorbed QC/TQC concepts, but also activity leaders at various levels of the group structure who, even though appointed from above, promoted diffusion of these activities.

- QC activities (and financial statements) followed the Japanese model. TQC activities transcended the Japanese model to reach a new stage of development in the Astra group.

Based on these findings, the rest of the chapter examines the nature of technology transfer, keys to success, and major characteristics of the Astra case, in light of the present situation in Indonesia and in comparison with the Japanese experience.

Evolutionary Nature of Technology Transfer

What is the nature of technology transfer in practice? The case study of Astra's QC and TQC activities tells us that the technology transfer process is evolutionary in nature, with repeated transplantations and diffusions, and with accumulated instances of adaptation and modification. 'Evolution' is not merely a sending–receiving process, in which the receiver passively receives what is being sent. In the evolutionary process, there must be changes on the receiver's side.

In Astra's case, the first point to note is who played the role of transmitter of the transplanted technology. Initially the teachers of QC courses were Japanese, who were later replaced by Indonesians as QC spread to other companies. TQC courses were conducted by Indonesian lecturers from the beginning. It was at this point, when Indonesians became the lecturers, that 'passive receivers' were transformed into 'active transmitters'. This was the first crucial change in the nature of the Indonesian receivers.

The change from passive receiver to transmitter gives the recipient scope to modify the technology received to suit local conditions. The second crucial change thus lies in the transplanted technology itself—in the adaptation of the technology to the receiver's environment. One example of such an adaptation was the broadening of QC to TQC at PT Gaya Motor after the company's leaders learned of TQC through courses. Another example is ATQC, in which company-wide TQC was extended to group-wide management. ATQC was further modified to become a unified group management system. As this example shows, adaptation may produce further modifications and improvements, including the creation of new technology.

Through such changes in the nature of receivers and in the technology itself, the initiative to lead the evolutionary process of technology transfer shifts from the sender to the receiver.

Key Factors in Technology Transfer

Why was Astra able to participate so effectively in this evolutionary technology transfer process, and what were the key factors in its success? In the QC/TQC activities of the Astra group, what the Japanese partners transplanted in the first stage was simply concepts, not activities. It took some time for the transplanted concepts to develop into activities, and required the presence of certain preconditions.

The first of these was an environment in which common information could be shared freely throughout the recipient organisation. As the lessons of the QC courses diffused, harmonious, communicative and information sharing relationships permeated, not only among managers but also at the plant level. This process took place first in some core companies, and then gradually spread to other companies in the Astra group. This indicates that lack of communication and 'divisionism' can be a major obstacle to the diffusion of technology, and consequently to the modification and creation of technology.

The second key condition was that there existed leaders at every stage of the evolution. In the second to fifth stages of development of QC/TQC activities, the initiative for further development was taken by the Indonesian side. In the diffusion of QC, supervisors and group leaders led activities in some core operating companies, while managers and supervisors acted as lecturers in other companies in the group. The managers of core operating companies, together with managers from headquarters and the group's top management, were involved in the diffusion of TQC. In this case, the multilevel structure of leaders and increasing numbers involved in the process are readily apparent. In producing financial statements, the leadership lay clearly with the Indonesian side, especially the top management of the group.

Astra's case thus demonstrates, first, that transmitters of technology transfer are not so hard to find among Indonesian partners. These are the activists and activity leaders (or human resources) who can transform the role of recipients from a passive to an active one. A second lesson is that leadership does not necessarily emerge naturally from the bottom levels of workers at production sites, as has been the Japanese experience, but can reside in the top levels of management. Appointed leaders and processes of technology transmission managed from above can function well. What is important is the existence of leaders; from where and how these leaders emerge is not a major issue.

To summarise the discussion so far, the case study of the Astra group has illustrated that technology transfer is an evolutionary process, needing an active response on the recipient side to adapt, modify and develop the transplanted technology and to create something new. The keys to success extracted from this are communicative and information sharing human relationships in the recipient society, and the existence of leaders as a locomotive in the evolving process of technology transfer.

Technology Diffusion in a Group of Companies

What are some of the general implications for the transfer of technology to Indonesia based on the Astra experience? This case study has demonstrated that Japanese management technology can be transferred to Indonesia. Indeed, the author's field survey indicates that the transfer of Japanese management technology within Japanese-affiliated companies has become more common in present day Indonesia. What renders the Astra case of special interest is that technology transfer is not confined to one Japanese-affiliated company, but has been extended to a group of affiliated companies, including locally owned firms.

In the case of a Japanese affiliate, the transplantation of management technology effectively takes place under the full control of Japanese managers. In most cases, there is little room for adaptation and modification by Indonesian personnel, which tend to be viewed by Japanese managers as deviations from preferred practice. Moreover, the sphere of diffusion is limited to the 'boundaries' of the company, and does not extend beyond the control of the Japanese managers.

This was also the case with the Astra group when QC was first introduced in some core Japanese-affiliated companies. Thereafter, the technology diffused through the Japanese-affiliated firms to non-affiliated group companies (first to Toyota and Honda-related companies, then to non-related ones such as PT Gaya Motor). Most of the diffusion of QC and TQC beyond the boundaries of the Japanese affiliates was undertaken by Indonesian transmitters of technology. There was considerable mobility among transmitters (the lecturers and QC/TQC activity leaders), which did not constitute a natural diffusion of information. As knowledge of QC and TQC spread, the technology was adapted and modified to ensure that the practices took root in each company. Because the Astra group contained an autonomous sphere of firms not affiliated to Japanese companies, this kind of active diffusion and modification—and even innovation—of Japanese management technology could occur. This indicates that a degree of autonomy is needed in the recipient society for the evolutionary process of technology transfer to proceed, at the recipient's initiative.

Furthermore, Astra's case proves that in Indonesia a group of affiliated companies can share a technology and a systematised pattern of corporate behaviour. In Japan, the *keiretsu* provide such an example of technology sharing among a group of companies, with member companies jointly participating in the development of new models, as well as process control optimisation under the *kanban* system. The group of companies in a *keiretsu* has no common ownership, unlike the Indonesian business groups. However, the experience of the Astra group indicates the possibility that a technology sharing group of companies, which has no common ownership but which is bound by a systematised corporate behaviour (for example, a smoothly functioning subcontracting system), can develop in Indonesia.

The Evolution Gap between Management Technology and Production Technology

The discussion so far has concentrated on the transfer of management technology. Based on the above analysis, the evolutionary process of management technology transfer can be formulated as follows:

learning, absorption, adaptation, modification/improvement, and creation/innovation. This process may run parallel to the transfer of production technology in general, which can be formulated as follows: operation, maintenance, component/subcomponent production, improvement, and development/innovation. In the history of technological development in Japan, the evolution of management technology has proceeded gradually, keeping pace with the evolution of production technology. This has not necessarily been the case in Indonesia, where the former has sometimes outstripped the latter. The reason is that production technology is increasingly embodied in machinery, while management technology is mostly embodied in persons.

Machine-embodied production technology can be transplanted easily in the short term by importing machinery, so that the machine-dependent processes of technology evolution—operation, maintenance and component production—can proceed relatively easily. The problem is that machine-dependent technology evolution tends to render recipients passive in nature, and is constrained by the specific embodied technology. This makes it hard for recipients to transform themselves into 'active' participants in their own technology improvement and development, which is a highly human-dependent process. A discontinuity therefore emerges between machine-dependent and human-dependent processes in the case of the transfer of production technology. The more easily the machine-embodied technology can be obtained, the larger the discontinuity becomes. This discontinuous progress may constitute one of the causes for Indonesia's frustration with the slow progress it is making in technology improvement and development.

Unlike production technology, the transfer of management technology can proceed only gradually, since it has to be mediated by persons. This gradual progress, however, is not accompanied by a discontinuity between processes, but by a continuous transformation of the recipients from passive to active participants; it is also free from the constraints of fixed embodied technology. Consequently, an 'evolution gap' between management technology and production technology sometimes occurs, in which recipients have become active in the modification of transplanted management technology, but are passive in regard to the transfer of production technology. This is the case in the Astra group: it has introduced innovations in management technology, but is still far from producing innovations in production technology, despite the fact that the Japanese partners began to transplant both technologies simultaneously. This evolution gap, which Japan has not experienced, is regarded as a feature of latecomers to industrialisation who receive an inflow of foreign direct investment with abundant machine-embodied production technology.

CONCLUSION

While discussions of technology transfer usually refer to production technology, the transfer of management technology is also worthy of study. In Indonesia, we found that Japanese management technology could be transferred, though the technology is allegedly Japan-specific and the Japanese partners are allegedly reluctant to transfer it. The Astra case study illustrates that the transfer of management technology is an evolutionary process, in which Indonesian partners have displayed a capacity to adapt and modify the transplanted technology, and subsequently to develop their own technology. The case study also indicates that management technology can diffuse beyond one company to form a technology sharing group of companies in Indonesia. Furthermore, an 'evolution gap' can emerge between a gradually evolving management technology and a discontinuously evolving production technology. With regard to management technology, the term 'technology transfer' may not be appropriate, because the transferred exogenous technology is merely an initial stimulant, while the endogenous (host country) initiative is a crucial factor in the subsequent evolutionary processes.

18

BATAM ISLAND AND INDONESIA'S HIGH-TECHNOLOGY STRATEGY

Shannon Luke Smith*

Led by its Minister for Research and Technology, Dr B.J. Habibie, the Indonesian government has pursued the promotion of high technology as an important component of its industrialisation plans. One of the key projects being developed under this framework is the transformation of Batam Island into a high-tech centre. The 1990s has witnessed the emergence of a not-insignificant electronics manufacturing industry on Batam, but there are questions as to its true character as well as its viability in the long term.

HABIBIE AND THE HIGH-TECHNOLOGY AGENDA

Since President Soeharto came to power in 1966, the Indonesian government has been characterised by two distinct and competing groups of economic advisers: the technocrats, and a diverse group of economic nationalists. While the technocrats are strongly committed to markets and competition, the nationalists have reservations about free market ideology and press for active government

---

*Shannon Luke Smith completed his Doctoral dissertation, 'Developing Batam: Indonesian Political Economy under the New Order', at the Department of Political and Social Change, Research School of Pacific and Asian Studies, Australian National University, in 1996.

intervention in and regulation of market behaviour. While both of these groups have influenced the policy making process, as well as policy choices, one of the most prominent individuals on either side in recent years has been the architect of Indonesia's high-technology strategy, Dr Habibie.

The economists, or 'technocrats', have been the chief source of macroeconomic policy under the New Order. They are philosophically committed to a market-oriented policy framework, a strategy emphasising labour-intensive industrialisation and regulatory reform. They have argued that, in view of Indonesia's labour surplus economy, the emphasis should be on export-oriented manufactures that can absorb cheap labour, rather than capital-intensive industries needing protected domestic markets in order to survive. While they have played a minor role in formulating industrial policy, their influence over macroeconomic policy formulation has won and maintained the confidence of foreign governments, financial institutions and investors.

The second broad category of economic advisers has been the economic nationalists, variously referred to as 'nationalists', 'interventionists', 'technologs', 'engineers' and 'technologists'. Made up of very diverse elements, they have generally been opposed to the doctrine of the free market, and committed instead to the notion that strong and focused state intervention in the allocation of resources can accelerate the process of industrial development. They argue that it is worth paying the short-term costs of protectionist policies in order to promote the development of state enterprises and *pribumi* [indigenous] entrepreneurs who cannot as yet compete in either domestic or world markets.

Both the economic nationalists and the technocrats recognise that the debate is part of an ongoing argument and that the positions of either side are not immoveable—in the late 1970s the technocrats acquiesced to the inward looking industrial policy advocated by then Minister of Industry Soehoed; and similarly, the 'broad spectrum', export-oriented industrial strategy developed by the traditionally 'nationalistic' Department of Industry from the mid 1980s shows shades of technocratic measures.

In the end, however, the decisive element in policy making and policy formulation is President Soeharto. Each side of the economic debate has to convince the President on an ad hoc, case-by-case basis. The President can be expected to listen to both sides of the argument and seek compromises, but in the end he calls the shots after the options have been presented to him.

One of the most forceful exponents of the 'nationalist' view that the government should maintain a large role in the economy has been Minister Habibie, who maintains that Indonesia can never hope to catch up economically with industrialised nations without a concerted government-led push to speed up the natural pace at which technology is transferred among nations. What the country needs, says Habibie, is a new comparative advantage in high-technology products. Believing that Indonesia's present export growth industries, such as textiles, clothing and footwear, have only a limited lifespan, Habibie sees Indonesia's future competitive advantage lying in investment in value added high technology and in the upgrading of human resource skills.[129]

Since entering cabinet in 1978, Habibie has been the chief architect of Indonesia's plan to master complex technologies, and with them leapfrog into the ranks of developed countries through medium-term generation of a self-sustaining high-tech manufacturing base. Habibie's economic concept is that high value added technology is the key to future economic success. He believes that Indonesia must focus on the 'competitive advantages' that only technology can provide, rather than relying on its traditional and 'comparative advantages' of abundant land and labour.

While the government has traditionally concentrated on basic infrastructure, such as roads, bridges and utilities, Habibie says that technology should be regarded as a crucial component of 'infrastructure' and invested in as such. In particular, he believes that Indonesia should not seek investment in labour-intensive industries, but pursue high-tech manufacturing in areas such as aircraft and ship production.

According to Habibie, the Indonesian economy would remain dominated by low-skill, labour-intensive and resource-based industries if its direction were left to market forces and comparative advantage, but a focus on technology would add value to domestic production and increase the productivity of Indonesian workers. Habibie believes that private companies will not on their own invest sufficiently in R&D or obtain transfers of technology from foreign firms, so the government must play a leading role in these areas. He therefore advocates government initiatives to establish

---

[129]For further discussion and analysis of Minister Habibie's approach, see Chapter 8 by Bishry and Hidayat and Chapter 9 by Rice in this volume.

high-technology industries and raise the skills of the workforce (Glassburner and Poffenberger 1983, pp. 19–20; Habibie 1993; *Far Eastern Economic Review*, 29 July 1993, pp. 58–60).

Habibie's focus on high technology has earned him plenty of critics, at home and abroad. The technocrats, and international institutions such as the World Bank, prefer a more traditional steady climb up the industrial ladder, with a priority on creating jobs; this would require investment in labour-intensive industries, and in raising labour productivity. Habibie himself associates his 'value added' strategy with the new generation of intellectuals and the engineering profession, and the comparative advantage approach with the economics profession, foreign lenders and international financial institutions (McLeod 1993, p. 4).

The major arguments against Habibie's ventures concern their limited effects on the economy as a whole. The technocrats, taking comparative advantage as their starting point, question the economic validity of high-technology production in a labour surplus economy. They point to the need to establish industries which provide jobs for a large labour force—there are over two million new entrants to the job market annually—and the need to acquire skills step by step. Fearing the creation of economic enclaves, the technocrats argue that industries which use sophisticated technology have limited linkages with an Indonesian domestic economy whose industrial capacity is very limited (*Far Eastern Economic Review*, 18 August 1983, pp. 55–57).

Other critics argue that Habibie's approach is costly for the economy, that Indonesia has little capital to invest in expensive projects, and that the high-technology strategy drains money that could be used for more productive purposes (*Kompas*, 4 March 1993, 11 March 1993). They also argue that his projects are aimed more at instilling national pride than at advancing the economy. The World Bank in particular has continued to press for an Indonesian economy which operates according to the principle of comparative advantage, and which is responsive to the free operation of the international division of labour.

More than any other minister, Habibie is untroubled by critics, for he has the best protection available—Soeharto—and in most ways Habibie's influence is derived from the President. As Hill (1994, p. 81 n. 26) noted, 'The Minister appears to be able to operate a personal fiefdom, immune to financial constraints and accountable only and directly to the President'.

HABIBIE'S BATAM ISLAND STRATEGY

Against this national backdrop, the debates surrounding the development of Batam Island have been little different. In the same way that Habibie has had strong presidential support for his plans to develop national aircraft and shipbuilding industries, both of which require heavy government subsidisation, he has received support for his plans to develop Batam Island as a high-technology industrial centre. President Soeharto has stated that 'Batam Island will be developed as a high-technology industry centre' (*Reuters News Service*, 11 December 1995).

Batam Island is said to be an integral part of Habibie's development strategy, although just what that entails has been poorly defined, and the strategy has been modified over the years. It is known that Habibie wants to promote technological industries on the island, but beyond that little is known. Habibie has never clearly outlined his high-tech aims for Batam. What exists are only vague references to 'high-technology industries', 'electronics' and engineering projects.

In a speech entitled 'Technology and the Singapore–Johor–Riau Growth Triangle', Habibie only once made reference to technology; judging by the numerous references to infrastructure projects, the speech may have been targeted towards his audience, who were engineers. Nonetheless, it is one of very few references Habibie has made about his technological aims for Batam Island. According to Habibie (1992):

> ... the technologies relevant to the future of the [Growth] Triangle are those which will help to increase the physical integration of the three corners of the [Singapore–Johor–Riau] Growth Triangle. They include technologies to build economic infrastructure; transportation—tunnels, bridges, harbours, airports and roads; communications; [and] clean water and energy. They also include the technologies relevant to the development of Batam as an industrial area ... Other than electronic components, other downstream industries adding value to raw materials produced nearby could be developed ...

Habibie's plan for Batam's development does not envisage the island as a place for relocating low value added and labour-intensive industries from Singapore or elsewhere. The favoured industries are non-polluting light, medium or heavy industries oriented towards exports, with low water consumption, and using skilled labour and medium to high technology (electrical, mechanical, optical, electronic), with the overall aim of making Batam a high-tech industrial centre employing highly skilled labour (interview with BIDA executive Gunawan Hadisusilo, 10 August 1994). So far several dozen domestic and foreign companies utilising low–medium technology, mainly in electronics, have decided to relocate to Batam Island.

Before assessing the current status of the high-technology strategy, it may be useful to provide a background of the evolution of Batam under Habibie. Habibie took over the project in 1978 and quickly stamped his mark on developments, overturning the open investment and regulatory environment created by the technocrats, and implementing interventionist and economic nationalist policies. The policies failed, however, and by the end of the 1980s a rethinking was necessary. Following regulatory changes in 1989, Batam Island experienced a boom in economic activity.

BATAM ISLAND IN THE 1980S: THE FAILURE OF ECONOMIC NATIONALIST POLICIES

Following the 1975 financial crisis in the national oil company, Pertamina, the Batam Island project, which it had pursued since the late 1960s, was handed over to Dr J.B. Sumarlin, a technocrat and then Minister for Administrative Reform and Deputy Chairman of Bappenas. With the support of several other technocrats, Sumarlin set about changing the nature of the Batam project as a whole, introducing wide-ranging regulatory reforms and reorienting development to complement that of Singapore.

The technocrats introduced a host of reforms for Batam Island, with the overall aim of encouraging development by stimulating private sector involvement and by aligning the regulatory environment to that of Singapore. Everything that Batam Island required to attract investment was provided—lower labour costs, land prices and harbour tariffs than in Singapore; new facilities and incentives for investors; and the bolstering of the Batam Island Development Authority (BIDA) to give it control over land allocations, investment applications and other areas that are usually the preserve of central government departments. The most important change was the decision to declare all of Batam Island a bonded zone, with the aim of encouraging the development of export-oriented industries and facilitating the importation of materials required by manufacturing industries located on the island.

The regulations introduced for Batam were an important step towards the liberalisation of trade and investment and the stimulation of foreign and private domestic sector involvement on Batam. More importantly, however, they were the key requirements for soliciting support in Singapore for the Batam Island project. Indonesia had made various efforts in the early 1970s to turn Batam Island into a competitor to Singapore, but the failure of those efforts, combined with the increasingly warm relations between Jakarta and Singapore, resulted in a turnaround in Indonesia's policy towards Batam. The technocrats realised that Singapore, located opposite Batam Island, was in a very favourable position to influence the island's development.

Instead of trying to develop Batam independently as a competitor to Singapore, the Indonesian government began to look to both the private sector and Singapore as the motor for Batam's development. The strategy was explained by Barli Halim, a technocrat and Chair of the Investment Coordinating Board (BKPM), who called for labour-intensive investments to move to Batam Island from Singapore and for a greater role for the private sector (BIDA 1980, p. 7; *Straits Times*, 10 June 1983).

When Habibie was handed responsibility for the development of Batam Island in late 1978, he inherited the economic agenda determined by the technocrats. They had created a framework for development based on a relatively open regulatory environment, complementarity with Singapore and a broad-based industrial make-up. Habibie—most likely guided by President Soeharto— initially pursued cooperation with Singapore, as witnessed by the signing of the Batam Cooperation Agreement in 1980.

However, Habibie quickly changed the focus of development away from broad-based industrialisation to intermediate, high-technology and capital-intensive industries, most notably by discouraging labour-intensive industries. Despite having negotiated the agreement with Singapore, Habibie was almost to shun new investment in favour of building up Batam's infrastructure, and where new investment was encouraged it was to be confined to technology-based industries. The result was limited economic activity for the decade ending in 1988.

Habibie argued that economic development and progress could take place only on a firm basis of investment in capital-intensive intermediate and high-tech manufactures. This could only be generated by deliberate state intervention, and would not be produced by free operation of an international market (Robison 1987,

p. 30, fn 30). Minister Ginandjar Kartasasmita, a Habibie supporter and Chair of BKPM in the mid 1980s, argued at the time that investment on Batam Island would improve Indonesian technical capabilities. According to Ginandjar (*Kaleidoscope International*, 1984 ix/1, p. 97):

> In Batam we want to develop high technology and machinery industries which hopefully will eventually serve other areas as well as being a staging point for export. This is its major purpose.

One of the main implications of this focus on high technology was that particular industries, notably those at the lower end of the technological scale, would not be welcome on Batam. The first indication that this would be the case was when Habibie began calling for capital-intensive investment on the island and urging labour-intensive investment to go to Java (*Kompas*, 21 May 1979). However, the restrictive investment climate became most apparent when, in cooperation with BIDA, the new BKPM Chair, Suhartoyo, issued an Investment Priority List (DSP) for Batam Island in 1982 (BKPM 1982a).

The DSP essentially restricted foreign investment to sectors requiring special technologies or skills, and those with export earning potential. It identified approximately 40 manufacturing and service industries as priority areas for investment on Batam, including food preservation, building components/materials, tools, electrical components, optical/photographic equipment, applied electronics and any oil industry activity. Most significantly, the DSP openly discouraged labour-intensive operations (BKPM 1982b; *Business News*, 2 August 1982; *Asia Research Bulletin*, 31 January 1983, pp. 1,002–1,003).

The restriction on labour-intensive industries severely damaged Batam's chances of attracting new foreign investment. During 1980, 12 Singapore companies had received approval for joint ventures and another 25 had made applications to invest on Batam Island, mainly in areas such as food processing, wood processing, brick making, garment making, biscuit making and warehousing (*New Nation*, 4 July 1980; *Straits Times*, 26 November, 6 December 1979). None of these investments was ever to be realised. For instance—and typically of prospective investors—textile manufacturers planning operations on Batam Island withdrew their investment applications because of the restrictions laid down in the DSP (*Straits Times*, 9 April 1981).

The restrictions on particular industries and labour-intensive operations were viewed with disdain in Singapore, for they were the antithesis of how the Singaporean government saw the role of Batam Island. Prime Minister Lee Kuan Yew expressed the opinion that attempts to leapfrog what should be a gradual industrialisation process would not be successful (*Straits Times*, 5 July 1980):

> There must be a gradual transformation—first labour-intensive, then gradually, skill-intensive. It is a process which cannot be short-circuited ... If Batam is to succeed, it must start off with labour-intensive industries. These are investments which will provide you with an opportunity to develop your core of skilled workers.

The results of the decade of development to 1988 were poor. In particular, BIDA made little headway in attracting new foreign investment. In the period 1978–88, only half a dozen or so new foreign investments were realised—one was a hotel first planned in the early 1970s and the rest were related to the oil industry, ironically as a result of 'nationalistic' policy measures introduced by Pertamina. By the end of 1988 only 13 foreign companies were operating on Batam, with export earnings amounting to only $44.2 million. The only new economic activity to make an impact was tourism, which accounted for 41% of total foreign exchange earnings (Ahmad 1993, pp. 96–97).

Potential investors continued to express reservations about the level of infrastructure, the regulatory environment and incentives. BIDA appeared to be responsive to existing investors' infrastructure needs—government investment accounted for around 50% of total investment on the island—yet it was still unable to supply sufficient electricity or water. But as far as the regulatory system was concerned, no real attempt was made to introduce policy changes or initiatives to make Batam Island a more attractive location for new investment.

Habibie had complete control over developments on Batam, and was largely unaffected by interference from the technocrats. This was clear from the speed with which he changed the focus of development, away from the technocrats' target of broad-based industrialisation, towards intermediate, high-technology and capital-intensive industries. Under Habibie, the state monopolised economic policy and activity on Batam—it was the main investor, and it controlled the activities of the private sector tightly through restrictive economic nationalist policies such as the DSP.

BIDA and Batam Island enjoyed special status. Neither the various sectoral ministries in Jakarta nor provincial authorities in Riau were willing or able to interfere. If the technocrats had wanted to introduce regulatory reforms for Batam they would not have had much success—Habibie was in firm control and apparently not to be budged from his policy of regulating investment behaviour there. However, considering the enormous changes taking place at the macroeconomic level, where the technocrats had greatest influence, it is likely that Batam was of little interest to them. In the mid 1980s, declining world oil prices necessitated a shift in economic policies towards progressive deregulation of the Indonesian economy, and it was at this level that the technocrats' efforts were most focused for a good part of the decade. As a result Batam Island remained a virtual enclave, unresponsive to outside influence, either from government or business.

In 1989, several crucial events made Batam Island a more attractive investment location. One was an agreement between Indonesia and Singapore to cooperate closely in its development. Another was a series of regulatory changes which allowed 100% foreign ownership of enterprises on Batam, as well as foreign investment in industrial estates. Coinciding with general trade and investment liberalisation in Indonesia and the beginnings of large Singaporean investment abroad, these changes led to an economic boom on Batam in the early 1990s. This boom appears sustainable for at least the rest of the decade.

BATAM ISLAND IN THE 1990S: REGULATORY CHANGE AND ECONOMIC TAKE-OFF

In 1989 the Indonesian and Singaporean visions for Batam Island finally converged. Earlier attempts at cooperation had received scant attention in either Indonesia or Singapore, mainly because of their differing goals and priorities. The 1980 Batam Cooperation Agreement, in particular, had achieved little in terms of cross-border movements of capital and goods. However, in 1989 it became clear that the leaders of both countries wished to cooperate in Batam's development.

At this time the Singapore government was becoming increasingly concerned about the growing number of Singapore-based multinational corporations (MNCs) relocating elsewhere in the region, and began looking at ways in which Singapore could maintain its multinational presence while remaining a competitive and attractive investment location. Two studies, one by the Monetary Authority of Singapore (MAS) and the other by the Economic Development Board (EDB), concluded that an industrial estate on Batam Island would assist in maintaining Singapore as a beachhead for direct investments in the region, while also creating significant investment and employment opportunities for Indonesia (*Singapore Business*, August 1991; *Straits Times*, 2 August 1991; interview with former Singapore Minister Goh Keng Swee, 29 September 1994).

During discussions with President Soeharto in August 1989, Prime Minister Lee Kuan Yew raised the possibility of MNCs, particularly in the electronics industry, obtaining terms and conditions for investments in Batam similar to those provided by Singapore, so that they could consider Batam when expanding (*Straits Times*, 3 August 1989; *Business Times*, 5–6 August 1989). Lee's idea was well received in Jakarta, where the initiative coincided with Indonesia's push for a more open and less regulated economy, in part to attract foreign investment.

Despite the fact that Indonesia had long concentrated on developing Batam in a 'nationalistic' manner, several factors led to a dramatic change in attitude towards cooperation with Singapore. In particular, there was a strong feeling that the government was 'running against time' in its efforts to attract investment to Batam Island and elsewhere in Indonesia, and that changes had to be made—both regulatory and in attitude. Furthermore, the Indonesian government, for long frustrated that limited funds and poor investor interest were hampering Batam's development, realised that great potential lay in cooperation with Singapore. At the same time, Singapore placed extra pressure on Soeharto and Habibie to make the regulatory changes the technocrats were urging.

Habibie was interested in attracting electronics MNCs to Batam Island. Already in early 1989, one such firm, French electronics company Thomson S.A., had established an assembly plant on Batam because of worker shortages and costs in Singapore (CIDB 1989, p. 5). The question for Habibie, however, was what kind of technology would be utilised by the relocating companies and what incentives would be needed. Following a visit to Singapore in September 1989, to discuss its experiences with multinationals as well as to visit several electronics MNCs, Habibie announced that he was open to having labour-intensive, technology-based industries, particularly electronics operations, on Batam Island (*Straits Times*, 6 September, 7 September 1989). Even though electronics assembly operations would be very low tech, with many simply supplying parts and components for assembly in Singapore, it appears that they conformed with Habibie's loosely defined aims for technology-based industries on Batam.

To facilitate the entry of foreign-owned electronics operations, the Indonesian government announced policy changes in October 1989 which provided for 100% foreign ownership on Batam Island, the only condition being a 5% divestment to an Indonesian partner within five years (BKPM 1989). The policy differed radically from that in the rest of Indonesia, where domestic ownership had to reach 51% within 15 years. It essentially made Batam Island a special economic zone, allowing foreign investors to set up companies without a local partner. The insistence that Indonesians hold at least 5% equity was a matter of contention, especially for electronics companies concerned about losing sensitive technology to their partners. However, the ruling appeared to be a compromise, catering to both Indonesian nationalist sentiments by maintaining an element of domestic ownership in foreign investment projects, while at the same time being designed to guard technology as demanded by electronics firms.

A further policy change, allowing both private foreign and domestic companies to set up industrial estates in Indonesia, was to be equally decisive for Batam Island. While the decision was aimed at circumventing infrastructure problems—industrial estates would provide investors with electricity, water, telecommunications, housing and factory space—it also paved the way for a Singaporean-led industrial park on Batam.

Specific details of joint cooperation on Batam Island were discussed by a joint working committee consisting of five high-level officials from both Indonesia and Singapore. At the first meeting, held in early December 1989, discussions about the EDB's proposal for an industrial estate on Batam Island only confirmed behind-the-scenes negotiations during the previous months. Having gained an investment waiver for electronics companies worried about intellectual property rights and 'piracy' in any relocation to Batam Island, having found a suitable land site, and having partnered Singaporean and Indonesian investors, all that was needed was to formalise existing agreements (*Straits Times*, 27 December 1989; interview with former Singapore Ambassador to Indonesia Barry Desker, 23 September 1994).

Batamindo Industrial Park was formalised by a Memorandum of Understanding and Joint Venture Agreement on 11 January 1990. This marked a turning point in Indonesian–Singaporean cooperation on Batam Island. According to Singapore Trade and Industry Minister, Lee Hsien Loong (*Indonesia Development News*, April 1990):

> Combining the resources, experience and expertise of Indonesia and Singapore will create a good investment environment for both countries, and facilitate efforts to attract large multinational corporations for their mutual benefit.

The major policy changes implemented by the Indonesian government, together with the interest and strong support of the Singapore government, were instrumental in overcoming the earlier reluctance of Singaporean and other foreign investors, and led to a surge of investment on Batam. It became both a privileged and attractive investment location in Indonesia, and for the first time received serious investor attention from Singapore and elsewhere.

The island experienced a boom in economic activity during the early to mid 1990s. Encouraged by changes to the investment regime, foreign investors rushed into Batam—approved foreign investment in 1990 was almost double that of 1989, and almost five times that of 1988. As shown in Table 18.1, foreign investment has remained strong since—total cumulative foreign investment increased from $428 million in 1989 to $1,916 million at the end of 1995 (BIDA 1996).

Equally important was that not only did the volume of economic activity change after 1989—the value of exports increased from $53 million in 1989 to $2,362 million in 1995 (Table 18.1)—but the nature of that economic activity changed the face of development. This is clear from the composition of exports. While in 1989 the main exports were oil equipment (64.2%), steel pipes and clothing, by 1995 the main export item was electronic products (83.2%), followed by jewellery and oil equipment (BIDA 1996).

More than 50% of private investment, and more than two-thirds of all foreign investment, on Batam has been poured into manufacturing industry. The industrial areas in Sekupang, Batu Ampar, Tanjung Uncang and Kabil have received their fair share of this investment, which has been dominated by manufactures, machinery and basic metals. Offshore oil drilling equipment remains important to Batam as the third biggest export item.

TABLE 18.1 Batam Island Economic Indicators, 1983–96

Year	Labour Force (no.)	Exports ($ million)	Foreign Investment ($ million)		Domestic Investment ($ million)	
			Annual	Cumulative	Annual	Cumulative
1983	5,500	3		206		43
1984			9	215		
1985	6,159		7	222		
1986	6,764	21	1	223		161
1987	8,672	27	–	223		
1988	9,478	44	66	289		
1989	11,041	53	139	428		
1990	16,085	152	256	684		1,515
1991	22,942	242	371	1,055	82	1,597
1992	31,644	565	33	1,088	436	2,033
1993	43,496	926	560	1,648	101	2,134
1994	69,630	1,339	225	1,873	162	2,296
1995	80,708	2,362	43	1,916	561	2,857
Jan–Jun 1996	84,667	1,513		2,033		2,549

Sources: BIDA, various publications.

However, since 1989 the majority of new investment in industry has been in electronics, with most electronics firms being located in Batamindo Industrial Park. This industrial estate—the keystone of Indonesia–Singapore cooperation on Batam—dominates the landscape to the extent that it is the most economically and politically important commercial venture on Batam. In fact, the transformation of Batam Island during the early to mid 1990s can largely be attributed to its establishment.

Batamindo Industrial Park is a S$600 million, 500-hectare joint venture between an Indonesian consortium with 60% ownership (majority-owned by the Salim and Bimantara Groups together with Habibie family interests) and two Singapore government-owned companies (Singapore Technologies Industrial Corporation with a 30% holding and Jurong Environmental Engineering with 10%)

(*Straits Times*, 12 January 1990; *Jakarta Post*, 12 January 1990). A self-sufficient, self-contained and integrated industrial park, the estate offers one-stop service and facilities—accommodation, transport and security services, immigration, recruitment, administrative support, estate management, factories, power supply, water supply, telecommunications facilities, commercial facilities—all of which are aimed at shielding investors from infrastructure inadequacies and bureaucratic inefficiencies (BIM 1994). The aim is to offer a Singapore-type environment with cheap labour and land costs in order to make investors feel 'at home'; that is, to feel just as if they were doing business in Singapore.

From its inception in January 1990, Batamindo Industrial Park has grown at a bristling pace. By the end of 1994 it had over 60 tenants with $250 million worth of investments, employing 35,000 workers and exporting $700 million worth of products. Of the 62 tenants signed in mid 1994, 25 were from Japan, 17 from Singapore, seven from the US, three from Germany, two from France, and one each from the Netherlands, Italy, the UK, Switzerland, Korea, Taiwan, Malaysia and Indonesia. The majority of tenants are involved in the manufacture or assembly of light industrial products such as telecommunications equipment, electronic components for audio and video products, printed circuit boards, integrated circuits and disk drives (*Media Indonesia*, 13 October 1995; *South China Morning Post*, 17 August 1995).

The estate's overall position and role in Batam Island's development is substantial. Average growth in tenants between 1991 and 1994 was 40%, outstripping that of Batam as a whole (*Business Times*, 10 November 1995). By 1994, just over three years after the first tenant began operations, the Park's tenants employed around 50% of the total formal workforce on Batam Island and around 70% of those involved in the industrial sector, in addition to providing other direct and indirect employment, such as in construction and services, and in the informal sector outside the park. Between them, the operators of Batamindo Industrial Park and their tenants had invested around $600 million, accounting for an eighth of total private investment on Batam. In addition, the tenants generated over half of Batam Island's total export value in 1994.[130]

---

[130]Exact figures are difficult to determine because they are based on details provided by investors in their BKPM submissions, and the actual values could be slightly higher or lower. However, because the tenants have shown high realisation rates, the figures are fairly accurate.

THE ELECTRONICS INDUSTRY ON BATAM ISLAND IN THE 1990S

The electronics industry has been one of the better performing industries on Batam in the 1990s, and is therefore a good candidate for investigating the success of Indonesia's strategy to move into high-technology industries. Only since deregulation in the late 1980s have exports of electronics increased rapidly on Batam Island and in Indonesia as a whole. These exports are composed mainly of consumer electronics, electronic components and industrial electronics.

Approximately 30% of total foreign investment on Batam is in electronics, ranging from sophisticated high-tech assembly and testing operations to simple low-tech and labour-intensive assembly operations, as well as skill-based supporting industries like precision machinery and contract manufacturing (*Bisnis Indonesia*, 27 December 1994).[131] Led by foreign firms, most of them based in Batamindo Industrial Park, exports of electronic products increased from nil in 1987 to $131 million in 1991, to comprise just over 50% of Batam's total export earnings (Table 18.2). The impact was felt nationally—in 1991 exports of electronic components such as integrated circuits and electronic parts from Batam Island amounted to more than half of Indonesia's total components and parts exports (Thee and Pangestu 1993, p. 131).[132] By 1995, exports of electronic products were approaching $2 billion and accounting for over 80% of Batam's total exports.

Despite impressive export figures, it is questionable whether the electronics industry has or will have a major impact on the Indonesian economy. One important issue is the development of an indigenous technological capability in the industry. According to Indonesian economists Thee Kian Wie and Mari Pangestu (1993, p. 133):

> ... there is very little connection as yet between building up of indigenous technological capability and export growth as in most cases the foreign partner is in control at the pre-investment and project execution stage; they import the majority of their components

---

[131] Of foreign investment on Batam, 53% is in industry; 58% of this is in electronics.

[132] Electronic components accounted for 34% of Indonesia's total electronics exports in 1991.

TABLE 18.2 Exports of Electronic Products
from Batam Island, 1987–96

Year	Total Electronic Products ($ million)	Total Exports ($ million)	Electronics as % of Total Exports
1987	–	26.8	–
1988	1.6	44.2	3.6
1989	8.0	52.9	15.1
1990	24.9	151.5	16.4
1991	131.2	242.0	54.2
1992	278.0	564.5	49.2
1993	689.6	925.8	74.5
1994	1,232.9	1,388.9	88.8
1995	1,965.4	2,362.0	83.2
Jan–Jun 1996	1,229.4	1,513.2	81.3

Sources: BIDA (1993a), *Realisasi Nilai Ekspor Pulau Batam, 1986–Okt 1993*, Jakarta; BIDA (1996).

so that there are [few] domestic linkages; obtain production, engineering and marketing capabilities through their headquarters or contractor; and there is no major change or design capability at the joint venture. Technological capability is only reflected in the workers trained, and the acquisition of minor change, and maintenance and repair capabilities.

That conclusion is consistent with the operation of foreign electronics companies on Batam Island where, in order to protect proprietary technology, Indonesia had to assure investors of 100% foreign ownership and management control. Virtually no effective linkages have been established by these firms with the science and technology infrastructure of the country. In fact, around two-thirds of Indonesian electronics exports in 1991 were accounted for by majority foreign-owned firms which had directly relocated their international operations to Java and Batam (Thee and Pangestu 1993, p. 133). Equally important is the fact that the majority of these electronics companies are second wave, medium-sized firms utilising low to medium-level technology; they are high tech only in the Indonesian context.

Typical is PT UIC Electronics, a 100% foreign-owned company which moved to Batam because of the tight labour and space situation in Singapore. Electrical components are shipped to the Batam factory, made into printed circuit boards, and sent back to Singapore for final assembly, packaging and export. It should be noted that the plan is eventually for most assembly to be done on Batam, with Singapore to be used only for logistic support and marketing (*Economic Bulletin*, May 1993).

A 100% foreign-owned contact lens manufacturer had to choose between relocating existing processes as part of planned obsolescence, or looking at the introduction of new processes and technology. The company decided to relocate to a low-wage centre in the Asia–Pacific region, the aim being to reduce production costs while waiting for new technology (interview with company executive, 4 October 1994).

In the case of joint ventures, foreign partners tend to transfer only enough technology to the local company to ensure efficient production operations—design and innovative work is carried out by foreign partners at the home base. There are few joint ventures in the electronics industry on Batam. Where they do exist, the role of the domestic partner is, more often than not, negligible, and since the June 1994 changes to foreign investment laws, an Indonesian joint venture partner is no longer required.

However, in the case of licensing arrangements, there tends to be a higher degree of innovation in the local firm: having bought the technology, it must adapt and modify it to suit local supply and demand conditions. PT Astra Microtronics Technology, a majority Indonesian-owned joint venture, manufactures semiconductors for export only. In 1995 it produced 236 million integrated circuits to earn $42 million (*Jakarta Post*, 17 January 1996). Astra offers a full range of packaging, assembly and testing services and has expanded its product lines by purchasing and utilising the latest technology in order to remain competitive. Semiconductors require a quick cycle time, and Astra decided to set up operations on Batam in order to reduce export–import processing time from what would have been two days in Jakarta to two hours on Batam. While Astra is currently air freighting through Changi airport in Singapore—the product arrives at its destination within an acceptable six days—it expects the completion of the Batam international airport and cargo centre to further cut delivery time (interview with Astra executive, 4 October 1994).

There were about a dozen domestic investments (PMDN) and half a dozen non-facility domestic investments in the electronics industry on Batam by the end of 1993 (BIDA 1993a, 1993b). For these domestic companies, not having a foreign link is perhaps a source of weakness: they may be less able to gain market access, export in sufficient volumes, or keep up with technology. This is particularly so for those that have been successful in the domestic market but are just beginning to export. According to Thee and Pangestu (1993, p. 134), the industries that develop can be expected to become more competitive as they attempt to meet international standards, produce at economic scales and build up backward linkages domestically. Only then will they become an important part of Indonesia's technological capability.

Public R&D institutes and government technology policies play a crucial role in the development of certain types of technology. This is no less so in Indonesia, where the private sector has traditionally shunned high-competition and high-risk activities in favour of safer activities such as building and property development. The role of the Agency for the Assessment and Application of Technology (BPPT) is to formulate technology policy; coordinate the application of technology; carry out technology research and assessment; and provide technology services (see Bishry and Hidayat, Chapter 8). However, in Indonesia R&D currently accounts for only 0.35% of government spending (interview with BPPT executive, 25 October 1994).

The government has set out to encourage value added knowledge and technology-intensive industries, often by providing incentives. BPPT recognises that assembly line operations and manufacturing under licence are low-technology activities offering little in the way of technology transfer. It is aware that, while Batam is attractive to investors because of its low wages, it will need to provide increased knowledge and technical capabilities if it is to attract added-value, high-technology industries to the island. Despite this, BPPT is yet to develop any significant or direct links with the private sector or electronics industries on Batam, or to become involved in Batam's overall development, even though the Chair of BPPT and BIDA is one and the same person, and even though over 50% of key BIDA personnel have BPPT links. On Batam, BPPT's activities have been confined to such projects as remote sensing, transportation and waste treatment systems, and more recently the building of the Barelang bridges (interview with BPPT executive, 25 October 1994).

It must be remembered, however, that the electronics industry is still relatively new to Indonesia, which failed to keep pace with other ASEAN countries in the race to develop medium and high-tech exports in the 1970s and 1980s. Two factors prevented Indonesia from deriving greater technological benefit during that period: first, a policy environment of strictly controlled foreign investment in the 1970s and early 1980s, which prevented the development of technology-based industries; and second, the low skill and education levels of Indonesian workers.

Although the present policy environment is more conducive to technology-based industries than that of the 1980s, Batam Island still suffers from the same shortages of skilled workers as the rest of Indonesia. Despite employment growth, a shortage of skilled labour and an abundance of unskilled labour pose a serious problem for Batam, where manufacturing industries require semi-skilled workers and trained technicians.

Current government efforts at basic skills education and training on Batam have largely been ineffective—while the government has supported such training, the content, quality and number of graduates has been disappointing (*Media Indonesia*, 1 February 1994; *Bisnis Indonesia*, 28 December 1994). This has led to the sourcing of workers from other parts of Indonesia by the Ministry of Manpower and private recruiting firms, often in cooperation with investors and industrial park operators. Under the EDB's Regionalisation Training Scheme, many companies in Batamindo Industrial Park provide on-the-job training in Singapore for Indonesian workers, who are then deployed on Batam Island. By September 1993, 31 such companies had participated in the scheme, training over 4,000 workers (*Straits Times*, 6 September 1993).

It is clear that the labour situation will be important in determining Batam's short and long-term prospects. While the advantage of Batam was once its cheap and abundant labour, it now risks becoming uncompetitive because of high labour costs and a shortage of skilled workers. Many see the moves to increase minimum wages on Batam—the highest in Indonesia—as an attempt by Habibie to move it up the technology ladder. However, the wage increases may have been sought by Minister of Manpower Abdul Latief, who has initiated minimum wage rises based on region, not industry, all over Indonesia.

Inadvertently, however, Habibie's restrictive industrial framework, in conjunction with labour market tightness, has ensured that wage costs on Batam continue to spiral upward, thereby eroding complementarities at a faster pace than previously thought possible. Perhaps recognising the restrictions placed on investors on Batam, as well as encouraged by early developments and successes, the Indonesian government began the development of neighbouring Bintan Island with a more open economic orientation. The government must consider carefully whether going high tech is the right move for Batam—does it have the necessary manpower, infrastructure, support industries and capital for the plan to turn it into a centre of excellence to succeed?

As things now stand, the nature of Batam's skills base is more conducive to large-scale, unskilled, labour-intensive production than to service and high-tech industries. For labour on Batam to remain competitive, particularly in light of recent wage increases, the Indonesian government will need to tackle human resource development. The development of scientific, engineering, technical and managerial manpower at all levels will be essential to the success of high-technology industrial development on Batam, particularly in the long term.

CONCLUSION

The appropriateness of Habibie's policy of promoting higher value added, high-technology, capital-intensive manufacturing on Batam Island is often questioned. The policy has restricted the outflow of lower value added manufacturing from Singapore, even though these are the very industries which could promote Batam and Riau's development at present. Indeed, it is possible that the boom that has occurred on Batam would have been bigger if the restriction to high-tech industries had not been there.

Habibie's aims have raised questions about the merits of challenging the wisdom of conventional development theory—that is, that countries should initially exploit their existing comparative advantage in labour-intensive manufactured goods. It can also be asked whether there is a preordained role for a country in the international division of labour. The question is whether Habibie can promote Indonesia to a particular position in the international division of labour, or perhaps how far he can proceed along this path.

Whatever Batam's high-technology future holds, the options will be conditioned by regulations set by Habibie and the Indonesian government, and success will rely on the continuation of selective resource allocation. The major irony of Batam Island is that the experience so far suggests that deregulation freed market forces only to steer export-oriented industries more in the direction of Indonesia's comparative advantage in low-tech, labour-intensive manufacturing.

19

DEVELOPING SCIENCE AND TECHNOLOGY COLLABORATION BETWEEN AUSTRALIA AND INDONESIA

Don Scott-Kemmis, Leslie O'Brien
and Remy Rohadian*

INTRODUCTION

This chapter focuses on interactions between the science and technology (S&T) 'systems' of Australia and Indonesia, from the perspective of one country seeking to promote S&T collaboration for mutual benefit. The discussion focuses on explicit government measures to facilitate bilateral collaboration in S&T. We recognise that the activities which will have the greatest long-term impact on technology collaboration are almost certainly trade and investment by Australian companies in Indonesia and by Indonesian firms in Australia, and the training of Indonesian students in Australia. The commercial relationship is developing strongly: by 1996 Indonesia had become Australia's sixth largest trading partner, and bilateral trade in that year grew by about 30% over the previous year. Manufactured goods account for almost 40% of Australia's

*The views expressed in this chapter are those of the authors and do not necessarily reflect those of the Australian government. Owing to space constraints, a number of detailed tables included in the original version of this paper have been omitted. These will be included in a longer version of the paper, which is in preparation.

exports to Indonesia, with elaborately transformed manufactures constituting one of the fastest growing categories of trade. Nevertheless, at this stage of the bilateral relationship and for some time to come, government-initiated mechanisms can play an important role in facilitating S&T collaboration.

Links that have developed between higher education institutions in the two countries through the Targeted Institutional Links Program of the Department of Employment, Education, Training and Youth Affairs (DEETYA, formerly DEET), Australian Research Council (ARC) grants and the initiative of individual institutions are an important additional dimension of cooperation in S&T, though they are not discussed here. However, in Indonesia universities account for only a small share of the country's total R&D activity, with just four or five undertaking the great majority of all university-level research.

S&T collaboration between Indonesia and Australia, like many other aspects of the bilateral relationship, grew slowly throughout the 1970s and 1980s and rapidly in the 1990s, particularly after 1992. Recent rapid development has been driven by several factors.

- From the mid 1980s in particular, Australian companies and organisations began to recognise the fundamental importance for Australia's future, and their own, of trade, investment and collaboration with the countries of East Asia. As a result, increasing numbers of companies have been seeking market entry mechanisms. Reducing the costs of market entry was particularly important because most Australian firms were small in size and relative novices in international trade and investment, while market opportunities were fragmented among the many growing national markets of the region.

- Australian governments have played an active role in the internationalisation of the Australian economy and the development of links with countries in the region. The role of government is, or is perceived to be, important in countries such as Indonesia where the national government's formal and informal intervention in the economy is substantial.

- Australian companies have recognised the opportunities generated in the short and long term by Indonesia's size, growth and proximity.

- Improvement in the bilateral relationship, since 1992 in particular, has promoted broader dialogue between the two countries and encouraged the desire—on Australia's part at least—to develop many dimensions of the relationship. The strong Indonesian government interest in S&T—more particularly the special role that the Minister for Research and Technology, Dr Habibie, has played in the Soeharto government and in Indonesian politics more generally—has given an additional impetus to strengthening the S&T dimension of the bilateral relationship.

- The increasing external earnings requirement for research and higher education institutions in Australia has made them responsive to opportunities provided by Australian government support for collaboration, multilateral agency funding for technical assistance and other projects, and recipient country funding. Research organisations in Indonesia, on the other hand, had long looked to bilateral and multilateral sources for a significant proportion of their 'research' funds.

- Indonesia requires assistance in the development and management of S&T organisations. Although funding by the Australian Agency for International Development (AusAID) has been important in areas such as agriculture, Australia—unlike many other OECD countries—has not been a major supporter of industrial technology transfer or R&D institutional development. The Viviani report into Australian priorities under the ASEAN–Australia Economic Cooperation Program (AAECP) did, however, recognise the important role of S&T collaboration in the ASEAN–Australia framework.

OBJECTIVES OF S&T COLLABORATION

The Australian and Indonesian governments recognise the potential complementarities between the two countries, in which S&T could play a significant role. However, there are many barriers to overcome before these complementarities can lead to trade and investment. Some of these arise from levels of protection and the complexities of market entry in Indonesia, others from a lack of awareness in Indonesia of Australian capabilities.

Barriers also arise from the characteristics of Australian manufacturing industry, the structure and capacities of which have developed over the past 50 years under an import substitution regime. The majority of Australia's high-technology sectors are, as a consequence, dominated by multinational corporations (MNCs); the recent opening of the economy and trends toward increasing concentration in high-technology global markets have tended to increase such dominance. The six high-technology sectors— information technology (IT), telecommunications, pharmaceuticals, the automotive industry, aerospace and energy generation—account for approximately 80% of industry R&D in OECD countries. The limited development of these sectors in Australia is the major reason for its relatively low level of business expenditure on R&D.

However, the dynamics of these sectors, including the role played by MNCs, is a great deal more complex in Australia than in industrialising countries. Some MNCs—General Motors and ICI, for example—carry out significant technological activity in Australia, interacting with Australian industry and research organisations. In addition, in most high-technology sectors there are many capable Australian firms undertaking (largely) non-core activities, from the supply of niche products to support services. Australia has a great depth and diversity of capabilities in sectors supporting industry and commerce, including in such fields as telecommunications and software services, engineering, instrumentation, business services, environmental management, and education and training. It has the capacity to contribute to solving many of the major challenges that Indonesia faces—in urban development, for example (where over 90 million people will become new urban residents in the next 25 years), the environment, agricultural productivity, diversification and commercialisation, and human resource development. Nevertheless, the lack of major national 'champions' capable of overcoming the high costs of market entry and 'pulling' the supply chain into markets impedes the development of linkages.

Research and higher education systems in Australia have been developed to international standards and maintain close relations with institutions in Europe and North America. The annual level of expenditure on R&D of over A\$6 billion underestimates performance in comparison with ASEAN countries. Research organisations in Australia have the benefit of capabilities accumulated over many decades of work, cumulative investments in infrastructure, and international recruitment and linkages.

Despite the barriers to trade and investment, there is great confidence that, as the Australia–Indonesia relationship expands and diversifies, and as understanding of opportunities increases, S&T cooperation will become increasingly important. The

Australian government has several objectives in actively encouraging such collaboration. These include:

- the marketing of Australian S&T services (Australia has invested heavily in its S&T base, with an annual investment about 10 times that of Indonesia);

- supporting Australian trade and investment, including by shifting perceptions about the Australian economy and capabilities;

- providing development assistance;

- investing in the building of relationships for long-term collaboration and diplomacy; and

- generating S&T knowledge that may be of benefit to Australia (examples include information on ocean currents, geological characteristics, or the migration of pests and diseases along the Indonesian archipelago).

DEVELOPMENT OF COLLABORATION: WHERE ARE WE?

The Role of Government

In the long run, government must assist and encourage Australian research organisations and S&T funding mechanisms (and companies) to support international S&T collaboration themselves as a means to achieve their objectives. While specific purpose government funding can facilitate the development of collaboration, it is essential that the availability of such funds promotes rather than retards the processes outlined above. It is also necessary to avoid a situation in which obtaining funds becomes an end in itself.

Government has an important funding role in the early stages of bilateral collaboration, in supporting both exploratory activities and pioneer ventures that may act as exemplars. But it is difficult to justify the commitment of further funds, which would require government to make an informed decision on the allocation of (limited) funding across a very wide range of fields, collaborators and types of activity. As collaboration progresses beyond the exchange of visits and trialing of projects, resource costs—salaries, travel, movement of equipment and materials, use of expensive infrastructure—rise sharply, and the role of government 'seed funding' declines. At this stage the case for government funding of substantial activities shifts from the promotion of bilateral collaboration to, in the case of Indonesia, either development assistance or support of R&D (or of other S&T-related activity) as an investment in achieving commercial or environmental objectives. In the latter case the 'investment' must be assessed in competition with alternative intranational investments—keeping in mind that assessing the appropriateness and costs and benefits of off-shore collaboration is challenging.

From this perspective the main roles of government are:

- signalling and promoting opportunities (obtaining assessments, identifying opportunities, supporting pioneers, influencing public S&T funding, seeking synergies between cooperation in, for example, S&T, and industry and development assistance);

- reducing search and learning costs (acquiring and disseminating information, supporting and organising visits and workshops);

- developing collaborative mechanisms (organising forums to exchange information, focusing the facilitation efforts of both governments); and

- raising the effectiveness of collaboration (identifying and addressing obstacles arising from government policies/ regulations or from the approach of organisations, diffusing knowledge about the management of cooperation).

Mechanisms to Promote Collaboration

The Australian government has invested heavily in the promotion of S&T cooperation with Indonesia. Expenditure over the past five years has probably been in excess of A$5 million, in addition to that provided by AusAID. This investment can certainly not be justified in terms of short-term returns.

There have been many agreements, memorandums of understanding and working groups, as well as an extensive series of major visits and substantial seminars to explore common interests. In 1994 Australia and Indonesia signed an Agreement on Cooperation on bilateral cooperation in S&T. This provides an umbrella framework for cooperation and an approach to dealing with intellectual property issues, and it lists areas of agreed mutual interest.

In 1992 Australia's Department of Industry, Science and Tourism (DIST) and Indonesia's Agency for the Assessment and Application of Technology (BPPT) established an organisation called Collaboration on Science and Technology between Australia and Indonesia (COSTAI) to promote direct collaboration in S&T. A steering committee was appointed, and meetings have been held in most years since 1992. In 1993 COSTAI held a workshop to raise awareness of the opportunities for collaboration in several fields. Of the wide range of projects identified, some went forward, but most did not lead to active collaboration. During Dr Habibie's visit to Australia in 1995, a memorandum of understanding agreeing to the restructuring of COSTAI was signed. This defined a clearer role for government and restated priority fields for collaboration: environment, energy, food and agriculture, biotechnology, marine S&T, telecommunications, automotive, aerospace and remote sensing. Under the new arrangements, an expanded steering committee comprising representatives of government, the scientific community and business in the two countries oversees COSTAI.

COSTAI meetings have usually attracted many expressions of interest, particularly from Indonesian researchers but also from Australian public sector organisations. However, COSTAI is a mechanism for facilitating the development of collaboration, not a funding institution. Neither the Australian nor the Indonesian government allocates funds to COSTAI to disperse for the support of collaborative projects.

COSTAI has thus far not played a strong proactive role in developing links between Australian companies and the Indonesian strategic industries under the Agency for Management of Strategic Industries (BPIS), or links with the 'national development projects' controlled by Minister Habibie (the Mamberamo Valley project, Natuna gas field, Maleo car project or N-2130 jet aircraft). There is no Australian liaison office in BPPT, whereas there are representatives from Germany, the US, the UK and Japan. It has often been suggested that the Australian government should appoint an appropriate person to join Dr Habibie's office. Recently, Dr Habibie invited former Australian Minister for Industry, Science and Technology John Button to be his personal adviser on developing links with Australian research organisations and companies.

Several of the working groups under the Indonesia–Australia Ministerial Forum conduct work programs involving aspects of technology cooperation or transfer: the Food and Agriculture Working Group, for example, is pursuing cooperation in technology issues related to livestock, horticulture and fisheries. Most substantial cooperative projects arising from the Ministerial Forum framework are dependent on AusAID funding, although some agencies, such as DIST and DEETYA, have allocated substantial funds to Ministerial Forum projects. COSTAI is functionally the Science and Technology Working Group of the Ministerial Forum.

Australian state governments have become active in facilitating cooperation with Indonesia, including S&T cooperation, often by building on sister state/province relationships. Thus far they have embraced such fields as development of technology parks, environmental cooperation, medical science, education and industrial projects.

The Range of S&T Collaboration

There is no coordination of Australia's international S&T cooperation, and there is no database that lists all—or even all significant—links. The list of collaborations given in Table 19.1 is by no means exhaustive: it focuses on the last few years; it does not include Australian Centre for International Agricultural Research (ACIAR) projects in agriculture, forestry and fisheries (ACIAR spends almost A$4 million in Indonesia each year); it mentions only some more recent AusAID projects; and it does not cover much of what happens in the higher education sector and mentions only a few multilateral (for example, AAECP) projects.

FACTORS PROMOTING OR IMPEDING COLLABORATION

The rapid growth of collaboration, expectations about the role of the Australian government in promoting and funding links, and differences in the scale and management of S&T between Australia and Indonesia have generated many challenges for managing the development of S&T collaboration. Several issues that present particular challenges are identified and discussed below.

Industrial Structure and Control

As discussed above, the structure of Australian industry and profile of Australian capabilities is not well understood in Indonesia. Major companies with global operations can plan and implement strategies for market entry, strategies which also provide a platform for related national firms. In the case of Indonesia, a good deal of technical assistance from MNCs is provided in association with strategic marketing—particularly in sectors such as aerospace, shipbuilding and telecommunications equipment where the government is

TABLE 19.1 Examples of Recent S&T Collaboration Projects

Technology/ Sector	Australian Organisation	Indonesian Organisation
Urban development		
Establishment of the Centre for Sustainable Urban & Regional Development	CSIRO (Building, Construction and Engineering), University of Melbourne	Lemtek, University of Indonesia
BCA Aider (software for standards)	CSIRO (Building, Construction and Engineering)	Dept of Public Works
Assessment of TOPAZ (software for spatial planning)	CSIRO (Building, Construction and Engineering)	Dept of Public Works
Sustainable housing & regional development	CSIRO (Building, Construction and Engineering)	Dept of Public Works
Heritage conservation	NSW Public Works	Dept of Public Works, West Java government
Driyoredjo Project	DIST, MBA, HIA, private sector	Dept of Housing, Dept of Public Works, private sector
Energy		
Low rank coal (assessment of coal)	Herman Research Labs Pty Ltd	BPPT
Testing of coal in pilot scale fluidised bed power station	CRC for Low Rank Coal	BPPT
Gasification combined cycle power generation (testing coal)	Herman Research Labs Pty Ltd	BPPT, PLN
Briquetting Indonesian coals for smokeless fuels	Herman Research Labs Pty Ltd	BPPT
Photovoltaic (village electrification) (AusAID/EFIC)	Solarex Pty Ltd	BPPT, LSDE
Collaboration in design of small-scale solar and other systems for Eastern Indonesia (GSLP)	Centre for Alternative and Solar Energy	LIPI
Civil nuclear technology	ANSTO	Batan

TABLE 19.1 (continued)

Technology/ Sector	Australian Organisation	Indonesian Organisation
Biotechnology		
Malaria vaccine (AusAID funds)	Walter and Eliza Hall Institute	Eijkman Institute
Database of organisations	Biotechnology Resources Australia	BPPT
Postgraduate diploma	University of Queensland	Eijkman Institute
Dengue virus detection and treatment	CSIRO (Biomolecular Engineering)	Gadjah Madah University
Human & animal reproductive technology	Monash University	Eijkman Institute
Biopesticides for insect control	CSIRO (Entomology)	
Environment		
Mamberamo Data Centre	CSIRO (Institute of Natural Resources and Environment)	BPPT
River catchment management	CSIRO (Institute of Natural Resources and Environment)	BPPT
East Java PCI Project (AusAID)	CMPS&F, Sinclair, Knight Merz, SAGRIC	Bapedal
PCI technology pilots (AusAID)	PCI (various suppliers)	
Waste treatment (NSW, DKI)	Aquatec, Maxcon Pty Ltd	DKI Sewerage Authority
Antarctic	CSIRO (Oceanography, Antarctic Research), DIST	BPPT
Meteorological cooperation	Bureau of Meteorology	Agency for Meteorology
Water supply technology	ACTEW	Bandung water service companies
Coastal Zone Environment and Resource Management (AAECP)	AMSAT	LIPI
Global change impacts on the terrestrial environment (AusAID)	CSIRO (Wildlife & Ecology)	BIOTROP

TABLE 19.1 (continued)

Technology/ Sector	Australian Organisation	Indonesian Organisation
Environmental laboratories (AusAID/DIFF)		
Hospital waste water treatment plant (AusAID/DIFF)		
Remote sensing		
Remote sensing applications	University of New South Wales	
AIVIS	Trippett-Sheddon Pty Ltd, CSIRO	BPPT
THEOS (multispectral scanner)	Comserve Pty Ltd, CSIRO	PT Ramatalindo
LADS (laser airborne depth sounder)	Visions Systems Pty Ltd, BHP	BPPT
	LADS Pty Ltd	Private sector company.
Remote sensing ground station cooperation project	ACRES (AusLIG)	Lapan
SAR (side aperture radar)	DSTO	BPPT, BPIS
Marine		
MREP (ADB)	SAGRIC, AMSAT	Bappenas: Bakosurtanal, LIPI, BPPT
CORRMAP (World Bank)	AMSAT	Bappenas: Bakosurtanal, LIPI, BPPT
Aquaculture in Eastern Indonesia (GSLP)	AMSAT	BPPT
Coral reef management (UNESCO)	AIMS	LIPI
Transport		
Maleo car design	Millard Design Pty Ltd	BPIS

TABLE 19.1 (continued)

Technology/ Sector	Australian Organisation	Indonesian Organisation
Aircraft structures (use of new materials for life extension.)	DSTO	BPIS, BPPT
Marine engineering feasibility study	DIST	Dept of Industry and Trade
Diverse aerospace cooperation	Royal Melbourne Institute of Technology	Institut Teknologi Bandung (Aerospace)
Marine engineering design	CRC for Marine Engineering	Hydrodynamics Lab, ITS
Other industrial technology		
Polymer chemistry	CRC for Polymers	BPPT
Information technology & telecommunications		
Microelectronic device design	University of South Australia	PT LEN
Telecommunications HRD and R&D cooperation	Telstra, TRL	PT Telkom, Telkom R&D
Microelectronics project (AAECP)	Various	LIPI, PT LEN
Mining		
Geological mapping of Irian Jaya & Kalimantan	AGSO	Geological R&D Centre, Dept of Mines and Energy
Mineral exploration technologies (GSLP)	CSIRO	
Forestry and agriculture		
Selecting varieties of *M. cajaputi* (ISTP, GSLP)	CSIRO	Forestry Research Centre, Dept of Forestry
Bovine disease diagnosis (GSLP)	CSIRO	
Acacia plantation production (GSLP)	CSIRO (Forestry)	

TABLE 19.1 (continued)

Technology/ Sector	Australian Organisation	Indonesian Organisation
Animal breeding technologies	Monash University	Eijkman Centre
Defence		
Propellents etc.	DSTO	AERIE
Aircraft structures	DSTO	BPPT, BPIS
High-frequency radio applications	DSTO	Lapan
Industrial standards		
Basic physical standards accreditation	National Measurement Laboratory	KIM, LIPI
Recognition of calibration network accreditation	National Measurement Laboratory	KIM, LIPI
National standards and calibration lab network (World Bank, government of Indonesia)	National Measurement Laboratory	KIM, LIPI
S&T management and awareness		
Scenario of future S&T cooperation (ISTP)	Centre for Research Policy, University of Wollongong	BPPT
Development of standards for accreditation of professional engineers (PSLP)	Institution of Engineers, Australia	Professional Engineers Association
Institutional strengthening (World Bank, government of Indonesia)	CSIRO	LIPI
S&T indicators (World Bank, government of Indonesia)	CRP, ACCI, SPRU	BPPT
S&T awareness (GSLP)	Questacon	Science Centre
Intellectual property management (AusAID, STP)	AIPO	Cabinet Secretariat

A selected list of acronyms is given in the glossary at the front of the book.

nurturing state-owned firms. Support of various types to BPIS ventures by British, American and German companies is returned to the providers through the sale of engines and other hardware to IPTN, the state-owned aircraft factory. In other cases such support is part of the usual investment by companies as they position themselves to compete for major infrastructure contracts (bridges, subways etc.).

Few of the major Australian-owned firms operating in core technology segments of high-technology sectors have a significant market share. Hence there are often no market entry leaders to facilitate the entry of the many Australian firms with specialised capabilities in high-technology sectors. Dr Habibie, anxious to support technology acquisition and accumulation in Indonesia, has sought to establish direct relationships with global firms, and has questioned the value to Indonesia of working with the Australian subsidiaries of MNCs.

Expectations and Mismatches in Approaches to Funding

Many proposals for collaboration have not succeeded because they have failed to attract funding. Australian government funds for international S&T collaboration are not large, and they are spread across many countries, fields of S&T and activities. Limited funds are available from DIST's International S&T Program (ISTP) to support the travel costs involved in collaboration projects. Projects requiring substantial funds must be financed by the participating organisation, or through national R&D programs or development assistance programs. The latter include AusAID's Government Sector Linkages Program (GSLP), Private Sector Linkages Program (PSLP) and other funding programs, and the programs of multilateral aid agencies. The ISTP is an open funding mechanism responding to applications from Australian public sector organisations, and using a committee of external advisers and peer review to select proposals. Indonesian government funding for researchers is, in principle, available only for projects consistent with priority national objectives. Many of these projects have not been of sufficiently high priority to Australian research organisations to attract their financial support. To some extent the requirement for each side to obtain funds from national sources has created a double hurdle to collaboration.

Many of the agencies involved in collaboration would prefer that the two governments channelled resources into a few selected priority areas. However, it is difficult to reconcile this approach with the more open, pluralist, bottom-up approach to funding within Australia. It could also lead to 'insiders' being positioned to capture funding rather than a mechanism that facilitated wider learning.

In promoting international S&T collaboration, the Australian government's central objective has been to generate mutual benefits, including those which accrue to Australia on a sustainable basis. In the case of Indonesia, aid and collaboration can become intertwined, shaping expectations and blurring objectives. Indonesian S&T agencies, with their limited funds, look to multilateral and bilateral development assistance as sources of financing. The Australian aid program is small relative to that of the US, Japan and Germany, and—with the two important exceptions of renewable energy applications for remote communities and the management of tropical reefs—most areas of priority to agencies under the Ministry of State for Research and Technology (Menristek) have not been high priorities for Australian aid.

The Australian government has not been prepared to be a 'risk partner'—a major funder of infrastructure or enabling investments—in major development projects such as the Mamberamo valley or the Maleo automotive project. Neither have the few Australia-based global firms participated as investors in these major projects.

Links between Private and Public Sector Research Organisations

Indonesian public sector R&D agencies have few links with the private sector; they have shown little interest in facilitating—or perhaps lack the capacity to facilitate—links between the Indonesian private sector and Australian public or private sector organisations. At this stage, many of the organisations under the scope of Menristek, including the strategic industries, form more of a technology island than a technology bridge. Consequently, few Australian firms see collaboration through government-to-government mechanisms as a significant route of market entry. The Indonesian private sector generally prefers to purchase technology inputs bundled with capital goods, or relies on the provider of capital to manage technology supply, adaptation and upgrading. There are important and increasing exceptions to this pattern, and therein lies a new level of opportunity for the public sector S&T organisations in Indonesia to develop links with industry.

Links between public sector research organisations and the private sector in Australia have improved markedly but remain uneven; they are strong in sectors such as mining but weaker in many manufacturing industries. Fostering industry-to-industry collaboration has been an important objective of the Australian government. Where collaboration remains locked in the public sector, or where collaboration in the public sector provides no particular advantages in developing effective collaboration with the private sector, its objective may not be achieved.

Coordination and Communication

Coordination and communication between Indonesian and Australian agencies is difficult and resource intensive. Different funding cycles and criteria create problems for managing interaction, and a lack of clarity about responsibilities causes delays. Whereas Australian S&T ministers are briefed by DIST and reinforce the agendas and priorities developed at the bureaucratic level, this is not the case in Indonesia. Two agendas—a bottom-up one devised by bureaucrats and a top-down one emanating from Dr Habibie—have evolved, further confusing the dialogue.

A lack of coordination among Australian agencies and states is a feature of Australia's interaction with countries in the region. The Australian Coordination Committee on S&T is examining ways to improve coordination of our international activities, but the problem has deep roots, in our pluralist S&T system and perhaps our culture. Those countries that pursue relatively coordinated approaches overseas generally have relatively cooperative approaches domestically. Few Australian public sector organisations pay more than lip service to developing a more coordinated approach.

The central partners in S&T collaboration have been DIST and BPPT. DIST has been pursuing a 'whole of government' approach, with ISTP funding collaboration in any sector or field of science. BPPT, on the other hand, has been pursuing firstly its own interests (in particular, seeking funds and collaborative activities for its—largely underemployed—researchers), and secondly the wider interests of the non-departmental agencies (LPND) coordinated by Menristek.

FUTURE APPROACHES TO S&T COLLABORATION

There are potentially great benefits in continuing to develop a strong S&T relationship based on close working relations at the enterprise and organisation level. Continuing growth in trade is one source of optimism about the future. Strong growth in Australian investment in Indonesia is another—Indonesia is now Australia's fifth most important destination for foreign investment, and there are well over 300 Australian companies with operations in Indonesia. The progressive opening of the Indonesian economy suggests that these trends in trade and investment will continue. As the economy opens, we can expect more Indonesian companies to source technology and technological services from a wider range of suppliers, thereby providing further opportunities for collaboration. Over the last few budget cycles Indonesian government investment in S&T has been increasing strongly, strengthening Indonesian capacities. The relatively large number of Indonesian students in Australia and the expanding links between our universities are particularly positive aspects of the current interaction. Of the more than 5,000 Indonesian students in higher education in Australia, about 50% are studying business administration or economics and about 30% engineering or science.

Information

The role of government in providing information has been greatly assisted by the development of the Internet. DIST, for example, is moving to ensure that extensive information is available on the Internet, and that the COSTAI home page has links to a great many other relevant databases. The focus of effort at present is to develop an extensive database and web of linkages through an APEC Industrial Science and Technology Working Group Project. Publications on Indonesian S&T activities and collaboration in various fields would also be useful for Australian organisations.

The Government Sector Linkages Program

This AusAID-funded program provides a mechanism for pursuing some of Indonesia's priorities for S&T collaboration that are more in the nature of development assistance than collaborative research for mutual benefit. Two DIST-managed projects (marine aquaculture in Eastern Indonesia and interfirm networking) and four projects proposed by the Commonwealth Scientific and Industrial Research Organisation (CSIRO) were supported by GSLP funds in 1997–98. The benefits of GSLP would be greatly enhanced if it was linked more closely to Ministerial Forum Working Group programs, and if Aus-AID consulted other departments over the allocation of funds.

National Goal-Oriented Research Programs

The development in Indonesia of major national research projects that are long term, top down and targeted provides a useful 'focusing device' for the communication of national priorities and the identification of institutional roles.

Bringing International Collaboration into the Mainstream of National Organisations

Major Australian public sector research organisations such as the CSIRO and the Defence Science Technology Organisation (DSTO), consulting organisations such as AMSAT, and several Cooperative Research Centres (CRCs) and universities are pursuing their own agenda for developing collaboration with Indonesia. We can expect to see an increasing level of activity by these organisations. However, this will most likely take the form of incremental and cumulative, rather than spectacular, growth. As discussed, several major Australian research organisations have now incorporated collaboration with Indonesia into their strategic plans and are allocating resources accordingly. In addition, the significance for Australia of effective international S&T links is increasingly being recognised at the core, rather than the margin, of research funding. The ARC now assigns international linkages to the first level of importance in its strategic objectives for Special Research Centres. Similarly, the CRC Program now has the development of international linkages as an explicit objective and selection criterion.

REFERENCES

Acero, L. (1984), *Technical Change in a Newly Industrializing Country: A Case Study of the Impacts on Employment and Skills in the Brazilian Textiles Industry*, SPRU Occasional Paper Series No. 22, University of Sussex, Sussex.

Ahmad, Mubariq (1993), 'Economic Cooperation in the Southern Growth Triangle: An Indonesian Perspective', in Toh Mun Heng and Linda Low (eds), *Regional Cooperation and Growth Triangles in ASEAN*, Times Academic Press, Singapore.

Akiyama, T. and A. Nishio (1997), 'Sulawesi's Cocoa Boom: Lessons of Smallholder Dynamism and a Hands-Off Policy', *Bulletin of Indonesian Economic Studies*, 33 (2), pp. 97–121.

Anderson, J.R. and J.B. Hardaker (1979), 'Economic Analysis in Design of New Technologies', in A. Valdes, G.M. Scobie and J.L. Dillon (eds), *Economics and the Design of Small Farmer Technology*, Iowa State University Press, Ames IA.

Archibugi, Daniele and Jonathan Michie (1995), 'The Globalisation of Technology: A New Taxonomy, *Cambridge Journal of Economics*, 19, pp. 121–140.

Arndt, H.W. (1978), 'Survey of Recent Developments', *Bulletin of Indonesian Economic Studies*, 14 (1), pp. 1–23.

ASEAN Secretariat (1997), *Science and Technology Indicators in ASEAN*, ASEAN Secretariat, Jakarta.

Astra (quarterly), Astra International, Jakarta.

Astra International (1997), Home page (astra.co.id).

—— (annual), *Financial Statements*, Jakarta.

Aswicahyono, H.H., K. Bird and H. Hill (1996), 'What Happens to Industrial Structure when Countries Liberalise? Indonesia since the Mid 1980s', *Journal of Development Studies*, 32 (3), pp. 340–363.

Australian International Education Foundation (1996), *The Indonesian Education System*, Commonwealth of Australia, Canberra.

Barker, R. and R.W. Herdt (1985), *The Rice Economy of Asia*, Resources for the Future, Washington DC.

Barlow, C. (1997), 'The Market for New Tree Crop Technology: A Sumatran Case', *Journal of Agricultural Economics*, 48 (2), pp. 193–210.

—— (forthcoming), 'Growth, Structural Change and Plantation Tree Crops: The Case of Rubber', *World Development*.

Barlow, C, S.K. Jayasuriya and S.C. Tan (1994), *The World Rubber Industry*, Routledge, London.

Basri, M.C. and H. Hill (1996), 'The Political Economy of Manufacturing Protection in LDCs: An Indonesian Case Study', *Oxford Development Studies*, 24 (3), pp. 241–259.

Becker, G.S. (1976), *The Economic Approach to Human Behaviour*, University of Chicago Press, Chicago IL.

Behrman, Jere R. (1990), *Human Resource Development*, ILO–ARTEP, New Delhi.

Berry, A. and B. Levy (1994), Indonesia's Small and Medium Industrial Exporters and Their Support Systems, Paper presented to the conference 'Can Intervention Work? The Role of Government in SME Success', World Bank, Washington DC.

BIDA (Batam Industrial Development Authority) (1980), *The Batam Development Program*, Jakarta.

—— (1993a), *Daftar Perusahaan PMDN di Pulau Batam Sampai Dengan 12/11/1993*, Jakarta.

—— (1993b), *Daftar Industri Non Fasilitas Batam, 27/10/1993*, Jakarta.

—— (1996), *Barelang: Development Data up to June 1996*, Jakarta.

BIM (Batamindo Industrial Management) (1994), *Batam Industrial Park: International Manufacturing Advantage*, Jakarta.

Birowo, A.T. and R. Gondowarsito (1990), 'Technology Transfer in Indonesian Agricultural Development', in P.G. Harrison (ed.), *Agricultural Technology Transfer: The Relevance of the Australian Experience*, Australian Institute of Agricultural Science, Darwin.

Bishry, R. (1992), An Overview of a Government Research Institute in Indonesia: BPPT, Unpublished paper, Jakarta.

BKPM (1982a), *Surat Keputusan Ketua BKPM No. 22/SK/1982*, Jakarta, 27 July.

—— (1982b), *Business Line List for Capital Investment in Batam Island, 1982/83*, Jakarta.

—— (1989), *Surat Keputusan Ketua BKPM No. 16*, Jakarta, 24 October.

Bleaney, M.F. and David Greenaway (1993), 'Long-Run Trends in the Relative Prices of Primary Commodities and in the Terms of Trade of Developing Countries', *Oxford Economic Papers*, 45 (3), pp. 349–363.

Boediono and Walter McMahon (1991), *Education and the Economy*, Education Policy and Planning Project, Department of Education and Culture, Jakarta.

Boeke, J.H. (1952), *Economics and Economic Policy of Dual Societies*, Tjeenk Willink, Haarlem.

Booth, A. (1988), *Agricultural Development in Indonesia*, Allen and Unwin, Sydney.

Borsuk, R. (1995), 'Suharto Pushes Plan to Build Jet in Indonesia', *Asian Wall Street Journal*, 11–12 August.

BPPT et al. (1995), *Indikator Ilmu Pengetahuan dan Teknologi Indonesia, 1994*, Second edition, BPPT, Ristek and Papiptek-LIPI, Jakarta.

BPS (1995), *Statistik Indonesia, 1994*, Jakarta.

—— (1996), *Statistik Indonesia, 1995*, Jakarta.

Braadbaart, Okke (1996), 'Machine Tools and the Indonesian Engineering Subsector: Consumption Trends and Localisation Efforts', *Bulletin of Indonesian Economic Studies*, 32 (2), pp. 75–104.

Campos, J.E. and H.L. Root (1996), *The Key to the Asian Miracle: Making Shared Growth Possible*, Brookings Institution, Washington DC.

Chapman, Ross (1992), 'Indonesian Trade Reform in Close-Up: The Steel and Footwear Experiences', *Bulletin of Indonesian Economic Studies*, 28 (1), pp. 67–84.

Chen, E.K.Y. (1997), 'The Total Factor Productivity Debate: Determinants of Economic Growth in East Asia', *Asian-Pacific Economic Literature*, 11 (1), pp. 18–38.

CIDB (Construction Industry Development Board) (1989), *Batam Outlook 1990*, Singapore.

Clark, David H. (1983), *How Secondary School Graduates Perform in the Labor Market: A Study of Indonesia*, World Bank Staff Working Paper No. 615, Washington DC.

Cline, William R. (1982), 'Can the East Asian Model of Development Be Generalized?', *World Development*, 10 (2), pp. 81–90.

Cole, David C. and Betty F. Slade (1996), *Building a Modern Financial System: The Indonesian Experience*, Cambridge University Press, Cambridge UK.

Collier, W., W.G. Wiradi and Soentoro (1973), 'Recent Changes in Rice Harvesting Methods: Some Serious Social Implications', *Bulletin of Indonesian Economic Studies*, 9 (2), pp. 36–45.

Creutzberg, P. (1975–80), *Changing Economy in Indonesia. A Selection of Statistical Source Material from the Early 19th Century up to 1940*, Nijhoff, The Hague.

Denison, E.F. (1967), *Why Growth Rates Differ: Postwar Experience in Nine Western Countries*, The Brookings Institution, Washington DC.

Departemen Perindustrian–Bank Indonesia (1989), *Buku Petunjuk Industri Tekstil Nasional 1987* [National Textile Guide 1987], Jakarta.

Dhanani, Shafiq (1994a), Training Needs in Indonesian Manufacturing, Paper presented to the DEPNAKER/World Bank Conference on Vocational Training, Labor Markets and Economic Development in Indonesia, Bali, 10–12 February.

―― (1994b), Unemployment and Underemployment in Indonesia in the 1980s, Unpublished paper, Jakarta, March.

Dick, H.W. (1981a), 'Urban Public Transport, Part I', *Bulletin of Indonesian Economic Studies*, 17 (1), pp. 66–82.

―― (1981b), 'Urban Public Transport, Part II', *Bulletin of Indonesian Economic Studies*, 17 (2), pp. 72–88.

Dijkman, M.J. (1951), *Hevea: Thirty Years of Research in the Far East*, University of Miami Press, Coral Gables FL.

Dinas Kesehatan (1997), Private communication, Kupang.

Djojonegoro, Wardiman (1991), Towards a Science and Technology Policy Approach: Supporting Industrial Development of Indonesia, Paper presented at the International Conference on Changing Technology Issues and Trends of Policy Research, Seoul, 30–31 October.

―― (1994), Education and Training for Business and Industry, Paper presented to the Australia Today Indonesia 1994 Education Conference, Jakarta, 6 June.

—— (1995), Education and Training for Industrial Growth, Paper presented to the Education and Training for Industry Growth Conference, Ministry of Education and Culture, Jakarta.

Dodgson, M. and J. Bessant (1996), *Effective Innovation Policy: A New Approach*, International Thomson Business Press, London.

Dosi, Giovanni, Keith Pavitt and Luc Soete (1990), *The Economics of Technical Change and International Trade*, Harvester Wheatsheaf, New York.

Eaton, Jonathan and Samuel Kortum (1996), 'International Technology Diffusion: Theory and Measurement', *Proceedings of the 25th Annual Conference of Economists*, Australian National University, Canberra.

Enos, John (1989), 'Transfer of Technology', *Asian-Pacific Economic Literature*, 3 (1), pp. 3–37.

Enos, John, Sanjaya Lall and Mikyung Yun (1997), 'Transfer of Technology: An Update', *Asian-Pacific Economic Literature*, 11 (1), pp. 56–66.

Evenson, Robert E. and Larry E. Westphal (1994), 'Technological Change and Technology Strategy', *Center Discussion Paper* No. 709, Economic Growth Center, Yale University, New Haven CT.

—— (1995), 'Technological Change and Technology Strategy', in Jere Behrman and T.N. Srinivasan (eds), *Handbook of Development Economics*, Vol. 3A, North Holland, New York, Ch. 37.

Fagerberg, J. (1994), 'Technology and International Differences in Growth Rates', *Journal of Economic Literature*, 32, pp. 1,147–1,175.

Fane, George and Timothy Condon (1997), 'Trade Reform in Indonesia, 1987–95', *Bulletin of Indonesian Economic Studies*, 32 (3), pp. 33–54.

Feder, G. et al. (1988), *Land Policies and Farm Productivity in Thailand*, Johns Hopkins University Press, Baltimore MA.

Feigenbaum, A.V. (1961), *Total Quality Control*, 1st edition, McGraw-Hill, New York NY.

Fox, J.J. et al. (1993), 'Groundwater Nitrate in East Java, Indonesia', *AGSI Journal of Australian Geology and Geophysics*, 14 (2/3), pp. 273–277.

Freeman, Chris and John Hagedoorn (1994), 'Catching Up or Falling Behind: Patterns in International Interfirm Technology Partnering', *World Development*, 22, pp. 771–780.

Garrett-Jones, Sam (1996), *The Development of Science and Technology Indicators in the ASEAN Region*, Report commissioned by the UNESCO Office, Jakarta, for the UNDP–ASEAN Technology and Environment Subprogram, Centre for Research Policy, University of Wollongong, Wollongong.

Gibson, J. (1966a), 'Production Sharing, Part I', *Bulletin of Indonesian Economic Studies*, 3, pp. 52–75.

—— (1966b), 'Production Sharing, Part II', *Bulletin of Indonesian Economic Studies*, 4, pp. 75–100.

Glassburner, Bruce and Mark Poffenberger (1983), 'Survey of Recent Developments', *Bulletin of Indonesian Economic Studies*, 19 (3), pp. 1–27.

Godfrey, Martin (1987), *Planning for Education, Training and Employment in Indonesia*, Summary Report, ILO/UNDP, Jakarta.

Green, Eric Marshall (1996), *Economic Security and High Technology Competition in an Age of Transition: The Case of the Semiconductor Industry*, Praeger, London.

Guerrieri, Paolo and Carlo Milana (1995), 'Changes and Trends in the World Trade in High-Technology Products', *Cambridge Journal of Economics*, 19, pp. 225–242.

Habibie, B.J. (1986), 'Industrialisasi, Transformasi, Teknologi dan Pembangunan Bangsa', *Prisma*, 15 (1), pp. 42–53.

—— (1992), 'Technology and the Singapore–Johor–Riau Growth Triangle', Speech delivered to the Tripartite Meeting and Seminar on Economic Development in the Growth Triangle and Its Environmental Impact, Batam Island, 8 May.

—— (1993), *Pembangunan Berdasarkan Nilai Tambah Dengan Orientasi Teknologi dan Industri*, Working Paper No. 1, CIDES, Jakarta.

—— (1995), *Ilmu Pengetahuan, Teknologi dan Pembangunan Bangsa: Menuju, Dimensi Baru Pembangunan Indonesia*, CIDES, Jakarta.

Harianto, F. (1993), 'Study of Subcontracting in Indonesian Domestic Firms', *Indonesian Quarterly*, 21 (3), pp. 331–343.

Hayami, Y. (1997), *Development Economics: From the Poverty to the Wealth of Nations*, Oxford University Press, Oxford.

Hayami, Y. and V.W. Ruttan (1985), *Agricultural Development: An International Perspective*, 2nd edition, Johns Hopkins University Press, Baltimore MD.

Helleiner, G.K. (1989), 'Conventional Foolishness and Overall Ignorance: Current Approaches to Global Transformation and Development', *Canadian Journal of Development Studies*, 10 (1), pp. 107–120.

—— (1990), *The New Global Economy and the Developing Countries*, Gower House, Aldershot.

Helpman, E. (1997), 'R&D and Productivity: The International Connection', *NBER Working Paper No. 6101*, National Bureau of Economic Research, New York NY.

Hicks, J.R. (1932), *The Theory of Wages*, Macmillan, London.

Higgins, B. (1956), 'The Dualistic Theory of Underdeveloped Areas', *Economic Development and Cultural Change*, 4 (2), pp. 99–112.

Hill, Hal (1983), 'Choice of Technique in the Indonesian Weaving Industry', *Economic Development and Cultural Change*, 31 (2), pp. 337–353.

—— (1992), *Indonesia's Textile and Garment Industries: Developments in an Asian Perspective*, Australian National University, Canberra, and Institute of Southeast Asian Studies, Singapore.

—— (1994), 'The Economy', in Hal Hill (ed.) *Indonesia's New Order: The Dynamics of Socio-economic Transformation*, Allen and Unwin, Sydney.

—— (1995), 'Indonesia's Great Leap Forward? Technology Development and Policy Issues', *Bulletin of Indonesian Economic Studies*, 31 (2), pp. 83–123.

—— (1996a), 'Indonesia's Industrial Policy and Performance: "Orthodoxy" Vindicated', *Economic Development and Cultural Change*, 45 (1), pp. 147–174.

—— (1996b), *The Indonesian Economy since 1966: Southeast Asia's Emerging Giant*, Cambridge University Press, Cambridge UK.

—— (1996c), 'Indonesia: From "Chronic Dropout" to "Miracle"?', *Journal of International Development*, 7 (6), pp. 775–790.

—— (1997), *Indonesia's Industrial Transformation*, Institute of Southeast Asian Studies, Singapore, and Allen and Unwin, Sydney.

Hill, Hal and Pang Eng Fong (1991), 'Technology Export from a Small, Very Open NIC: The Case of Singapore', *World Development*, 19, pp. 553–568.

Hobday, Michael (1995), *Innovation in East Asia: The Challenge to Japan*, Edward Elgar, Aldershot.

—— (1997), East vs South East Asian Innovation Systems: Comparing OEM with TNC-Led Growth in Electronics, Paper presented at the International Symposium on Innovation and Competitiveness in Newly Industrialising Economies, Seoul, 26–27 May.

—— (forthcoming), 'Understanding Innovation in Southeast Asia: Malaysia's Experience in Electronics', in K.S. Jomo (ed.), *Malaysia's Industrial Technology*.

Hu, Liu (1994), 'An Account of China's Technology Import and Export in 1992', *Almanac of China's Foreign Economic Relations and Trade 1993/94*, Beijing.

Hughes, Helen (ed.) (1992), *Dangers of Export Pessimism: Developing Countries and Industrial Markets*, ICS Press, San Francisco CA.

IMF (International Monetary Fund) (1996), *Balance of Payments Statistics: 1996 Yearbook*, Washington DC.

Ishida, T. (1991), *An Introduction to Textile Technology* (revised edition), Osaka Senken Ltd, Tokyo.

James, Ted (1996), Indonesia: Non-oil and Non-gas Export Performance in 1995, Jakarta, mimeo.

Johnson, Harry (1964), 'Development as a Generalized Process of Capital Accumulation', in G.M. Meier (ed.), *Leading Issues in Economic Development*, 4th edition, Oxford University Press, New York NY, pp. 631–636.

Keyfitz, Nathan (1989), 'Putting Trained Labour Power to Work: The Dilemma of Education and Employment' *Bulletin of Indonesian Economic Studies*, 25 (3), pp. 35–56.

Kim, Linsu (1997), *Imitation to Innovation: The Dynamics of Korea's Technological Learning*, Harvard Business School Press, Boston MA.

Koesmawan (1996), *Textile Export Marketing and Technology Acquisition Efforts in the Indonesian Textile Industry: A Case Study*, University of Twente, Enschede.

Koike Kazuo and Inoki Takenori (1987), *Jinzai Keisei no Kokusai Hikaku* [International Comparison of Human Resource Formation], Toyo Keizai Shinposha, Tokyo.

Kokko, Ari and Magnus Blomstrom (1995), 'Policies to Encourage Inflows of Technology through Foreign Multinationals', *World Development*, 23, pp. 459–468.

Krugman, Paul R. (1991), *Geography and Trade*, MIT Press, Cambridge MA.

Kumar, Nagesh (1996), 'Intellectual Property Protection, Market Orientation and Location of Overseas R&D Activities by Multinational Enterprises', *World Development*, 24, pp. 673–688.

Lall, Sanjaya (1990), *Building Industrial Competitiveness in Developing Countries*, Development Centre Studies, OECD, Paris.

—— (1993), 'Understanding Technology Development', *Development and Change*, 24 (4), pp. 719–753.

—— (1996), *Learning from the Asian Tigers*, Macmillan, London.

Lall, Sanjaya and Kishore Rao (1995), *Indonesia: Sustaining Manufactured Export Growth*, Vol. 1, Main Report, Revised draft, Jakarta, August.

Lin, O.C.C. (1998), 'Science and Technology Policy and Its Influence on Economic Development in Taiwan', in H.S. Rowen (ed.) *Behind East Asian Growth: The Political and Social Foundations of Prosperity*, Routledge, London, pp. 185–208.

LIPI (1994), *Overview of the Final Report: STAID Project Element No. 10–13, R&D Management System*, Vol. 1, STAID and LIPI, Jakarta.

Lipsey, R.G. (1997), 'Globalization and National Government Policies: An Economist's View', in J.H. Dunning (ed.), *Governments, Globalization and International Business*, Oxford University Press, Oxford, pp. 73–113.

Lok, H.P. (1993), 'Labour in the Garment Industry: An Employer's Perspective', in Chris Manning and Joan Hardjono (eds), *Indonesia Assessment 1993. Labour: Sharing in the Benefits of Growth?* Political and Social Change Monograph, Research School of Pacific Studies, Australian National University, pp.155–172.

Long, Simon (1997), 'A Survey of Indonesia: Suharto's End Game', *The Economist*, 26 July.

LPEM (1996), *Studi Kemitraan Usaha Besar dan Kecil* [A Study of Partnerships between Large and Small Enterprise], Faculty of Economics, University of Indonesia, Jakarta.

Luhulima, Cornelis P.F. (1996), 'Management of Technology: The Indonesian Case', in Karen Minden and Wong Poh-Kam (eds), *Developing Technology Managers in the Pacific Rim: Comparative Strategies*, M.E. Sharpe, Armonk NY, Chapter 9.

Lynn, Leonard H. (1985), 'Technology Transfer to Japan: What We Know, What We Need to Know, and What We Know that May

Not Be So', in Nathan Rosenberg and Claudio Frischtak (eds), *International Technology Transfer: Concepts, Measures, and Comparisons*, Praeger, New York NY.

Manning, Chris (1979), Wage Differentials and Labour Market Segmentation in Indonesian Manufacturing, Unpublished PhD thesis, Australian National University, Canberra.

—— (forthcoming), Indonesian Labour in Transition: An East Asian Success Story? Cambridge University Press, Cambridge UK.

Manning, Chris and Sisira Jayasuriya (1996), 'Survey of Recent Developments', *Bulletin of Indonesian Economic Studies*, 35 (2), pp. 3–45.

Manning, Chris and P.N. Junankar (forthcoming), 'Choosy Youth or Unwanted Youth? A Survey of Unemployment', *Bulletin of Indonesian Economic Studies*.

Mathews, John A. (1996), 'High Technology Industrialisation in East Asia', *Journal of Industry Studies*, 3, pp. 1–77.

McKendrick, David (1992), 'Obstacles to "Catch-Up": The Case of the Indonesian Aircraft Industry', *Bulletin of Indonesian Economic Studies*, 28 (1), pp. 39–66.

McLeod, Ross H. (1993), 'Survey of Recent Developments' *Bulletin of Indonesian Economic Studies*, 29 (2), pp. 3–42.

—— (1997), 'Survey of Recent Developments', *Bulletin of Indonesian Economic Studies*, 33 (1), pp. 3–43.

Menristek (1996), *Kebijaksanaan Strategis IPTEK Nasional Pada PJP-2* [Strategic Policies for National Science and Technology Development during the Second Long-Term Development Plan], Jakarta, December.

—— (1997), *Petunjuk Penyusunan Proposal Riset Unggulan Kemitraan* [Guidelines on Drawing up Priority Partnership Research Proposals], Jakarta.

Menteri Perdagangan (1995), Peranan Standardisasi dalam Era Perdagangan Global [The Role of Standardisation in the Era of Global Trade], Paper presented at the Seminar on Standards and the Assessment of Conformance, Jakarta, 7 November.

Middelton, John, Adrian Ziderman and Arvil van Adams (1993), *Skills for Productivity: Vocational Education and Training in Developing Countries*, Oxford University Press, New York NY.

Ministry of Education and Culture (1995), *Skills toward 2020*, Report of the Taskforce on Development of Vocational Education and Training in Indonesia, Jakarta, December.

Moedjiono (1995), Tantangan dan Peluang dalam Mewujudkan Jasa Sertifikasi Sistem Mutu Sebagai Komoditas [Challenges and Opportunities in Promoting the Services of a Quality Certification System as a Commodity], Paper presented at the Seminar on Standards and the Assessment of Conformance, Jakarta, 7 November.

Mowery, David C. and Joanne E. Oxley (1995), 'Inward Technology Transfer and Competitiveness: The Role of National Innovation Systems', *Cambridge Journal of Economics*, 19, pp. 67–93.

Myint, H. (1985), 'Organizational Dualism and Economic Development', *Asian Development Review*, 3 (1), pp. 24–42.

Najmabadi, Farrokh and Sanjaya Lall (1995), *Developing Industrial Technology: Lessons for Policy and Practice*, World Bank, Washington DC.

Nelson, R.R. (ed.) (1993), *National Innovations Systems: A Comparative Analysis*, Oxford University Press, Oxford.

OECD (1995), *Industry and Technology: Scoreboard of Indicators 1995*, Paris.

—— (1996a), *Technology and Industrial Performance*, Paris.

—— (1996b), *Main Science and Technology Indicators 1996/2*, Paris.

—— (1997), *Technology and Industrial Performance*, Paris.

Ostry, J.D. (1996), Current Account Imbalances in ASEAN: Are They a Problem?, Paper presented to the conference on Macroeconomic Issues Facing ASEAN Countries, Jakarta, November.

Ostry, Sylvia and Richard R. Nelson (1995), *Techno-nationalism and Techno-globalism: Conflict and Cooperation*, The Brookings Institution, Washington DC.

Otsuka, K., G. Ranis and G. Saxonhouse (1988), *Comparative Technology Choice in Development: The Indian and Japanese Cotton Textile Industries*, Macmillan, London.

Pacey, A. (1980), *Rural Sanitation: Planning and Appraisal*, Intermediate Technology Publications, London.

Pack, Howard (1987), *Productivity, Technology and Industrial Development: A Case Study in Textiles*, Oxford University Press, New York NY.

Pack, Howard and L.E. Westphal (1986), 'Industrial Strategy and Technological Change: Theory Versus Reality', *Journal of Development Economics*, 22 (1), pp. 87–128.

Pangestu, M. (ed.) (1996), *Small-Scale Business Development and Competition Policy*, Centre for Strategic and International Studies, Jakarta.

Pavitt, Keith (1985), 'Technology Transfer among the Industrially Advanced Countries: An Overview', in Nathan Rosenberg and Claudio Frischtak (eds), *International Technology Transfer: Concepts, Measures, and Comparisons*, Praeger, New York NY.

Penny, D.H. (1969), 'Indonesia', in R.T. Shand (ed.), *Agricultural Development in Asia*, Australian National University Press, Canberra, pp. 251–279.

Porter, A. et al. (1996), 'Indicators of High Technology Competitiveness of 28 Countries', *International Journal of Technology Management*, 12 (1), pp. 1–32.

Porter, Michael E. (1990), *The Competitive Advantage of Nations*, Macmillan Press, London.

Raillon, F. (1990), *Indonesia 2000, The Industrial and Technological Challenge*, CNPF-ETP and Cipta Kreatif, Paris and Jakarta.

Rakasima, Mahmud F. and Tamsil Linrung (eds) (1995), *Wawancara Habibie*, Penerbit Amanah Putra Nusantara, Jakarta.

Ramelan, Rahardi (1997), Visi Industri dan Teknologi Indonesia dalam Era Ekonomi Informasi, Paper presented to the panel discussion, 'Membangun Visi Industri dan Teknologi Indonesia dalam Era Ekonomi Informasi', Faculty of Economics, University of Indonesia, Jakarta, 29 April.

Ray, David (1996), 'Measuring Indonesia's Technology Capacity: New and Old Approaches', *Kelola: Gadjah Mada University Business Review*, 5 (13), pp. 75–102.

Repelita VI, 1994/95–1998/99 (1994), Koperasi Pegawai Bappenas, Jakarta.

Rice, Robert C. (1990), 'Indonesian Approaches to Technology Policy during the Soeharto Era: Habibie, Sumitro and Others', in Robert C. Rice (ed.), *Indonesian Economic Development: Approaches, Technology, Small-Scale Textiles, Urban Infrastructure and NGOs*, Centre of Southeast Asian Studies, Clayton.

—— (ed.) (1990), *Indonesian Economic Development: Approaches, Technology, Small-Scale Textiles, Urban Infrastructure and NGOs*, Centre of Southeast Asian Studies, Clayton.

Riddell, R.C. and M. Robinson (1995), *Non-governmental Organizations and Rural Poverty Alleviation*, Clarendon Press, Oxford.

Robinson, Richard D. (1988), *The International Transfer of Technology: Theory, Issues, and Practice*, Ballinger Publishing Company, Cambridge MA.

Robison, Richard (1987), 'After the Goldrush: The Politics of Economic Restructuring in Indonesia in the 1980s', in Richard Robison, Kevin J. Hewison and Richard A. Higgott (eds), *Southeast Asia in the 1980s: The Politics of Economic Crisis*, Allen and Unwin, Sydney.

Rohdewohld, R. (1995), *Public Administration in Indonesia*, Montech, Melbourne.

Rosenberg, Nathan (1994), *Exploring the Black Box: Technology, Economics and History*, Cambridge University Press, Cambridge UK.

Rosenberg, Nathan and Claudio Frischtak (eds) (1985), *International Technology Transfer: Concepts, Measures, and Comparisons*, Praeger, New York NY.

Rowen, H.S. (ed.) (1998), *Behind East Asian Growth: The Political and Social Foundations of Prosperity*, Routledge, London.

Safioen, H. (1990), *Prospek Perkembangan Industri Tekstil dan Kebutuhan Tenaga Kerja*, Asosiasi Pertekstilan Indonesia, Himateksi, 17 November.

Sandee, H. (1995), Innovation Adoption in Rural Industry: Technological Change in Roof Tile Clusters in Central Java, Indonesia, PhD thesis, Vrije Universiteit, Amsterdam.

Sarwono, B. (1990), *Peternak Kambing Unggul* [Improved Goats], Pembar Swadaya, Surabaya.

Sato Yuri (1996), 'The Astra Group: A Pioneer of Management Modernization in Indonesia', *Developing Economies*, 34 (3).

Schiller, J. and B. Martin-Schiller (1997), 'Market, Culture and State in the Emergence of an Indonesian Export Furniture Industry', *Journal of Asian Business*, 13 (1), pp. 1–23.

Schmoch, U. (1996), 'International Patenting Strategies of Multinational Concerns: The Example of Telecommunications Manufactures', in *Innovation, Patents and Technological Strategies*, OECD, Paris.

Siahaan, B. (1996), *Industrialisasi di Indonesia sejak Hutang Kehormatan sampai Banting Stir* [Industrialization in Indonesia from the Debt of Honour to the Change of Course], Pustaka Data, Jakarta.

Simon, Denis F. (1991), 'International Business and the Transborder Movement of Technology: A Dialectic Perspective', in Tamir

Agmon and Mary Ann von Glinow (eds), *Technology Transfer in International Business*, Oxford University Press, New York NY.

—— (1997), 'Techno-security in an Age of Globalisation', in Denis F. Simon (ed.), *Techno-security in an Age of Globalisation: Perspectives from the Pacific Rim*, ME Sharpe, Armonk NY.

Sjaifudian, Hetifah (1997), 'Graft and the Small Business', *Far Eastern Economic Review*, 16 October.

Soesastro, H. and M.C. Basri (1998), 'Survey of Recent Developments', *Bulletin of Indonesian Economic Studies*, 34 (1).

STAID (Science and Technology for Industrial Development) (1993), *Science and Technology Indicators of Indonesia 1993*, Jakarta.

Stiglitz, J.E. (1996), 'Some Lessons from the East Asian Miracle', *The World Bank Research Observer*, 11 (2), pp. 151–177.

Strudwick, J. and A.M Cresswell (1994), *Secondary Education Outcomes in Indonesia*, USAID, Jakarta.

Sudradjat, A. (1996), *Permasalahan Pengembangan Industri dan Perdagangan Teskstil dan Produk Tekstil*, Working paper, Asosiasi Pertekstilan Indonesia, Bandung, 26 March.

Suryodiningrat, Meidyatama (1995), 'Indonesia Offers Thailand a Rice-for-Planes Deal', *Jakarta Post*, 7 January.

Szirmai, A. (1993), *Comparative Performance in Indonesian Manufacturing, 1975–1990*, Research memorandum 538 GD-3, Groningen Growth and Development Centre, University of Groningen, Groningen.

—— (1994), 'Real Output and Labour Productivity in Indonesian Manufacturing, 1975–1990', *Bulletin of Indonesian Economic Studies*, 30 (2), pp. 49–90.

Tarmidi, L. (1996), 'Changing Structure and Competition in the *Kretek* Cigarette Industry', *Bulletin of Indonesian Economic Studies*, 32 (3), pp. 85–107.

Thee Kian Wie (1990), 'Indonesia: Technology Transfer in the Manufacturing Industry', in H. Soesastro and M. Pangestu (eds), *Technological Challenge in the Pacific*, Allen and Unwin, Sydney, pp. 200–232.

—— (1991), Technological Developments and Its Implications for Indonesia's Garment Industry, Paper delivered to the Economics of Trade and Development Seminar, Australian National University, Canberra, 9 July.

—— (1993), 'Industrial Structure and Small and Medium Enterprise Development in Indonesia', *EDI Working Papers*, World Bank, Washington DC.

—— (1996), Raising Indonesia's Industrial Competitiveness, Unpublished paper presented at the Fifth East Asian Economic Association Convention, Bangkok, 25–26 October.

—— (1997), *Pengembangan Kemampuan Teknologi Industri di Indonesia* [The Development of Industrial Technology Capability in Indonesia], Penerbit Universitas Indonesia, Jakarta.

Thee Kian Wie and Mari Pangestu (1993), 'Technological Dynamism in Indonesia's Exports of Electronic Products', in *Dialog Teknologi dan Industri: Pemacuan Teknologi Menuju Terbentuknya Industri Nasional Yang Kuat dan Berdaya Saing Tinggi*, BPPT, Jakarta, pp. 125–148.

—— (1994), *Technological Capabilities and Indonesia's Manufactured Exports*, Final draft report to UNCTAD's Technology Program, Jakarta, January.

Timmer, C.P. (1973), 'Choice of Technique in Rice Milling in Java', *Bulletin of Indonesian Economic Studies*, 9 (2), pp. 57–76.

Tran, Van Tho and Shujiro Urata (1995), 'Emerging Technology Transfer Patterns in the Pacific Asia', in Denis F. Simon (ed.), *The Emerging Technological Trajectory of the Pacific Rim*, ME Sharpe, Armonk NY.

Tubagus Feridhanusetyawan (1997), 'Survey of Recent Developments', *Bulletin of Indonesian Economic Studies*, 33 (2), pp. 3–39.

UN (1989), *An Overview of the Framework for Technology-Based Development*, Bangalore.

UNIDO (1979), *Appropriate Industrial Technology for Textiles*, New York NY.

US Department of Commerce (1992), *U.S. Direct Investment Abroad, 1989 Benchmark Survey*, US Government Printing Office, Washington DC.

van der Eng, P. (1996), *Agricultural Growth in Indonesia: Productivity Change and Policy Impact since 1880*, Macmillan Press, London.

van der Kamp, R. (1997), Technology and Human Resources in the Indonesian Textile Industry: The Role of Technological Progress, Education and HRD in Economic Performance, MSc theses series 97.15, Eindhoven University of Technology, Eindhoven.

van Diermen, P. (1997), *Small Business in Indonesia*, Ashgate, Aldershot.

Westphal, Larry E., Kim Linsu and Carl J. Dahlman (1985), 'Reflections on the Republic of Korea's Acquisition of Technology Capability', in Nathan Rosenberg and Claudio Frischtak (eds), *International Technology Transfer: Concepts, Measures, and Comparisons*, Praeger, New York NY.

Williamson, J. (ed.) (1994), *The Political Economy of Policy Reform*, Institute for International Economics, Washington DC.

Wong Poh-Kam (1997), 'The Proliferation of Technology-Based International Strategic Alliances: Contrasting Perspectives', in Denis F. Simon (ed.), *Techno-security in an Age of Globalisation: Perspectives from the Pacific Rim*, ME Sharpe, Armonk NY.

World Bank (1991), *Indonesia: Employment and Training— Foundations for Industrialization in the 1990s*, Report No. 9350-IND, Washington DC.

—— (1992), *Indonesia: A Private Sector Assessment*, Report No. 10901-IND, Washington DC.

—— (1996a), *Indonesia: Dimensions of Growth*, Report No. 15383-IND, 7 May.

—— (1996b), *World Development Report 1996: From Plan to Market*, Oxford University Press, Washington DC.

—— (1996c), *Industrial Technology Development for a Competitive Edge*, Report No. 15451-IND, Washington DC, 29 May.

—— (1997a), *Indonesia: Sustaining High Growth with Equity*, Report No. 16433-IND, Washington DC.

—— (1997b), *Training and the Labor Market in Indonesia: Productivity Gains and Employment Growth*, Report No. 16990-IND, Washington DC.

Yoshihara Kunio (1988), *The Rise of Ersatz Capitalism in Southeast Asia*, Oxford University Press, Singapore.

Zhao, Hongxin (1995), 'Technology Imports and Their Impacts on the Enhancement of China's Indigenous Technological Capability', *Journal of Development Studies*, 31, pp. 585–602.

AUTHOR INDEX

SUBJECT INDEX